29.95
68I

Fundamentals of Historical Geology and Stratigraphy of India

Ravindra Kumar
Centre of Advanced Study in Geology
Panjab University, Chandigarh
India

A HALSTED PRESS BOOK

JOHN WILEY & SONS
New York Chichester Brisbane Toronto Singapore

Copyright © 1985, Wiley Eastern Limited
New Delhi

Published in the Western Hemisphere
by Halsted Press, A Division of
John Wiley & Sons, Inc., New York

Library of Congress Cataloging in Publication Data

Ravindra Kumar, Dr.
Fundamentals of historical geology and stratigraphy of
India.

"A Halsted Press book"
  Bibliography: p.
  Includes index.
  1. Geology—India.   I. Title.
QE295. R34   1984   551.7'0954   84-19640

ISBN 0-470-20124-X

Printed in India at Modern Printers, K-30, Naveen Shahdara, Delhi-110032.

Foreword

Historical Geology is a very important part of geology embraces as it does the entire history of the earth. It is a journey through space and time; as such the scope is vast and all the physical transformations of the earth have to be taken into account. The movement of the oceans, rise and decay of mountains, igneous activities, development of life, climatic changes such as freezing episodes, aridity conditions, crustal movements etc. form the subject matter of Historical Geology.

We had excellent books on Indian Stratigraphy by renowned authors like D.N. Wadia and M.S. Krishnan, but with the new concepts available now, a book with a modern approach has become absolutely necessary. We have now fairly large data on geochronological dating of rocks, better knowledge of fossil record and new techniques of stratigraphical studies.

The book on Historical Geology and Stratigraphy of India by Dr. Ravindra Kumar who has been a successful teacher at the Tribhuban University, Nepal and Panjab University, Chandigarh, and has considerable field experience, is most welcome.

I am confident that the book will be extremely useful for all levels of University students.

3rd June, 1982
Ram Krishna Marg
Lucknow, India

R.C. MISRA
F.N.A.
Formerly Professor & Head
Department of Geology
Lucknow University
Lucknow

Preface

The Indian subcontinent presents a marked diversity in its landforms and climate. While in the east, vast swampy regions of Sunderban are frequented by marine waters, up in the north, we have the world's tallest mountains, topped by the Mount Everest. Likewise, there exists a sharp contrast in the climate. We have the world's wettest spot, Chirapunji, in the east and dry deserts of Rajasthan in the west. Whereas the northern mountaineous region of the Himalaya experiences sub-zero temperatures in the winter, in summers the mercury rises above 50°C in some parts of the Gangetic plains. In the southern coastal regions, the temperature is almost 30°C throughout the year. These regions of climatic contrasts also differ from each other in terms of relief, geological structures and rock formations.

This beauteous subcontinent is but a momentous incidence in the everchanging shape of the earth's surface. The diversity in the landforms and the contrasts in their climate, which fascinate us most, are merely a passage from uninterrupted past to unknown future. The swamps and deserts, the mountains and lowlands, the land and the sea have been exchanging their locales on the surface of the earth as a consequence of the pulsating nature of the mother earth. The migration of the shore lines in recent past is interpreted in the recorded history and archeology. Vast areas, which were under marine water in the geological past, form parts of the present topographic contrasts of the Indian subcontinent. Their history is recorded in terms of the rock formations and geological structures. The areas, which were above the marine waters in the geological past, have in some cases completely disappeared leaving behind no traces of their existence. Their presence is interpreted by indirect methods. Reconstruction of the past shore lines, distribution of diverse type of landforms on the continents and the topography of the oceanic bottoms during the geological past form the subject-matter of Historical Geology.

The term "Historical Geology" may, in special cases, be considered as synonym for "Stratigraphy". However, the latter term has often restricted application to a sequential arrangement of stratified rocks which are laid down on the surface of the earth. Nevertheless, the concept of sequential arrangement is based on dynamic and physical activity operating both on the surface and within the earth. The restoration of the stratified rocks in sequential order of their deposition is the only means to restore the paleogeography during successive periods of the geological history of the earth. The physiographic characters of the surface of the earth is an expression of complex changes which have been continuously taking place in the composition and the structure of the earth's crust. The Historical Geology thus helps in restoration of the evolution of the earth's crust during the geological past.

A graduate student of Geology is often given to understand that Stratigraphy deals with certain set of names of rock formations and fossils which are memorised until the examinations were over. It is never realised that the memories put to experience become part of our lives which energises to think for ourselves and to explore the never ending mysteries of the earth. It was this need to correlate the 'meaningless' names in stratigraphy with the past events that led me to reorient the content of a text book on historical geology.

The subject-matter of Historical Geology (Stratigraphy) is introduced to the undergraduate students at their second year of study. Such a course has generally two aspects, the first part dealing with the principles and methods of stratigraphy and the second part systematically covering the descriptions of rock formations of various geological ages. Both these aspects are available separately in the respective text books dealing with them in rather a specialised manner. These advanced books are not of much use to the undergraduate students and thus a need was felt to cover these two aspects in a single text book. The present book will also be useful for the engineering students who opt for geology as one of their subsidiary subjects. The book has been written in a simple language with less of technical jargon so that it may interest those wishing to learn the earth sciences in a general way.

Some of the excellent books on the stratigraphy of India were written a few decades ago. Although new editions were made available, these books remained based largely on earlier views on the stratigraphy of India. Some of these earlier views and concepts have become outdated on account of many new data which have been added since these books were written. For certain stratigraphical problems latest techniques of geology and paleontology have been employed and the results have been encouraging. Geochronological dating of rocks based on quantitative analyses of radio-isotopes included in the composition of these rocks have further helped in solving some of the intricate problems of stratigraphy, specially of the Precambrian rocks.

In recent years, a number of excellent monographs, review papers and research papers on the stratigraphy of India have been published. These may be used as a reference for those who wish to go deeper in the problem.

A graduate student is primarily concerned with the comprehension of the subject-matter. With this aim in mind, certain generalisations have been made in the present book regardless of the fact that contrary views exist. To experts, these generalisations may appear to be hasty conclusions but they will bear me out that for the wood the forest has not been lost. It will take several decades or even whole of our life time when a consensus may emerge on these complex problems.

The book comprises eleven chapters. In an introductory chapter, aims and objects of Historical Geology and a brief historical account of the development of the science are given. The second chapter elaborates on the basic methods and concepts of Historical Geology (Elements of Stratigraphy). Paleogeographic and paleotectonic reconstructions and related concepts are introduced chapter three. The fourth chapter briefly outlines the tectonic framework of continents and oceans. The tectonic divisions of India and their physiographic expressions are dealt with in greater detail. Precambrian rock formations of India and their special problems of correlation are described in chapters five to seven. Chapter eight deals with Paleozoic formations whereas chapters nine and eleven describe the rock formations of Mesozoic and Cainozoic Eras respectively. The Gondwana Sequence of India that was deposited in continental basins during the Late Palaeozoic-Mesozoic time has been dealt in chapter ten.

There has been a spurt of research publications on Indian Geology during the last over two decades. I have, as far as possible, tried to incorporate most of the recent contributions on the subject. Accordingly some papers are quoted in the text while others are included in the Bibliography of Selected References. But, in view of obvious limitations, it has not been possible for me to include all the relevant research publications in this work. As regards the shortcomings, I crave the indulgence at the hands of the readers who are requested to bear in mind the immensity of the subject-matter capsuled in a brief introductory course such as the present book.

I have received active encouragement from my colleagues and students during the preparation of the materials for the present work. I am specially grateful to Professor V.J. Gupta and Professor Ashok Sahni for critical appraisal of parts of the manuscript and for suggesting many improvements. I had some fruitful discussions with Professor S.B. Bhatia on general aspects of stratigraphical problems that helped in arranging the subject-matter for the present book. Professor A.K. Prasad very kindly reviewed the contents of the book and made useful suggestions for its improvement. Thanks are due to Shri Piyush Panwar, Sh. S.K. Saigal and my other research students, for inserting the corrections in the typed manuscript. The present work was given a preliminary shape during the tenure of my deputation to the Tribhuban University (Nepal) through the Indian Cooperation Mission. Fruitful discussions with Professor B.M. Pradhan, Dr. M.P. Sharma and other faculty members and students of the

Department of Geology, Trichandra Campus, Kathmandu are gratefully acknowledged. Mr. G.D. Sharma took great pains in typing the manuscript with utmost care. Line diagrams were redrawn in India ink by Mr. M.C. Mankoo and Mr. Gian Chand. Last but not least, I am to express my sincere gratitude to my wife, Smt. Punam Srivastava for constant encouragement and consistent devotion by playing a positive role in completing this work in its present form.

October 1984　　　　　　　　　　　　　　　　　　　RAVINDRA KUMAR
Chandigarh, India

Contents

Foreword   *iii*
Preface   *v*

PART I : ELEMENTS OF HISTORICAL GEOLOGY   1-56

Chapter 1. Introduction   3
    Brief Historical Review   *4*
    Geological Investigations in India   *6*

Chapter 2. Elements of Stratigraphy   9
    Criteria for Stratigraphic Classification and Correlation   *9*
      Non-palaeontological Criteria
      Palaeontological Criteria
    Standard Stratigraphic and Time Scales   *17*
      Litho-stratigraphic Classification
    Absolute Age of Rocks   *23*
      Geochronology

Chapter 3. Principles of Palaeogeography and Palaeotectonics   26
    Contemporary Marine Basins   *27*
      Marine Sediments
      Marine Life
      Bionomical Zones
    Contemporary Continental Basins   *31*
    Palaeogeographic Reconstructions   *33*
      Palaeogeographic Maps
    Palaeotectonic Reconstructions   *36*
      Tectonic Analysis of Facies Maps
      Tectonic Analysis of Isopach Maps
      Diastems and Angular Unconformities

Chapter 4. Earth's Crustal Structure and Tectonic Divisions of India   40
    Tectonic Elements of Continents   *41*
      Cratons
      Folded Mountain Belts
    Tectonic Elements of Oceans   *45*
    Tectonic Divisions of India   *46*
      Peninsular India
      Extra-Peninsular India
      Indo-Gangetic Plain

x *Contents*

PART II : STRATIGRAPHY OF INDIA 57-225

Chapter 5. Precambrian Basement of Indian Peninsula 59
- Precambrian History of the Earth *60*
- Precambrian Basement *61*
  - Dharwar Province
  - Eastern Ghats Province
  - Central Indian Province
  - Singhbhum-Orissa Province
  - Aravalli-Bundelkhand Province

Chapter 6. Proterozoic Formations of Indian Peninsula 81
- Proterozoic History *82*
- Basement Cover Transition *83*
- Proterozoic Succession *84*
- Lower Purana Succession *85*
  - Delhi Supergroup
  - Bijawar and Gwalior Groups
  - Kolhan Group
  - Cuddapah Supergroup
  - Kaladgi and Pakhal Groups
- Upper Purana Succession *94*
  - Vindhyan Supergroup
  - Kurnool Group
  - Northern Extension of Kurnool Group

Chapter 7. Precambrians of Extra-Peninsula 103
- Precambrians of the Tethyan Basement *104*
  - Salkhala Group
  - Vaikrita Group
  - Bhimphedi Group
  - Jutogh Group
  - Daling Group
- Precambrians of the Lesser Himalaya *112*
  - Western Sector
  - Central Sector
  - Nepal Himalaya
  - Eastern Himalaya

Chapter 8. Palaeozoic History 121
- Tectonic History *123*
- Palaeozoic Life *124*
- Precambrian Cambrian Boundary *126*
- Marine Palaeozoic Formations of India *130*
  - Tethyan Regions
  - Lesser Himalayan Regions

Chapter 9. Mesozoic History 147
- Tectonic History *149*
- History of Mesozoic Life *150*
  - Marine Forms

          Land Forms
      Permian-Triassic Boundary *151*
      Marine Mesozoic Formations of India *155*
          Tethyan Himalaya
          Lesser Himalaya (Krol Belt)
          Indian Peninsula

Chapter 10. Gondwana Sequence of India   **176**
      Sedimentation and Palaeoclimates *177*
      Lower Gondwana Sequence *178*
          Talchir Formation
          Marine Intercalations
          Bap and Badhaura Formations
          Damuda Group
          Lower Gondwana of Eastern Himalaya
      Upper Gondwana Sequence *186*
          Damodar Valley Basin
          Satpura Basin
          Rajmahal Hills
          Mahanadi-Son Valley Basin
          Pranhita-Godavari Basin

Chapter 11. Cenozoic History   **194**
      Tectonic History *195*
          Rise of the Tertiary Mountains
      History of Cenozoic Life *199*
      Boundary Problems *200*
      Indian Cenozoic Formations *204*
          Himalayan Palaeogene Succession
          Himalayan Neogene Succession
          Indus Belt
          Deccan Traps
          Assam-Arakan Region
          Andman-Nicobar Islands
          Northwestern Peninsula
          Cauveri and Godavari Basins

Bibliography of Selected References   **227**
Subject Index   **243**

Part I

ELEMENTS OF
HISTORICAL GEOLOGY

Chapter 1

Introduction

Historical geology is a science which deals with the historical development of the earth. It aims at reconstruction of the earth's evolutionary history and formulation of general laws governing the evolution. The reconstruction is based on the study of rock successions exposed on and below the surface of the earth. The rocks are studied not only in the field in their natural setting but also in the laboratory for their textural and mineralogical characters. The organic remains preserved in certain rocks have special significance for their immense utility in assigning the age for the rocks.

The reconstruction of the historical evolution of the earth's crust during the geological past is based on four different aspects of historical geology. The first aspect is to establish the age of rock formation exposed on different segments of the earth's surface. The second aspect pertains to palaeogeography which describes the distribution of landforms and sea in the geological past. The third aspect covers past tectonic movements inferred on the bases of palaeogeography and structure of the rocks. The fourth aspect deals with a synthesis of palaeogeography and palaeotectonics of different segments of the earth's crust.

Establishing the age of rock formations and their correlations are the basic tasks of the historical geologists. The history of the earth's evolution can be reconstructed only when the succession of rock formations has been worked out. The succession is generally established in terms of 'older' and 'younger' ones, i.e., their relative time of formation. Sometimes it is possible to establish the age of rock formations in terms of absolute time or in terms of some relatively broad geological time range.

Palaeontology, the science which deals with organic remains preserved in rocks, has been most widely used for determining the relative age of rock formations formed during the last about 570 million years. The type of organic remains preserved in rocks indicate the kind of life which existed on the earth at the time of formation of these rocks. This form of life is assigned a certain stage in the evolutionary history of the organic world. The relative ages of rocks are worked out in terms of a particular stage of evolution of life. The age of remaining rock formations are established according to their relationship with the sedimentary rocks containing dateable palaeontological records. Techniques of **Geochronology** help in determining the age of rocks in absolute terms with the help of radio-active elements present in some rocks. **Stratigraphy** deals with the mutual relationship and succession of rocks. It aims at the grouping of natural association of rocks, establishing their ages and correlating them with the rock formations of different areas.

Natural association of various rock types and their structures and textures reflect the physico-chemical conditions under which the rocks were laid down. Such an interpretation of past conditions of deposition of rocks helps in reconstruction of past relief of the earth's surface at the time of formation of these rocks. Absence of record of any rock succession from a segment of the earth's surface for a given geological period may indicate that this segment constituted the land area at that time. It is also possible that this segment may have been under sea and the sediments were deposited at that time. Subsequently, the segment was uplifted and eroded leaving behind no trace of sedimentary record of that time. The distribution of land and sea, climate and animals and plants over the earth's surface of the geological past are displayed with the help of **palaeogeographic maps**.

The relief of the earth's surface and the distribution of land and sea have been constantly changing during the successive geological epochs, which are, as a rule, governed by movements and transfer of matter within the earth's crust and mantle. Such movements are known as **tectonic movements**. Thus the sedimentary record of the geological past serves, at the same time, as an evidence of past tectonic movements. Subsequent deformation, metamorphism and igneous intrusions in rock successions are also evidences of tectonic movements. These tectonic movements reflect the pattern of evolution of the earth's crust and are often complex. The science of **Tectonics** deals with the specific problems of origin and evolution of the structure of earth's interior.

Historical geology with all its basic data on the succession of rock formations, palaeogeography and palaeotectonics helps in establishing a generalised theory of the evolution of the earth's crust. In spite of a fairly good knowledge of geology of any segment of the earth's crust, it is difficult to state the generalised theory of the earth's evolution in simple language. Elaboration of such a generalised theory is inherent in unfolding the geological history contained in the rock successions. As a generalised theoretical science, historical geology does not solve the specific problems of prospecting and exploration of mineral deposits. Nevertheless, it arms the economic geologists with the theoretical bases for mineral exploration. Historical geology explains the laws of distribution of rock formations and associated mineral deposits in terms of comprehensive and generalised theory.

BRIEF HISTORICAL REVIEW

Historical geology came into being as a science during the 18th and 19th centuries when palaeontological methods were introduced in geology. These methods were introduced by W. Smith in England and J. Kiev and A. Bronnairre in France. These scientists compiled the first stratigraphical columns, geological maps and cross-sections with the help of palaeontological data. Most of the Geological Systems of the International Stratigraphic Scale based on palaeontology were introduced during the first half of the 19th century. During this period, geological maps and stratigraphical columns for a number of European countries were compiled. Gradually a vast amount of geological data was accumulated for a meaningful theoretical generalisation.

The first generalisation was put forward by J. Kiev in 1812 in his **theory of catastrophy** which received wide acceptance during the first half of the 19th century. According to this theory, catastrophic events took place leading to sudden changes in the relief of the earth's surface and annihilation of life. This theory, however, was questioned by J. B. Lamark, Charles Lyell and Charles Darwin. Charles Lyell in his book "Fundamental of Geology" (1830-33) showed that the great changes on the earth's surface took place not as a result of catastrophy but due to slow and steady geological processes which are operating even today.

Charles Lyell proposed the concept of **uniformitarianism** in his generalised theory of the evolution of the earth's surface. According to this concept, the past geological processes are explained in terms of the contemporary processes which can be observed and recorded at present. The concept of uniformitarianism was a big leap forward in the advancement of geological science. Later, the treatise of Charles Darwin on the "Origin of Species" (1859) significantly added to our knowledge of organic evolution in terms of historical science. Geological history of various parts of the globe which had been published by the end of 19th century was synthesised by an eminent Austrian scientist, Edward Suess, in his work entitled "The Face of the Earth" (1904).

In the beginning of the 20th century, the French scientist, E. Huag, dealt with, at length, the contemporary geological processes and their manifestations in the geological pasts. He demarcated the differing characters of the evolutionary history of the platform and geosynclinal areas which were later elaborated in the work of A.W. Grabau, G. Stille, M. Kay, S. Bubnov and N.S. Shatsky. The palaeogeography of the North America was worked out by Charles Schuchert in 1910 and the first text book on Historical Geology of America by L.V. Pirsson and Charles Schuchert appeared in 1924.

During the later part of the present century, many new data on the geology of different segments of the earth's crust were collected. Several new techniques of the bordering sciences such as geophysics and geochemistry helped in formulating a more comprehensive theory on the historical evolution of the earth. Special mention should be made of the concept of **Plate Tectonics** which evolved during the last over two decades. As a generalised tectonic theory on the evolution of the earth's uppermost layer, the Lithosphere (i.e., earth's crust and parts of Upper Mantle), it has greatly influenced the contents and methods of study of historical geology. Recent advances in theoretical and experimental petrology have also greatly influenced the bases of palaeogeographic and palaeotectonic reconstructions from the observed successions of rock formations.

GEOLOGICAL INVESTIGATIONS IN INDIA

Systematic geological investigations in India began only about a century ago with the establishment of the Geological Survey of India. The geological mapping during the 19th century was carried out only in a few isolated areas. These investigations, though of preliminary nature, helped in establishing the basic features of the geology of India. During the earlier half of the present century, extensive and intensive pioneer investigations laid the foundation of the Indian geology. The later part of the present century has been a period of intensive detailed work in independent India. The Central and State Governments, universities and other research institutions of India have tapped resources and provided facilities for these intensive detailed investigations on the geology of India. New organisations, such as Oil and Natural Gas Commission, Atomic Energy Commission, Mineral Exploration Corporation, Coal Authority of India Ltd., Central Ground Water Board, etc., were set up thereby opening venues of employment to thousands of geologists in the country. Besides, the Geological Survey of India and the State Geological Surveys were greatly expanded during the last three decades. As a result of the team work and the individual efforts of geoscientists working in research institutions and universities, the progress in the science of geology of India has undergone a revolutionary change.

The major elements of the Indian peninsula were already known by the end of the previous century. The oldest element of the peninsula, i.e., the Archaean foundation, consists of rocks formed more than 2500 million years ago. These rocks are the main source of metallic deposits of India. They are overlain by rock formations referred to as Puranas and Vindhyans ranging in age from about 2500 million years to about 570 million years. These rocks are specially known for the Panna diamond mines in the north India and the Kurnool diamond mines in the south India. The next in succession are the Gondwana rocks formed some 320 to 136 million years ago. These rocks are the chief source of coal of India. The Deccan Traps succeeding the Gondwana formations cover a vast expanse of the central-western India. These volcanic rocks were laid on the surface of the earth about 70 million years ago and they were exposed to chemical weathering under tropical conditions for several million years. Consequently, the rocks over extensive areas were converted into rich bauxite deposits, the main source of aluminium ore of India. The marine transgressions inundated a major part of north-western and north-eastern India and coastal regions of the south and south-eastern India during the time interval of 200 million years to about 1 million years before present. The rock formations that were deposited in these marine basins are the only source of petroleum in India.

The Archaean basement is beset with extremely complex geological history. To unravel the complex geology, earlier efforts relied mostly on broad lithology and major structures. Stratigraphic successions were established with a fair degree of certainty in regions of simpler structures. However, uncertainty remained in most of the intensely deformed and metamorphosed successions. For example, in Singhbhum-Orissa region, the structure has been alternatively explained time and again in terms of anticlinorium or synclinorium. Thus the Iron Ore Series have alternatively been regarded older and younger than the Gangpur Series. In Karnataka, the Dharwar sediments were alternatively considered to have been deposited over or intruded by the Peninsular Gneiss. In recent years, advanced techniques of geochemistry, petrology, structural geology and geochronology have been used to settle some of the century old problems of the Archaean geology of India. Yet much remains to be done to authenticate the earliest epoch of the geological history of India.

The Purana succession overlying the Archaean basement of the Peninsula has a relatively less complex structure. The basic stratigraphic divisions of Cuddapahs and Vindhyans established in the pioneer work done in the first half of the present century have undergone very little change. In recent years, emphasis has been laid on the faunal and floral records, geochronology and palaeogeography of these rock formations.

The terrestrial deposits of the Indian Gondwana rocks were widely studied and classified during the twenties and thirties of the 20th century. More recent studies have been made on the classification, age and palaeogeography. Latest techniques of sedimentology have been employed to interpret the environmental and depositional history of Gondwana sediments. Some recent records of fossil fauna and flora have further added to our knowledge of the palaeobiogeography.

A wealth of data on marine Mesozoic and Tertiary rock formations fringing the Precambrian basement of the Indian Peninsula have accumulated over two decades of intensive oil exploration in India. Subsurface geology of the region has been worked out on the basis of drill hole data and geophysical surveys. Synthesis of these investigations have improved upon our knowledge regarding the Mesozoic and Tertiary geological history of the region.

The Deccan Traps are horizontal volcanic flows interlayered with the Intertrappean Beds. The stratigraphy of the Deccan Traps was established in the pioneer work carried out on the basis of fossil records in the Intertrappean Beds in the first half of the century. In recent years, similar volcanic successions were identified in Kathiawar, Kutch, NW Himalaya,

Orissa, Bengal and South India. The biostratigraphy of the fossiliferous beds intervening the volcanic layers have been better established as a result of recent fossil findings. A wealth of data is now available on the geochemical and petrographical characters of the volcanic rocks.

The geology of the Himalayan sector was sketchy till the end of 19th century. The region which is one of the most difficult terrains still remains largely inaccessible. In the pioneer work of the first half of the present century, the broad structural-geomorphic divisions of the Himalaya were demarcated. During the last about two decades, the Himalayas have attracted more and more geologists in comparison to other regions of India. It being the youngest and the loftiest mountain chain of the world, the Himalaya occupies a prominent place in any concept dealing with the Plate Tectonics. In addition to immense work done by the geologists of various Government agencies, significant contributions on the Himalayan Geology have also been made by the scientists of the Centre of Advanced Study in Geology, Chandigarh; the Wadia Institute of Himalayan Geology, Dehradun and score of other universities of India. The new data that are still pouring in have upset the concepts established by the pioneer workers in the first half of the present century.

The Precambrian formations of the Himalaya were demarcated primarily on negative evidence of the absence of fossil records and sometimes on lithological similarities with the Precambrian formations of the Indian Peninsula. The more metamorphosed rocks of the Himalaya were assigned the Archaean age whereas the less metamorphosed ones the Middle and the Upper Precambrian ages. Supported by the new data on geochronology, structural and metamorphic history and palaeontology, some successions have now been assigned younger ages. The rocks of the Lesser Himalaya are largely unfossiliferous and unmetamorphosed. Their stratigraphy was worked out on the basis of indirect correlations. New fossil findings and data on structure and stratigraphy have led the geologists to modify the earlier stratigraphic divisions. Most of the fossiliferous rocks of the Phanerozoic age (less than 570 million years) are located in the Tethyan region in the north of the snowy peaks of the Himalaya. Systematic mapping of these areas in recent years has led to establish a more detailed stratigraphic classification of this region. The northern limit of the Himalaya is marked by a lineament along the upper Indus river in Ladakh. This region has, of late, been considered as the key region for the understanding of the Himalayan tectonics. Though some new data on the geology of this region have been published yet much more remains to be done in revealing the complex geological history of the same.

Chapter 2

Elements of Stratigraphy

Stratigraphy deals with the sequential arrangement of layered rocks according to the time of their formation on the surface of the earth. An account of the sequence of deposition of sedimentary rocks in any area is depicted in stratigraphical column prepared on the basis of geological mapping of that area. The comparison and generalisation of stratigraphical columns in the neighbouring areas help in establishing a common stratigraphical column or the whole region in which all the rock formations are graphically represented according to their chronological order of deposition. Stratigraphic columns of neighbouring regions are synthesised to describe the geological history of that continent.

Stratigraphic classification helps in grouping the sedimentary and volcanic rocks into rocks formations on the basis of their lithological characters and fossil contents if any. The chronological sequence of rock formations established in terms of their relative ages is graphically represented in vertical columns. The older formations are shown below the younger ones. The place of intrusive rocks of the area is determined by their relationship with the layered rocks of the column and, accordingly, they could also be shown on the graphic representations.

Correlation of rock formations of different regions form an important theme of stratigraphy. The correlation aims at matching the rock formations of distant areas deposited at the same stage of the earth's evolutionary history. The two formations thus correlated may not have been deposited at the same time, or in other words, they may not be synchronous in their origin. These formations deposited at the same stage of the earth's evolutionary history are referred to as **homotaxial**. A standard stratigraphic scale has been established for correlation of rock successions of different regions.

CRITERIA FOR STRATIGRAPHIC CLASSIFICATION AND CORRELATION

Relative age of rock formations refers to the relative time of their formation with respect to certain other groups of rocks formed at different times. The rocks that deposited later will generally lie over the ones deposited earlier. Thus the older rocks are overlain by the younger ones under normal conditions of stratification. The relative younger ages are also indicated in terms of different and generally advanced forms of organic remains enclosed in the younger rock formations. A study of these fossils helps in determining the relative ages of the fossiliferous rock formations. However, many rock formations, especially those deposited more than 570 million years ago, are generally devoid of any fossil record. For determining the relative ages of fossiliferous as well as unfossiliferous rocks, certain set of non-palaeontological criteria are used.

Non-Palaeontological Criteria

Order of Superposition: Since each overlying bed is younger than the underlying bed under the normal conditions of stratification, this principle can be easily applied to determine the relative ages of layers of sedimentary and volcanic rocks in the areas of simpler structures (Fig. 2.1 a). In the case of deformed rocks (Fig. 2.1 b), the top and bottom of beds at the time of their deposition are determined with the help of sedimentary structures like current bedding, graded bedding and current ripple marks. Relative ages of succeeding beds determined on each outcrop give the chronological succession of the whole sequence.

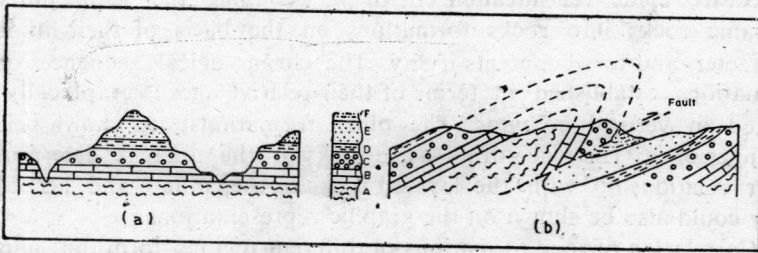

Fig. 2.1: Geological cross sections of regions of (a) horizontally bedded, and (b) deformed rocks.

Petrographical Characters: Petrographical study of rocks helps in dividing the rock sequences into formations as well as members which differ from one another in their mineral composition, textures and structures. Correlation of rock formations of the neighbouring areas is often carried out by comparing the petrographical characters of the formations in the two areas. The correlation is accomplished by tracing the formations having some distinct lithology. Such beds are known as **marker horizons**. The rock formations situated in between two marker horizons in the neighbouring areas are considered to be of the same relative age provided there is no trace of erosion between the two marker horizons. In rock succession of uniform lithology, the stratigraphic classification and correlation are often carried out with the help of minerals of high specific gravity such as zircon, epidote, garnet, magnetite, etc. These heavy minerals tend to vary in proportion at various stratigraphic levels of a sequence having uniform lithology. This variation is attributed to different sources of sediments at successive stages of the deposition.

The petrographical-mineralogical characters of rock succession enable to classify and correlate the rock formations deposited in a single basin of sedimentation. Rock formations of the same relative age but deposited in different sedimentary basins may greatly differ from one another in their lithological characters. For example, the Permian rock formation of the Salt Range in Pakistan (*Productus* Limestone) is lithologically different from that of the Spiti area (*Productus* Shale).

Structure and Tectonics: The basic premise of this criteria for stratigraphic classification and correlation is that the tectonic movements are simultaneously active over a large part of the earth's crust. The sedimentary rocks deposited on the floors of basins are uplifted above the surface of water at different times in the earth's history. As a result of these tectonic movements, they are deformed and partly eroded. During the succeeding subsidence, these rocks are overlain by younger beds. The surface separating these two successions of rock formations is known as the **surface of unconformity**. A sequence of rock formations of an area can be divided into major rock groups separated from one another by such surfaces of unconformity. Formations lying between the two successive surfaces of unconformity in neighbouring areas are considered to be of the same age (Fig. 2.2).

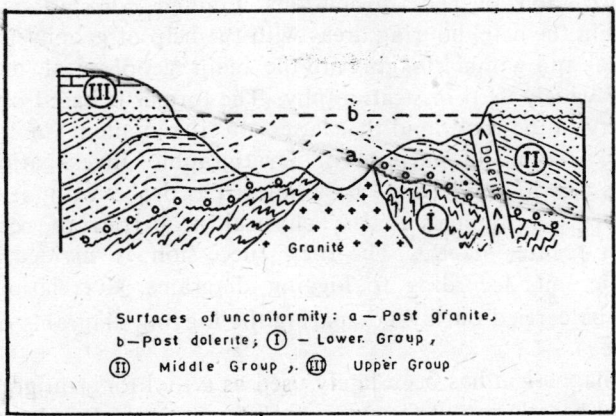

Fig. 2.2: Stratigraphical classification based on unconformities.

However, uplift and subsidence of large regions do not begin simultaneously everywhere, neither do they have the same rate of movement everywhere. Thus, it is possible that in certain parts of the region the rock succession are still in the process of erosion while the other parts have undergone subsidence and started receiving sediments. Thus, the rock formations occupying a position between two surfaces of unconformity may not be strictly synchronous in detail everywhere. Nevertheless, for large stratigraphical groupings, the surface of unconformities constitute important natural boundaries between successive groups of rock formations.

Tectonic movements are also recorded within a single group of rock succession in the form of rhythmicity of different lithologic types. The basin of deposition may undergo successive phases of deepening or shallowing which are reflected in the form of deposition of sediments alternatively of deeper and shallow basins. The deepening of marine basins is related to a **marine transgression** over the parts of continents and the shallowing of basins corresponds to a **marine regression** from the land areas. Thus, the phase of marine transgression is recorded by the presence of pure limestone among a relatively shallow water marly sediments. On the other hand, marls among the clays or shales, shales among the sandstones and marine sediments among the continental deposits indicate a phase of marine regression. Relative subsidence or uplift within a sedimentary basin can be used as a reliable criterion of correlation within a single basin of sedimentation.

Geophysics: Geophysical instruments help in recording changes in the physical properties of subsurface rocks, such as, specific gravity, magnetism, electrical resistivity, seismicity etc. These physical properties reflect the lithological characters of rock formations. The geophysical data thus can be used in determining the successive order of superposition of a subsurface succession of rock formations. Limited correlations can also be carried out in the neighbouring areas with the help of geophysical data.

Electrical and gamma logging are the main geophysical methods that have been widely used in stratigraphy. The former is based on the electrical resistivity of the rocks and the latter on the content of radioactivity of the rocks. For the purpose of stratigraphic investigations, special instruments are lowered in the bore holes. The probe with self recording apparatus registers changes in the physical properties of rocks as it is lowered to greater depths. The rock succession is divided into various stratigraphic units according to logging diagrams. Correlations of these units can be carried out by comparing the logging diagrams of the neighbouring areas.

Palaeomagnetism has been lately used as a tool for stratigraphic classification and correlation of Cenozoic rock formations. It is known that the earth's magnetism has been changing its polarity periodically throughout the geological time. The earth's magnetism at the time of deposition of rocks controls the orientation of ferromagnetic minerals under conditions of free settling. This orientation is preserved with the lithification. With a careful technique, this orientation can be recorded in rock samples collected from different parts of the rock succession. The polarity is recorded in terms of normal or reversed with respect to the present pole orientations of the earth's magnetism. For example, in a succession of volcanic rocks, four successive magnetic normals and reverses have been recognised. These rocks have been dated with the help of K-Ar radiometric dating method and reversals in polarity have been recorded for the duration of 4 million years (Fig. 2.3).

The velocities of propagation of seismic waves through the earths' crust is controlled by the physical behaviour of the rocks. Contrasting physical properties of different rock layers are often well recorded in seismic reflections. Changes in rock types produce changes in their reflectivity which is observed in seismic data. Such seismic data are useful in inferring stratigraphic changes. The methods and concepts of such interpretations have been evolved in to a new emerging science of **seismic stratigraphy**. Artificial seismic waves are produced at specific shot points. The waves, while traversing through the subsurface rock layers, are reflected and refracted from different interfaces of rock sequences. P-wave reflections are the most widely used data in seismic stratigraphy. The seismic data are analysed along with core logs and other drilled well data.

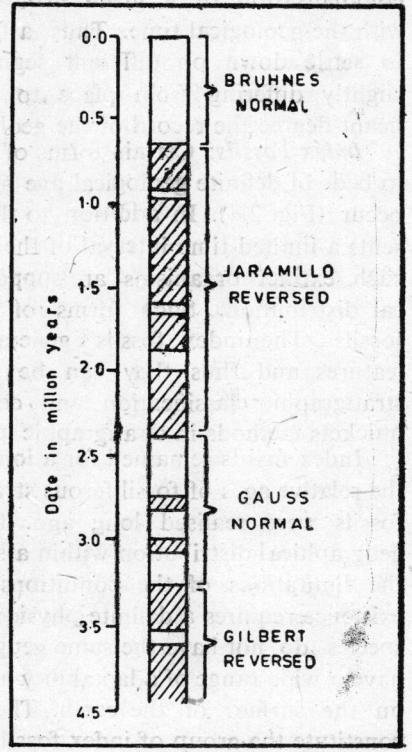

Fig. 2.3 : Sequence of magnetic reversals and radiometric dates in volcanic rocks.

Palaeontological Criteria

The palaeontological records are more reliable for determining the relative ages of rock formations and their correlation over large areas. This method is based on the premise that a given faunal or floral assemblage is never repeated again in the geological time and that the organisms continuously undergo changes in their morphology. The law of irreversibility of biological evolution states that the organisms cannot evolve to a preceding form even if they find themselves in exactly similar conditions in which their predecessors lived. Thus, each assemblage of past life preserved in a rock reflects a specific stage of the evolution of life. Thus, the fossil record can be used for determining the relative ages of different beds in a rock succession.

Wide areal distribution of the living organism helps to carry out correlation of rock formations over a larger region. The time taken for colonialisation of a given group of organisms is insignificant compared with the geological time. Thus, a form of organic life has a large scope to settle down on different segments of the earth's surface at a time slightly differing from place to place and yet not affecting to any significant degree the record of the geological time.

Index Fossils: Certain forms of extinct animals and plants are restricted to beds of definite geological age beyond which they are not known to occur (Fig. 2.4). In addition to their limited vertical spread, which represents a limited time interval of their existence on the surface of the earth, such extinct organisms are supposed to have a relatively wide geographical distribution. Such forms of organic remains are known as index fossils. The index fossils generally have characteristic morphological features and thus they can be easily identified. Use of index fossils for stratigraphic classification and correlation is considered as one of the quickets methods in stratigraphic palaeontology.

Index fossils remained for a long time the main tools for determining the relative ages of fossiliferous strata. However, the limitations of these fossils were realised long ago. The premise that index fossils have wide geographical distribution within a short interval of geological time, ignores the limitations of the conditions of habitat. Any organism for its own existence requires a definite physico-chemical conditions and, therefore, all species do not have the same geographical distribution. Some groups may have a wide range of adaptability and thus they may be widely distributed on the surface of the earth. These groups of organisms do not generally constitute the group of index fossils as they generally have a wider range of their existence on the geological time scale. On the other hand, others may be confined to certain favourable zones of environment and any change in environment may cause either their migration or, in extreme cases, extinction of these highly susceptible organisms. After a lapse of certain geological time, the favourable conditions of environment may return and these organisms may reappear higher up in younger beds.

Fossil Assemblage: In view of the limitations of index fossils, the total fossil record of the bed is examined and compared with those of other beds. Inference regarding the relative ages of rock formations is thus based on the total number of fossils preserved in those rock formations and not on one or a few index fossils which might have been found accidently. The geological dates inferred from one group of fossils can be cross checked with the other group of fossils preserved in the same strata, thus, minimising the possibility of any error.

The components of any fossil assemblage which occur in any rock stratigraphic unit may have three different derivations. Some of the species of the fossil assemblage inhabited the area of deposition. Such a community of inter-related organisms inhabiting an area is known as **biocoenose** and the area inhabited by the community adapted to its environment is known as **biotope,** Study of those species of the fossil assemblage which constituted the biocoenose of the basin of deposition helps in reconstruction of ecological conditions at the time of deposition. The second group of species of the fossil assemblage may be derived from neighbouring or distant biotopes. The total fossil assemblage in a rock unit of organisms which existed during certain phase of geological history and which thrived in different biotopes is referred to as **thenatocoenose.** Such species which inhabited on the surface of the earth during the time of deposition of a particular rock stratigraphic unit and which are preserved as fossils in that unit are useful tools of stratigraphic classification and correlation. The fossil assemblage may contain a third group of fossils which are derived from an older rock formation along with other sediments by the usual sedimentary processes. **These derived fossils** can be usually identified and separated from the indigenous material. Such derived fossils do not have any stratigraphic significance except for that they indicate the upper age limit of the rock formations in which they now occur.

The species of the fossil assemblage which constitute the thenatocoenose of the rock formation usually differ from each other in their evolutionary history. Certain species existed on the surface of the earth for a longer duration of time than the other. The time of appearance and extinction and the duration of their existence of organisms are recorded in the succession of rock formation (Fig. 2.4). Thus a number of species appeared for the first time in a particular unit but ranging on into the overlying strata. Yet another group of species ranged through the underlying strata

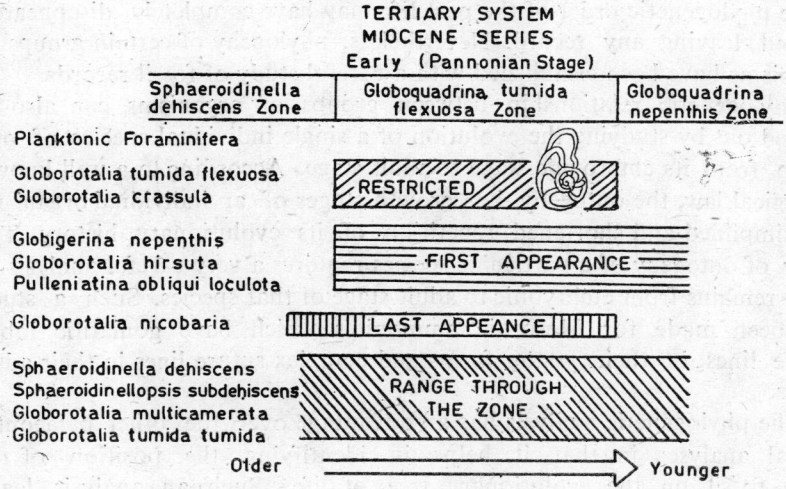

Fig 2.4: A biostratigraphic zone defined by assemblage of planktonic foraminifera (based on Srinivasan and Srivastava, 1975).

but disappeared during the time of deposition of the given unit. Some fossils appear and become extinct during the course of deposition of this unit (Index Fossil) and some ranged throughout the duration of the deposition of this unit.

Any change in the assemblage of fossils over a certain period of geological time depends primarily on the external conditions. Certain groups of organisms readily adapt to the changed physical conditions others may undergo certain changes in their external morphology. Such changes in morphological characters of certain species of fossils across a stratigraphical section has been successfully used for stratigraphic classification and correlation.

For the classification of the Cainozoic rocks of West Europe, Charles Lyell analysed the relative proportion of vorious groups of fossils in the total assemblage preserved in rock successions. Many species of living mulluscs are also known from older formations as fossils. However, their proportion in the fossil assemblage shows a gradual increase in younger formations. Thus Eocene (Series) contains less than 3.5%, Miocene up to 17%, Lower Pliocene up to 35% to 50% and Upper Pliocene up to 95% fossil forms of the living mulluscs in their fossil assemblages.

Phylogeny: Phylogeny of any group of organism describes the duration of its life and pedigree. The duration of existence of any form of life on the surface of the earth can be established on the basis of their fossil records in the rock formations. However, to establish the predecessors and successors of organisms, a careful biological study is required which is difficult to carry out from fossil records. The biological basis for establishing the pedigree of any organic form takes into account the total physiology and not only the hard parts preserved as fossils. In some cases, links in the phylogenetic order of the past life may have completely disappeared without leaving any record. Nevertheless, phylogeny of certain groups of organisms have been established with a careful study of fossil records.

Phylogenetic relationship between groups of organisms can also be worked out by studying the evolution of a single individual species of one group from its embroynic stage to adult stage. According to a well known biological law, the ontogeny, i.e., growth stages of an individual organism, is a simplified and shortened repetition of its evolutionary history. The study of ontogeny of a particular species requires a very careful analysis of fossil remains from embryonic to adult stage of that species. Such a study has been made for Mesozoic ammonites which have goniatitic lobed suture lines in their young forms and complex suture lines in their adult forms.

The phylogenetic analysis has an advantage over the other palaeontological analyses in that it helps in identifying the position of the given fossil on the evolutionary tree of life. Such an analysis leads in establishing the biostratigraphy of the area. The phylogenetic method of analysis of fossil records has a greater degree of flexibility even in the cases of correlation of rock formations situated at far distances from one another.

Micropalaeontology: Study of fossil records of micro-organisms for stratigraphical classification and correlation were begun over half a century ago. Micro-organisms, such as, foraminifera, radiolaria, ostrocods, diatoms, pollens and spores, etc., require special methods of collection and analysis. Because of these special techniques, micropalaeontology is considered as an independent science which has a wide application especially in the exploration of petroleum and natural gas. The micro-fossils, because of their small sizes, can be collected even from the cores of the bore holes. A large number of microfossils can be collected from a relatively small number of rock samples thus giving a much wider spectrum of data. These microfossils are identified in the laboratory with the help of stereomicroscopes, petrological microscopes and in special cases electrons can microscopes.

Foraminifera are the most important microfauna for classification and correlation of marine formations. These organisms have a wide geographical distribution and are present in large numbers in many marine rocks. Ostrocods (crustacea) are found both in marine and continental formations and they are known to exist since Palaeozoic till today. They have been found quite useful for stratigraphic purpose because of their wide distribution and much longer range in geological time.

Spores and pollens are better preserved because of their hard resistent parts. They appear annually in large numbers and they tend to spread over large area on continents, lagoons and marine waters due to their agility to flying and floating. However, their application in stratigraphy has many limitations. In spite of a large amount of work on the study of pollens and spores from different parts of the world, sometimes it is difficult to identify them even at the generic level. In some cases, the spores and pollens of later geological periods creep into an already deposited sedimentary rocks through cracks and other opening in rocks. These materials get mixed up with the *in situ* fossils. Since the preservation of earlier spores and pollens is as good as the ones which were introduced much later, it becomes difficult to isolate the contaminated forms.

STANDARD STRATIGRAPHIC AND TIME SCALES

The concept of time is realised in an uninterrupted flow of events and phenomena. The **geological time**, encompassing a total span of over 4.6 billion years since the earth was born, is represented by past geological events and phenomena. These events and phenomena are recorded in rock sequences as the pages of the Earth's history. A certain amount of time is involved in deposition of any sequence of sedimentary rocks. During this time interval the organisms also undergo evolutionary changes.

The rock sequence and its fossil contents are the records of that duration on the geological time scale.

The rock sequences which were deposited during a certain duration of geological time and which contain characteristic fossils constitute **Chronostratigraphic Units**. Thus, each chronostratigraphic unit represents a certain course in the Earth's history and for each chronostratigraphic unit there is a corresponding **Time Unit**. The chronostratigraphic units and time units have separate sets of terminology (Table 2.1). An erathum (chronostratigraphic unit) is divided into systems, system into series, series into stages and stage into zones. Likewise an Era (time unit) is divided into periods, period into epochs, epoch into ages and age into phases.

Table 2.1 : Terminology of chronostratigraphic and time units

Chronostratigraphic Units	Example	Time Units	Example
Erathem	....	Era	Cenozoic Era
System	Tertiary System	Period	Tertiary Period
Series	Miocene Series	Epoch	Miocene Epoch
Stage	Panonian Stage	Age	Panonian Age
Zone	*Globorotalia tumuda flexuosa* Zone	Phase	*Globorotalia tumuda flexuosa* Phase

Successive arrangements of stratigraphic and time units as standard reference for world-wide correlation are known as **Standard Stratigraphic Scale** and **Geological Time Scale**. These scales are based on past geological processes. Each division in the standard scale represents a natural stage in the earth's history and the evolution of life. Passages from one stage of evolution to another define the natural boundaries between the subdivisions of the standard stratigraphic and time scales. Such passages in the history of the earth are often recorded by a number of crises, such as, orogeny, extinction of a particular group of organisms, sudden climatic changes, etc. These crises in the history of the earth are not likely to be recorded everywhere on the surface of the earth. Thus, only some of the standard stratigraphical units are recognised in an area or a region.

The wider stratigraphical units, i.e., Erathems, Systems and Series, are readily recognised in most parts of the world and hence a single set of nomenclatures for these units is used by the stratigraphers all over the world. However, many of the stages are difficult to trace on all continents. Zones are established only for a particular biogeographical region. There are only a few zones of global significance, e.g., Graptolite zones for Ordovician and Silurian systems and zones of Palaeogene and Neogene based on characteristic planktonic foraminifera.

An **Erathem**, represented by a succession of rock formations deposited during an **Era** is a record of the broadest division in the earth's history.

Erathem boundaries are characterised by the boundaries of most pronounced discontinuity in the earth's history. An erathem is generally made up of three or more systems.

Palaeontology has been the most important basis for determining the relative ages of rock sequences. Thus all the rock successions of the earth were grouped into five broad chronostratigraphic units with reference to the type of life that existed on the surface of the earth (Fig. 2.5). The **Palaeozoic, Mesozoic** and **Cainozoic** successions contain remains of past life because the organisms living during these eras had developed hard parts. The group of these eras is referred to as **Phanerozoic** denoting the presence of forms of life with preservable hard parts. The primitive life of **Archaeozoic** and **Proterozoic** eras did not contain hard parts and therefore the rocks deposited during this time are generally unfossiliferous. These sequences were earlier referred to as **Azoic** denoting absence of life. The classification of the Phanerozoic successions is based on sound palaeontological data. Such subdivisions have been assigned suitable names for international correlation. The classification of Azoic rocks is not yet based on reliable systems of classification. Subdivisions of the Azoic successions have generally local or regional significance.

The names of five erathems have been derived from Greek words. Thus, 'Archeos' means oldest, 'Proteros' older, 'Paleo' early, 'Mesoz' intermediate, 'Cainos' new and the endings of each name, 'Zoicos' means life. The Archaeozoic and Proterozoic Eras are also combined into one unit, named as **Precambrian Eon**. The Precambrian Eon represents that part of the geological

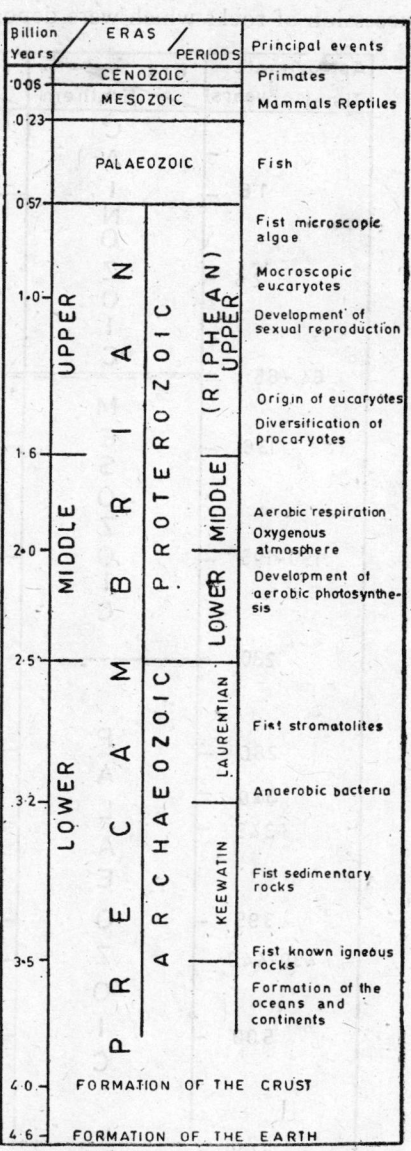

Fig. 2.5: Major divisions of Geological Time Scale and principal events.

20 *Elements of Historical Geology*

time which is older than the Cambrian Period, the oldest period of the Palaeozoic Era. The Precambrian Era is divided into three units, the Lower corresponds to the Archaeozoic, the Middle and the Upper to Proterozoic (Fig. 2.5).

A **System** constitutes a part of an erathem and it is represented by a succession of rocks which were deposited during a **Period**. In Phanerozoic

Approximate Time(10^6 years)	Era / Erathem	Period / System		Epochs / Series
	C A I N O Z O I C	Quaternary		Holocene (Recent)
1·8				Pleistocene
		Tertiary	Neogene	Pliocene
2·6				Miocene
			Palaeogene	Oligocene
				Eocene
64-65				Palaeocene
	M E S O Z O I C	Cretaceous		Late (Senonian)
136				Early (Neocomian)
		Jurassic		Late (Malm)
				Middle (Dogger)
190-195				Early (Lias)
		Triassic		Late
				Middle
230				Early
	P A L A E O Z O I C	Permian		Late
				Middle
280				Early
320		Carboniferous		Late (Pennsylvanian)
345				Early (Mississipain)
		Devonian		Late
				Middle
395				Early
430-440		Silurian		
		Ordovician		Late
500				Early
		Cambrian		Late
				Middle
570				Early
	PRECAMBRIAN			

Fig. 2.6: Standard Stratigraphic and Geological Time Scales of the Phanerozoic Eon.

rock formations, a system is characterised by a typical fossil assemblage. The successive systems are demarcated on the basis of substantial renewal of fauna and flora. The system boundaries are also often demarcated by sudden change in sedimentary facies or, in some cases, angular unconformity. A system is made up of two or three series of rock formations.

A large number of systems of Phanerozoic rock formations were established by the British and French geologists during the first half of the 19th century (between 1822 and 1841). The second International Geological Congress held at Bolon in 1881 approved names of ten systems, viz., Cambrian, Silurian, Devonian, Carboniferous, Permian, Triassic, Jurassic, Cretaceous, Tertiary and Quaternary. Later on the upper part of Cambrian and lower part of Silurian were grouped into an independent system named as Ordovician. The Tertiary Period has also been divided into Palaeogene and Neogene which are given independent status of period (Fig. 2.6).

The names of the most of the systems were derived from their type areas from where the rock successions of each system were first described. The 'Cambrian' is derived from the old name of Welsh province of Great Britain (Cambria), 'Devonian' from Devenshire in the southwest England, 'Permian' from a province of that name in Russia, 'Jurassic' from the Jura mountains of Switzerland. The names of 'Ordovician' and 'Silurian' were derived from the old tribes (Ordovics and Silurs) living in England. In case of Carboniferous and Cretaceous, the names were derived from their typical lithology in the type areas, coal (carbon) in the case of former and chalk (creta) in the case of the latter. The name of Triassic signifies its three-fold division in its type locality in Germany. Tertiary and Quaternary as well as Palaeogene and Neogene represent the stages of organic evolution.

A **Series** constitutes a part of the system which were deposited during an **Epoch**. In marine basins, these epochs often correspond to a duration of either marine transgression or regression. For Phanerozoic formations, a series is characterised by a typical assemblage of fossils (Subfamily, genus and species). Each series is made up of two or more stages of rock formations.

A series is sometimes assigned a name according to its relative position in the respective system. In a two-fold division of system, they are referred to as Lower and Upper and in three-fold division as Lower, Middle and Upper. The time units are accordingly known as Early and Late in two-fold division and Early, Middle and Late in three-fold division of the period. In case of Jurassic, Palaeogene, Neogene and Quaternary systems, series have been assigned special names (Fig. 2.6).

A **Stage** is a part of series which was deposited during a geological **Age**. Stages of the Phanerozoic formations are characterised by a typical assemblage of fossils (genus, sabgenus and species). These stages are established on the basis of organic remains of wide geographical distribution, e.g., foraminifera, brachiopods, graptolites, ammonites, etc. Each stage is established with the help of a detailed description of the lithotypes and fossil records in the type area and then assigned a name after its type area e.g., Oxfordian Stage of the Upper Jurassic (Malm) Series. In some cases, the stage names are based on ethnic nomenclature, e.g. Sarmatian Stage of the Miocene Series.

A **Zone** is the smallest unit of the stratigraphical scale. Its distribution is highly restricted to certain biogeographical areas. A zone is characterised by an assemblage of fossils referred to as 'Zonal Assemblage'. The constituents of the zonal assemblages of deposits formed during the same geological time may differ from area to area. However, their general character remains the same. A zone is assigned a name according to its most characteristic fossil form. This name is retained even for those areas which may not contain this typical fossil form although the zonal assemblages of the two areas are comparable. The name of the author of the species of fossil being used as a zonal name is not used, e.g., *Globorotalia tumida flexuosa* Zone named after a planktonic foraminifera, *Globorotalia tumida flexuosa* (Koch).

Litho-Stratigraphic Classification

Stratigraphic divisions of the western Europe were established earlier than for other countries. Most of these divisions were adopted for the standard stratigraphic scale. With the expansion of the geological investigations in Asia, Africa and America, it has been realised that the standard stratigraphic divisions cannot be defined exactly with the same boundaries because of the differing characters of the earth's evolution in its different segments. In many regions, larger stratigraphic divisions (Erathems and Systems) could be delineated but the smaller divisions (Series and Stages) were frequently difficult and often impossible to delineate on the suggested criteria. Moreover, for stratigraphic divisions of unfossiliferous strata, the standard stratigraphic scale cannot be readily used.

In view of these limitations of standard stratigraphic scale, a new scheme of classification of rock successions known as **Lithostratigraphic Classification** has been proposed. Such a scheme of classification is based primarily on the lithological criteria which are easily recognisable in the field. The lithostratigraphic units are assigned regional names and they are used for describing the geology, structure and economic resources of that region. Lithostratigraphic classification follows the basic principles of stratigraphic classification and correlation.

Group, the broadest unit of the lithostratigraphic classification, comprises a thick succession of rock formations spread over a large area. The successive groups are demarcated on the basis of well marked unconformities. A group is divided into more than one formations. Each group is assigned a name after its type area where all the subdivisions or most of the subdivisions are well exposed. An association of mutually related groups is known as **Supergroup**. Thus a supergroup and a group roughly corresponds to the Erathem and System of the standard stratigraphic scale respectively.

Formation, the fundamental unit of the lithostratigraphical classification, is a mappable lithological unit on an intermediate scale (1 : 50,000) of geological mapping. It is represented by a rock succession deposited under a relatively uniform physicochemical condition. A formation may comprise a predominant rock type or an association of two or more rock types. Successive formations are separated from one another by a change in lithology which may be gradual (transitional) or sudden (sharp). The formational boundaries are, in some cases, characterised by a break in sedimentation. A formation, roughly corresponding to a series of the standard scale, is assigned a name after its type area (e.g., Kulikhani Formation) or after the name of the type area and the dominent lithology of the formation (e.g., Dhading Dolomite).

A part of the formation, which has some remarkable lithology or some characteristic fossil assemblage, is demarcated as **Member**. A member is given a name on the same principle as a formation. It is also a mappable unit on an intermediate scale of geological mapping, however, a member has lesser vertical thickness and lateral extent as compared to a formation. A formation and a member can be further subdivided into **Beds** of still smaller thickness and lateral extent. The thickness of a bed is generally too small and, therefore, it cannot be shown on geological map of intermediate scale.

Sometimes, the term **Complex** may be used for an association of groups or formations for which it is difficult to assign any rank of formal stratigraphical classification. A complex includes a number of groups or formations without any associational relationship. Boundaries between successive complexes is generally characterised by tectonic dislocations.

ABSOLUTE AGE OF ROCKS

Absolute ages of the earth and its rock formations have been determined by various methods since the beginning of the eighteenth century. These methods relied on, for example, the rate of sedimentation, rate of increase in the salinity of sea and rate of biological evolution. These methods, however, remained unsatisfactory because of the fact that these natural processes do not have a constant rate throughout the earth's history. A reliable method of determining absolute age of rocks, developed at the beginning of the 20th century, is based on the principle of radioactive decay. The radiometric dating of rocks and its interpretation in terms of geological phenomena have since been evolved into an independent science known as **Geochronology**.

The radioactive elements constantly undergo a process of decay giving rise to the formation of their stable isotopes known as daughter elements. The process of decay does not depend on temperature, pressure or any other physical parameters. For each radioactive element, its half life period is defined as the time taken for the decay of half the quantity of that element. The half life period for each radioactive element can be determined experimentally in the laboratory.

The radiometric dating of rocks is carried out with the help of minerals containing radioactive elements. From the moment these minerals are formed, the radioactive elements entrapped in the mineral begins to decay. The daughter element of this element gets accumulated within the mineral. At any given time after the formation of the mineral, the quantities of the radioactive element (mother element) and the end product of decay (daughter element) can be determined in laboratory with the help of sophisticated instruments. Since the half life period of that radioactive element is known, the time taken for the formation of the daughter element can be calculated with the above data. This time corresponds to the formation of the given mineral and hence the rock.

Radioactive elements with long period of half life (in millions and billions of years) are used in geochronology. The most commonly used mother elements are Uranium, Thorium, Potassium and Rubidium. Carbon is used for determining the age of quaternary rock formations or rocks and minerals of archaeological interest. Half life period of these elements and their decay products are given in table 2.2. The half life period of some of the radioactive elements is very large, e.g., In^{115} (10^{21} years), Te^{130} (10^{21} years) and Sm^{147} (10^{16} years). These elements could be very useful for determining the absolute ages of very old rocks. However, these elements occur in very small quantities in rocks. Laboratory techniques for determining such small quantities are very complex.

Table 2.2: Half-life periods and daughter elements of some widely used radioactive elements in geochronology.

Mother Element	Daughter Element	Half-Life Period
U^{238}	Pb^{206}—8 He^4	4.468×10^9 years
U^{235}	Pb^{207}—7 He^4	0.704×10^9 years
Th^{232}	Pb^{208}—6 He^4	14.01×10^9 years
Rb^{87}	Sr^{87}	50×10^9 years
K^{40}	Ar^{40}, Ca^{40}	11.9×10^9 years
C^{14}	N^{14}	5570 years

Uranium and Thorium decay to form stable lead isotopes with the expulsion of alpha particles (helium). Minerals containing more than 1% U and Th, such as uranite, monzonite, zircon and allinite are used for isochron dating. Zircon dates often refer to the time of formation of rocks since the zircon is quite resistant to any subsequent change in rocks.

The Potassium-Argon method is a very versatile method applicable to many igneous and metamorphic rocks and sedimentary rocks containing glauconite, illite, etc. Potash felspars and mica are the suitable minerals for the determination of ratios of the mother and the daughter elements. The whole rock, low in potassium, can also be used for radiometric dating. Sometimes, low apparent ages may be obtained due to loss of argon during a thermal event subsequent to the formation of the rock. Anomalously high ages may also result due to accumulation of argon in rocks from other sources.

Rubidium-strontium method is also very widely used because of common occurrence of rubidium as trace element in alkali felspars and mica. The Radio-carbon method is used for determination of the absolute ages of rocks and minerals which were formed less than about 60,000 years ago. The isotope ratios are calculated from the plant remains enclosed in young sedimentary deposits.

Although geochronology as a science has highly developed during the last few decades yet the techniques for determining the ratios of mother and daughter elements are not foolproof. Even an error of one per cent in case of rocks formed more than one billion years ago can lead to an error of over 10 million years. The radioactive methods of dating the age of rocks are still complex and expensive which prohibit their wide application. It is also quite probable that the minerals used for the radiometric dating may have recrystallised during metamorphism of rocks thus destroying the daughter elements. The clock is then said to have been re-set. The dates arrived at in such cases record the age of metamorphic event and not the time of formation of the rock.

The radiometric methods of dating the rocks have become indispensable tool in Precambrian Stratigraphy. The Precambrian rocks in most cases were affected by repeated high grade metamorphism during the course of geological history. The ages determined by many of the existing methods give the time of last strong transformation of rocks. Some elements, such as samarium (Sm) and its daughter element niodymium (Nd), however, are not affected by the metamorphic processes. Thus, the Sm-Nd method of radiometric dating helps in observing the earliest geological events through the veil of the later events.

The K-Ar method usually gives the period of time when the pressure and temperature of deeper parts of the earth's crust cease to influence the transformation of rocks by uplifting the crustal block above a critical level. The K-Ar method, thus, roughly shows the time of folding, metamorphism and plutonism (endogenic geological process). In a few cases, it is possible to determine the time of sedimentation (exogenic geological process) in cases of slightly altered or unaltered rocks. Glauconite, a syngentic mineral of sedimentary origin, is commonly used in K-Ar and Rb-Sr methods of dating of sedimentary rocks. The age of unaltered carbonate rocks can be determined by lead isotope method.

Chapter 3
Principles of Palaeogeography and Palaeotectonics

Physical geography deals with the relief of the earth's surface, climate and distribution of organisms. The physiographical conditions, which existed on the earth's surface in the geological past are dealt with in **palaeogeography**. Whereas a landscape of our time is a part of the earth's surface where physical environment is on an average constant, the palaeogeography reconstructs past landscapes through various intervals of geological time. Palaeogeographic reconstructions are based on direct analogy of the contemporary dynamic processes with those of the geological past. Distribution of various types of sediments and organisms on the earth's surface are carefully studied in their association with the existing relief and climate. Similar association of sedimentary rocks and remains of organic life of various stratigraphic units is the evidence of past relief and past climates.

Chief types of sediments, which are deposited under different conditions of relief and climate, are grouped into glacial, fluvial, lacustrine, paludal, eolian, lagoonal and marine. They are differentiated from one another on the basis of their textural, structural and mineralogical characters as well as associational characters of various subtypes of each group of sediments. Distribution of these genetic types of sediments in rock successions of various geological periods reflects the characters of landscape of that time.

Lateral variation of one genetic type of sediment into another on the surface of the earth is known as facies variation. Such facies variations are recorded in the rock successions which were deposited during the same geological period. A litho-facies is defined on the basis of its petrographical and associational features, and a characteristic fossil assemblage. Facies analysis is a methodical study of variations in lithofacies of the same age. Such an analysis helps in reconstruction of the past landscapes and at the same time elucidates the genesis of the related deposits. For example, a clay bed which is devoid of any fossil is laterally replaced in all directions by shelly limestone of shallow marine conditions of deposition. From such an association, it can be safely inferred that the clay bed was also deposited under shallow marine conditions.

CONTEMPORARY MARINE BASINS

The part of earth's surface covered under marine water is divided into four oceans, namely, the Pacific, the Atlantic, the Indian and the Arctic. These areas are underlain by oceanic crust which differs in structure and composition from the continental crust. Parts of continents are also submerged under marine water which constitute various seas of the world. Some of them are open to oceans (e.g., Bay of Bengal, Arabian Sea, etc.), others are separated from the oceans by a chain of islands (e.g., Japan sea, Bering sea, etc.) and still others are either completely cut off or connected by very narrow straits to the oceans (e.g., Mediterranean sea, Black sea, Baltic sea, etc.). Some inland large lacustrine areas have increased salinity of their water and they are also known as seas, e.g., Caspian sea and Arab sea.

The part of continents under marine water is known as **Continental Shelf** (Fig 3.1). It is gently sloping (slope angle less than 1°) and reaches up to a depth of 100 to 200 m from the mean sea level. The slope abruptly becomes higher (up to 7°) on the outermost margin of continents under the marine water. This part of the continent is known as **Continental Slope**. The depth of the oceanic bottom in the abyssal sea ranges from 2000 to 3000 m, occasionally reaching up to 6000 m. The ocean floor is dissected by deep trenches where the depth is more than 6000 m and sometimes reaches up to 10000 m. Such trenches are characteristic for the western Pacific on the outer side of the island arc system. The ocean floor is also characterised by mid-oceanic ridges which often emerge out of water in the form of a series of islands.

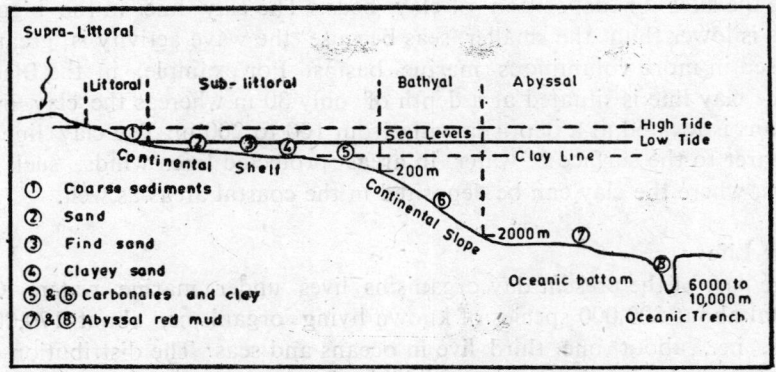

Fig. 3.1: A profile (idealised) of marine basin showing bionomical zones and distribution of characteristic sediment groups.

The average salinity of marine water is 35 ppt (parts per thousand), whereas the average salinity of fresh water is less than 0.5 ppt. The salinity of surface marine water in the equatorial and circum polar zones ranges from 33 to 34 ppt. The salinity of tropical marine areas reaches up to 38 ppt.

Some inland seas, e.g., Black Sea, have reduced salinity (18 ppt.) while others have enriched salinity, e.g., Caspian Sea (41 ppt.). The salts of marine water contain, on average, chlorides 88.7%, sulphates 10.8%, carbonates 0.3% and other salts 0.2%.

Marine Sediments

The chief source of sediments of marine basins is the continental area raised above the mean sea level. Rivers are the most important agents of erosion, transportation and discharge of a great quantity of clastic and dissolved sediments into the marine basins. For example, the average annual discharge of sediments by the Mississippi river is estimated to be 600 km^3, Amazon river 3787 km^3, Congo river 1260 km^3 and Volga river 225 km^3. The other agents for the supply of sediments from the land areas to ocean basins are glaciers, winds and erosional activity of sea waves in the coastal regions. On an average, the seas and the oceans of the world receive about 12.5 billion tons of clastic sediments and about 5 billion tons of dissolved sediments annually. Submarine volcanism further adds to the total bulk of deposits of marine basins.

In a generalised profile of marine basins (Fig. 3.1) the coarser sediments are laid down nearer to the coast and the finer sediments are dispersed far away from it. The finest clastic sediments are deposited in quiet water of abyssal ocean along with calcareous and siliceous organic sediments where the waves cannot effect displacing the sediments. The lower boundary up to which the waves are capable of keeping the finest sediments in suspension is known as **Clay Line** (a line in the profile but a surface in three dimension). Only the oceanic bottoms which are deeper than the clay line can be the sites of deposition of clay beds. The clay line in the bigger oceans is lower than the smaller seas because the wave activity is greatly increased in more voluminous marine basins. For example, in the Black Sea, the clay line is situated at a depth of only 30 m whereas the clay line in oceans is located at a depth ranging from 100 to 200 m. The clay line is still nearer to the surface of water in areas protected from winds, such as in bays, where the clay can be deposited in the coastal areas as well.

Marine Life

A large part of the present day organisms lives under marine waters. Of the total about 500,000 species of known living organisms, about 170,000 species, i.e., about one third live in oceans and seas. The distribution of life in marine basins is, however, highly irregular depending upon the physical and chemical conditions such as sun light, temperature, salinity, gaseous regimes and mobility of the medium. In warm waters, the marine life is most varied. For example, in Mexico Bay, about 1500 species are known whereas in Barnets Sea only 200 species are known to live. A rich community of life thrives up to shallow depths of the continental shelf and upper parts of continental slope as in this region the oxygen supply is plentiful. Any change in the average salinity of marine basins leads to impoverishment of organic life.

The aquatic plants, which have to depend upon the sun light for their growth, thrive up to an average depth of 200 m. The plants are also the main source of food for the marine fauna. This explains as to why the most of the organisms of sea are known to occur up to a depth of 150 m. At greater depths, anaerobic conditions prevail due to complete depletion of oxygen and non-availability of the sun rays. Under these conditions, only those forms of life are known to survive which depend upon sulphides and sulphates for their existence.

A group of organisms which has adapted a given set of physico-chemical conditions is referred to as **biocoenose** and the place of their habitat is known as **biotope**. A number of bionomical zones are demarcated on different parts of the marine basins depending upon the character of their biocoenoses. The marine fauna is classed into a number of groups depending upon their habits. The biocoenosis of each bionomical zone includes a combination of marine forms having differing habits.

The forms which live on the marine floors constitute **benthonic** fauna. Among the benthonic forms, three subgroups are identified; the **sessile** forms which are fixed to the seabottom, the **vagrant** forms which are locomotive type, and the forms having **burrowing** habit. There is another group of organisms which thrives near the surface of water. They are known as **pelagic** fauna. The pelagic forms, which are passive and which float in or near the surface of water, are known as **planktons**; and active swimmers are known as **nektons**.

The benthonic fauna include many inverteberates, sedentary marine plants and some fishes. These forms tend to be thickly populated in shallow marine basins where, with other favourable conditions, the density of population may reach up to several kilograms of forms per square meter. The density of benthonic forms is greatly diminished at greater depths to about 50 gm per square meter at a depth of 1000 m. Benthonic forms of life constitutes only about 1/2 of the number of nektonic and about 1/4 of the number of planktonic forms. The nektons include fishes, some crustacea, marine mammals, reptiles, cephalopods, etc. The planktons include all unicellular plants (diatoms, dinoflagellates and algae), protozoans (foraminifers, radiolarians, etc.), fine ratchets, jelly fishes, some worms and larva of many inverteberates.

Bionomical Zones

Distribution of benthonic forms of organisms in relation to depths of marine basins enables to distinguish four bionomical zones, viz., littoral, neritic, bathyal and abyssal (Fig. 3.1). The lagoons and brackish water lakes of the coastal regions are distinguished as independent environment having partly marine and partly terrestrial conditions.

Littoral zone is a part of the continental shelf between the marks of high and low tide. The marine organisms living in this zone are adapted to repeated withdrawal of marine water during low tide period. The composition of the biocoenosis of the zone depends upon the type of soil in the littoral zone. Rocky substratum supports numerous sedentary shells of crustacea, such as barnacles, oysters, some bryozoans and boring mulluscs. Burrowing bivalves, gastropods, worms, decapod cray fishes, crabs and colonies of mussels thrive on sandy soil. The softer soil also supports the growth of aquatic plants.

Sediments of the littoral zone normally include beach gravels and beach sands. Finer sediments are often laid in the bay regions which are protected from the waves. Along the steeper shores, coarse sediments, often of the boulder sizes or even bigger rock blocks, are deposited along with the fine sediments. The littoral sediments are repeatedly eroded and deposited due to advancing and receding shore lines and the wave activity. These sediments are, therefore, well sorted and show sedimentary structures such as cross-bedding and ripple marks. These sediments contain the remains of organic forms, both which inhabited this zone and those which have been transported from inland as well as from oceanic areas. Sometimes, the remains of the organic forms of other regions exceed in number than those forms which inhabited the littoral zone.

Neritic zone, which is also known as **sub-littoral zone**, reaches up to 200 m depth on the continental shelf. This zone is characterised by plentiful supply of oxygen and sun rays. The most varied marine life is observed up to a depth of about 40 to 80 m. Even the plants are most abundant giving rise to what may be referred to as "submarine meadows". The deeper parts of the neritic zone become darker as the sun rays are not able to penetrate through greater depths. As a result, the population of the organic life decreases rapidly towards the outer parts of the neritic zone. The organic life of this zone include colonies of bivalves, mulluscs, brachiopods, corals, calcareous sponges, marine dactyles, bigger foraminifera, bryozoa, etc.

Sediments of the neritic zone are well sorted and they range in grains size from sand to fine clays. Ferrugenous sand and shelly detrital oolites are the most predominant sediments. Clay and silt are characteristic of upper parts of neritic zone which also contains, sometimes, glauconitic sands. The outermost parts of the zone often contains phosphoritic sediments.

Bathyal zone corresponds to the continental slope of the ocean bottoms. The depth of the water ranges from 200 m to about 3000 m. This zone is characterised by a temperature of about 10° C, almost complete absence of current activity and weak sun light. Only blue rays of the light spectrum reaches up to these great depths. The benthonic forms, which is greatly reduced as compared to the neritic zone, include some bigger barnacles, ostocorallia, marine dactyles, etc.

Sediments of the bathyal zone are mainly composed of clays and marls.

The coarser grades of sediments (sand and higher grades) normally constitute less than about 30% of the total sediments. The sand grade sediments are known from depths up to 1000-2500 m. Glauconitic and phosphoritic sediments are commonly known from depths of 1000 to 1500 m. The sediments of bathyal zone contain some remains of organisms foreign to this biotope.

Abyssal zone is an area of complete darkness with a very low temperature of about 1° to 5° C. Relief of the ocean bottom is very complex characterised by deep oceanic trenches and mid-oceanic ridges. The life is almost non-existent except for some anaerobic bacteria and similar organisms which receive their energy either from reducing reactions of sulphides and sulphates or from decomposition of protein matter of organic remains which fall on the bottom of abyssal oceans.

Sediments of abyssal zone are extremely fine grained comprising mainly clay and calcareous and siliceous oozes. **Red clay** is a typical very fine grained sediment of Pacific ocean found at depths of about 4000 m. The clay is coloured by manganese and ferric compounds and is possibly derived from the continental areas. The rate of sedimentation in abyssal zone is extremely slow which suggests that the sedimentary successions of a particular geological period deposited in abyssal zone would normally be much thinner as compared to the successions of the same period deposited in other regions. Such rock successions with reduced thicknesses are known as **condensed horizons**. Some parts of the abyssal zone are completely devoid of sediments while some are known to contain nodules of ferromagnesium compounds among the floor sediments.

Lagoons and **brackish water lakes** are partly or completely disconnected from open seas. The salinity in some cases is more and in others less than the normal salinity of the marine water. As a result of an increase in salinity, the organic life is greatly reduced. A decrease in salinity also reduces the population of marine forms. In addition to the population, the sizes of the organic forms also decrease in these basins of abnormal salinity. The forms which are susceptible to a change in salinity are known as **stenohaline** forms. The resistant forms, known as **euryhaline** forms, generally have luxurious growth in these basins. The sediments of these coastal basins are similar to those of littoral and sub-littoral zones (i.e., from gravels to fine clays). However, the environment is quieter due to almost nonexistent wave action. The sediments in this respect tend to resemble in their sorting and other textural characters with the sediments of the terrestrial lakes.

CONTEMPORARY CONTINENTAL BASINS

Land is primarily an area of erosion and denudation. The processes of sedimentation are confined to only certain depressions in the relief of the continents, such as lakes, river basin and flood plains. These sediments rarely preserve organic remains because of their intermittent erosion and redeposition. The continental sediments are highly variable and they are grouped into a number of genetic types, viz., elluvial, deluvial, proluvial, alluvial, paludal, lacustrine, gravitational (landslide, creep and scree), glacial and fluvio-glacial, eolian and cave deposits. Each genetic type is formed by a set of dynamic processes characterised by climatic and relief conditions. A careful mapping of these genetic types in stratigraphic successions of various geological periods helps in identifying palaeo-climates and palaeo-relief on the surface of the earth during the geological past. Some typical associations of these genetic types are briefly described below.

Glacial zone comprises glacial (moraines), fluvioglacial and glacio-lacustrine sediments. The glaciated regions show structures such as slickensides and striations which are formed due to the abrasive action of the glacier movements. These structures are also seen on the pebbles and boulders transported by the glaciers. The typical glacial moraines contain unsorted sediments ranging in grain size from large boulders to fine clays. Varved clay beds, which show characteristic fine laminations, are also characteristic of deposits of glacial zones.

Humid plains are characterised by sediments classed as alluvial, deluvial, eluvial, peludal and lacustrine. These sediments are laid in river valleys, flood plains, swamps and lakes of humid regions. In temperate regions, eluvium, a product of mechanical weathering, predominates. These sediments are generally devoid of any plant remains. The tropical and subtropical regions contain sediments of predominantly chemical weathering. These regions are characterised by a zone of weathering which reaches up to a depth of 100 to 120 m from the surface of the earth. Such chemical weathering leads to the formation of laterites, bauxites, kaolinite and oolitic iron ores at the sites of weathering. The detritus is often completely decomposed. The sediments contain pure quartz sand and ores of manganese and iron. Peat is formed from decomposition of carbonaceous matter from a luxurious plant growth in the humid plains of tropical and subtropical regions which gives rise to the formation of typical coal bearing facies.

Arid plains consist of eolian and proluvial sediments which give characteristic landforms to arid plains, such as, sand dunes, alluvial fans and proluvial floats. Variegated and red coloured clastic sediments with evaporites such as halides and sulphates constitute the typical sediments of the arid plains. Gravels and pebbles of this region are known to carry imprints of "desert varnish". Vegetation is rare. Loess, a yellow white fine grained loam containing quartz, felspar, calcite and other minerals, is a characteristic sediment of the peripheral parts of deserts especially in the vicinity of glaciated regions.

Foredeep and **intermontane basin** are sedimentary traps formed in front of and within a rising chain of mountains. Characteristic sediments of these basins include alluvial, proluvial, gravitational, lacustrine and fluvio-glacial genetic types. The detrital sediments, which are derived from the weathering and erosion of the rising mountains, are sorted and laid in these basins. Coarse rock fragments are deposited in the neighbourhood of the mountains, whereas the finer sediments are laid away from the mountains. The source area of the sediments. i.e., the high mountains, are characterised by a vertical climatic zonation which is reflected in the sediments representing humid to dry and warm to cold areas of erosion. The rate of accumulation of sediments in these basins is generally large and often it has a sposmadic character reflecting the various phases of the upheaval of the mountains. As a result of tectonically active region of deposition, the sedimentary facies of the foredeep and intermontaine basins quite often show lateral and vertical facies variation typical of what is known as **molassic deposits**.

Volcanic areas are characterised by the accumulation of magmatic lava and pyroclastic materials which are often spread over the earth's surface around the volcanic centres giving rise to typical volcanic relief. Volcanic cones are formed around the volcanoes which primarily eject lava of acidic composition having higher viscosities than those of basic lavas. The latter spreads over much wider regions giving rise to flat or table land topography typical of the plateau flood basalts. Deccan plateau of western India is a typical example of terrestrial basic volcanism.

PALAEOGEOGRAPHIC RECONSTRUCTIONS

Palaeogeographic reconstructions are based on the facies analysis of the stratigraphic successions of the area. A facies is represented by a typical association of rock types and, for Phanerozoic rock formations, by a typical assemblage of fossils. A facies at the same time represents an environment of deposition of layered rocks. However, the environment in which the organisms accumulate after their death need not always coincide with the environment in which they lived. For correlation and determination of relative ages, the total fossil assemblage of rocks are taken into consideration. However, for palaeogeographic reconstructions, only a part of fossil assemblage is considered which represents organisms inhabitating the basin of deposition.

The total assemblage of fossilised organisms, which is found in any stratigraphic unit, is referred to as **thanatocoenose**. In a bionomic analysis, **biocoenosis** of the basin of deposition is worked out from the thanatocenosis on the basis of certain known empirical biological characters. For example, the sessile and burrowing organisms are in most cases fossilised at the places of their habitat. The litho facies of the rocks enclosing the fossil forms also help in determining whether the organisms lived in those basins or they were transported from shallower or deeper water areas.

34 Elements of Historical Geology

The mineral composition of the detritus is a record of the composition of rocks of source area (provenance). Degree of chemical decomposition of detritus indicate the type of weathering which helps in suggesting the climate operating in the source area. The provenance of detritus is often established on the basis of **palaeocurrent** analysis with the help of sedimentary structures such as current bedding and current ripple marks.

The non-detrital components (chemical precipitates) of sediments, on the other hand, indicate the physio-chemical conditions of the area of deposition. Abundant calcium carbonate (limestone) indicates warmer climate and alkaline conditions. Ratio of ferrous and ferric compounds help in estimating the oxidising or reducing environments of the basins. Rock salt, gypsum and red coloured rocks indicate dry environments. On the other hand, coal, kaolin, laterites and bauxite, etc., indicate humid climate. Glauconite and phosphorite are formed in neritic areas of normal salinity. Abundance of pyrite and marcasite indicates hydrogen-sulphide contamination of the basin which often can be traced to decay of organic matter under anaerobic conditions. Presence of volcanic components indicate a volcanic phase in the neighbourhood of the basin.

Dynamic environment of the basin is inferred on the basis of textures and structures of sedimentary rocks. Thin parallel stratification indicates absence of waves at the site of deposition. Different types of cross stratifications are characteristic of littoral, eolian, fluvial and deltaic deposits. The inclination angles and directions of stratifications enable us to infer the current directions (palaeocurrents) or palaeowind directions. Grain sizes and shapes are used to infer the proximity of erosional areas to the basins of deposition.

Palaeogeographic Maps

The palaeogeographic analysis are summarised and graphically represented by a set of palaeographic maps. A palaeogeographic map shows the broad characters of physiography and climates which existed on the surface of the earth at a particular stage of the earth's evolution. A palaeogeographic map is prepared for an interval of geological time for which average physiographic and climatic conditions are inferred on the basis of facies and bionomical analyses of stratigraphic unit deposited during that time interval. For each set of palaeogeographic maps of a given region, the time intervals are chosen depending upon the details of the litho-and biostratigraphic observations from that region. Palaeogeographic maps covering regions of continental sizes are prepared for different period of the standard time scale. For detailed basin analysis of an area or region of any continent, epochs and ages are normally the time interval chosen for palaeogeographic maps (Fig. 3.2).

Preparation of palaeogeographic maps are based on field and laboratory data. Four successive stages in their preparation may be identified. First, the synchronous stratigraphic complexes of the region are delineated with detailed descriptions of their lithological and palaeontological characters. These data give the bases of litho-facies maps of specific time intervals for the region. At the second stage, the litho-facies are traced and extrapolated for those areas for which data are not available either due to cover of younger formations or due to erosion of the given stratigraphic unit.

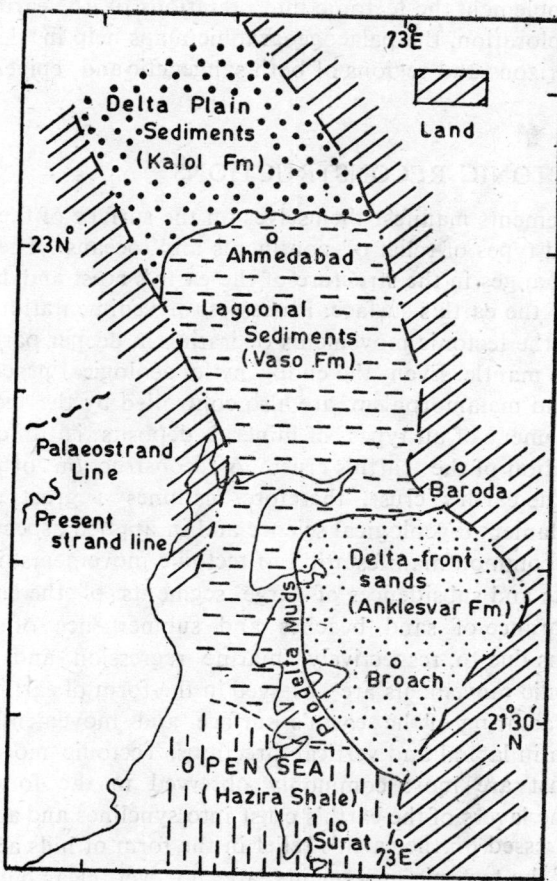

Fig. 3.2: Palaeogeographic and facies map of the Cambay Basion for the Eocene Epoch (based on Sudhakar and Basu, 1973).

The palaeo-relief and the palaeo-climate for the given time interval are inferred on the bases of the areal distribution of litho-facies and fossil records at the third stage. Lastly, with suitable cartographic expressions, major elements of the landscape which existed on the surface of the earth during that time interval are shown on an areal map of the region. On this map, the present geographical elements are also shown for reference of outcrops.

Palaeogeographic maps have great significance both for fundamental research in geology as well as for exploration of mineral deposits. A set of palaeogeographic maps for successive intervals of geological time helps in inferring the past geological processes and their migration through various times on different parts of the earth's surface. So far as the surface processes and relief are the expressions of the processes of the earth's interior, these maps supplement the tectonic interpretations of the earth's evolution. In mineral exploration, the palaeogeographic maps help in delineating the favourable horizons and regions of both syngenetic and epigenetic mineral deposits.

PALAEOTECTONIC RECONSTRUCTIONS

Tectonic movements manifest themselves on the surface of the earth in the form of varied types of relief of continents and oceans. These movement bring about changes in the structure of the earth's crust and hence changes in the relief of the earth's surface. Processes of sedimentation are directly controlled by the tectonic movements operating in deeper parts of the crust and the upper mantle. Even the endogenetic geological processes, such as magmatism and metamorphism, are also controlled by the tectonics of the crust. Emplacement of all types of mineral deposits is influenced by the tectonic evolution of the earth's crust. A reconstruction of palaeotectonic evolution of the earth's crust, therefore, assumes a great significance in both the fundamental geological science and in applied aspects of geology.

The most common manifestation of tectonic movements is observed in gradual uplifts and subsidences of large segments of the earth's surface, such as appearance of sand beaches and submergence of some parts of coastal regions due to, respectively, marine regression and transgression. Quicker tectonic movements are observed in the form of earthquakes which often cause fracturing of the earth's surface and movements across these fractures both in lateral and vertical directions. Tectonic movements of the geological past are most commonly observed in the form of general warping of the layers of the earth's crust into synclines and anticlines which are often expressed on the earth's relief in the form of hills and valleys.

Causes of the tectonic movements are not yet clear but the observed results of the movements obviously depend on the depths of the processes with which they are related. Primarily two types of tectonic movements are distinguished. These are horizontal and vertical tectonic movements. Among the vertical movements, uplifts are considered positive whereas the subsidences are regarded as negative movements. Magnitude of the tectonic movements, which can be observed today on different parts of the earth's surface, is recorded experimentally in terms of few millimeters per year. Tectonic movements of the geological past are inferred with the help of deformed structures of the rock successions which form the subject-matter of Structural Geology and Tectonics.

On the basis of rates of the tectonic movements, their two broad divisions are recognised. **Epierogenic movements** are spread in time over a few hundred million years. On the other hand, **orogenic movements** take place in a shorter span of geological time, i.e., a few tens of million years. The orogenic movements are confined to relatively narrow and long tracts of the earth's surface which are known as **orogens**. Orogeny literally means the processes of mountain building. Orogenic movements are accompanied by intense deformation and regional metamorphism of rocks, igneous activity and emplacement of granitic plutons. The epierogenic movements, on the contrary, have a relatively quiet character.

Tectonic Analysis of Facies Maps
Since the earth's relief is a reflection of tectonic movements within the earth's interior, and since the earth's relief controls the character of the litho-facies of various stratigraphic units, the past tectonic movements can be inferred with the help of facies maps which show the distribution of various litho-facies deposited during a particular interval of geological time. Based on these interpretations, **palaeotectonic maps** are prepared which give tectonic zonations of the earth's surface for various intervals of geological time.

Regions with negative vertical movements are represented on the palaeotectonic maps by the basins of deposition. Within this negative region of first order, various sub-basins of second order are distinguished which are separated by relatively uplifted areas. Likewise, uplifted regions of first order are delineated into various second order positive areas which are separated by areas of relative subsidences. Sometimes, an area may undergo alternatively positive and negative vertical movements during that specific interval of geological time. Such movements are recorded in terms of shallow water and deeper water facies of sedimentary rocks. Rock succession of a particular stratigraphic unit often have uniform lithological characters which reflect stable tectonic environment during the period of their deposition. The stable tectonic environment is often inferred as a representative of epierogenic movements whereas the fluctuating tectonic environment are related with the orogenic movements.

Tectonic Analysis of Isopach Maps
Thickness of stratigraphic units vary in accordance with variable relief of the basin of deposition and varying rates of sedimentation. In areas of faster subsidence, greater thickness of sediments are laid down as compared to basins with slower rates of subsidence. Analysis of stratigraphic thicknesses is carried out with the help of **isopach maps** (Fig. 3.3) which are prepared by joining the points of equal thickness of a particular stratigraphic units on different parts of a sedimentary basins. Such lines are known as **isopach lines.**

The sedimentation rates in certain sedimentary basins are compensated with an equal amount of subsidence of the bottom of the basin. For example, deposition of say 100 m thick shallow water oyster limestone can take place only if the bottom of the basin had been subsiding at the same rate as the rate of sedimentation. Thus, this succession indicates a subsidence of the basin by 100 m during its deposition. Such sedimentary successions are known as **compensated sequences**.

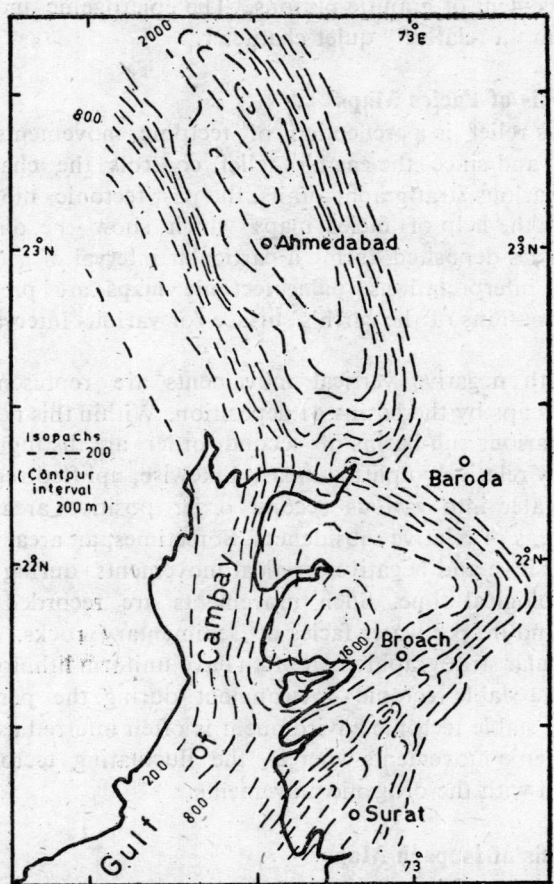

Fig. 3.3: Isopach map of Eocene rocks of the Cambay Basin (based on Mathur et al., 1968).

In another case, the subsidence rate may exceed the rate of deposition. A succession with 5 m thick beach sandstone followed by 95 m thick shale with benthonic fauna indicates that the bottom of the basin during the deposition of beach sandstone was near to the surface of water while during the deposition of the succeeding shale, the basin bottom had subsided to a depth of about 200 m. Thus, a total 100 m thick beach sandstone-shale succession indicates the subsidence of the basin by more than 300 m. Such Rock sequences are known as **non-compensated sequences**.

Tectonic analysis of isopach maps is carried out with the help of facies map of the same stratigraphic units. The basin configuration of a stratigraphic unit can be represented by drawing geological cross sections. In compensated sequences, the upper horizontal line corresponds to a common datum level at the end of the period of deposition. Thickness of each lithofacies are shown from top to bottom. The lower irregular line, thus, shows the configuration of the bottom of the sedimentary basin.

The thicknesses in non-compensated sequences indicate variable magnitude of subsidence in different parts of the basin. The basin relief is inferred on the basis of the litho-facies. The thickness of each lithofacies are shown from the basin bottom line downwards. The lower line, in this case, is a record of form and magnitude of the subsidence of the basin during the period of deposition of the given stratigraphic unit.

Diastems and Angular Unconformities

Positive vertical movements are inferred from stratigraphic records of diastems and angular unconformities which are at the same time natural boundaries of litho-stratigraphic units. An unconformity is a record of break in sedimentational history of the area which is caused by an uplift of the area of deposition. A break in deposition of lesser time magnitude is represented by diastems whereas the angular unconformities normally represent a big hiatus in deposition.

In diastems, the overlying and underlying successions are laid with an apparent conformity, i.e., the bedding planes of the two successions are nearly parallel. Such breaks are characteristic of epierogenic movements. In angular unconformity, the bedding planes across the surface of unconformity are not in parallelism. Quite often, the underlying older successions are more deformed as compared to the overlying succession. Such angular unconformities indicate that the underlying successions were deformed during a phase of mountain building (orogeny), and that the mountains were later peneplained and subsided to become the site of deposition of the younger succession of rocks.

As one moves away from the region of vertical movements to a neighbouring region, it is observed that the magnitude of these movements gradually diminishes. A regional unconformity, thus, gradually paves way to a local unconformity and ultimately no unconformity at all. In these neighbouring regions, stratigraphical units, representing the time gap of deposition in another area, are represented by a succession of rocks. Such rock successions define structural stages in the evolution of the earth's crust of those regions.

Chapter 4

Earth's Crustal Structure and Tectonic Divisions of India

The earth's surface is divided into land areas and areas under the marine water. The oceanic areas constitute about 51.5% (i.e., 263 million square kilometres) of the total earth's surface. These two primary units have different geological and geomorphic features. A great variety of landscape exists both on the continents and the areas under marine water. If the marine water were removed from the oceanic areas, the great contrast between the physiographic characters of oceans and the continents would be immediately evident. A major part of the continents is characterised by plain areas raised only a little above the mean sea level. A major part of oceanic areas are also plain areas, which are at depths greater than three kilometres from the mean sea level. The peneplaination on continents is brought about by the interaction of endogenic tectonic processes and exogenic processes of erosion and deposition. However, denudational and erosional processes are non-existent in the case of oceanic areas. Oceanic plains have been formed by the endogenic tectonic processes alone.

Mountain chains of continents are relatively narrow belts raised above the average heights of the continental plains. The youngest mountain chains are composed of peaks often more than seven kilometres high. The older mountain chains are, to a great extent, denuded to much lower heights. Similar belts of high relief raised above the average 'ground' level are also known in oceanic areas. Since these 'mountains' tend to occupy the middle portions of oceans, they are known as **mid-oceanic ridges**. They are often raised above the sea level in the form of chain of islands. In other parts of the oceans, specially near the continental margins, the oceanic bottoms are depressed in narrow and arcuate oceanic trenches where the depth reaches 7 to 11 kilometres. Island arcs characterised by profuse volcanism are situated on the continental seas between the oceanic trenches and the continental areas.

The great contrast in the relief of oceanic and continental areas is directly related with the differing structure and composition of the earth's crust beneath the continents and oceans. The crustal structure is inferred

Earth's Crustal Structure and Tectonic Divisions of India 41

from the study of seismic waves which are produced by earthquakes. These seismic waves traverse through different layers of the earth's interior. The velocities of their travel through these different layers depend upon the physical properties of each layer. The velocities are recorded from the seismographs which is an important aid in interpreting the composition and structure of the earth's interior.

Earth's crust beneath the continents comprises three layers (Fig. 4.1). The upper layer is made up of sedimentary rocks having an average density of 2.2 gm/cm^3 and velocities of seismic waves varying from 1.8 to 5 km/sec. Thickness of this upper layer is variable from 2 to 10 km. The middle layer has a density of 2.4 to 2.6 gm/cm^3 corresponding to seismic velocities of 5 to 6.2 km/sec with thickness varying from 15 to 20 km. Since the physical properties of this layer correspond to those of the granites and metamorphic rocks, this layer is known as **granitic layer**. The lower layer has a density of 2.8 to 3.3 gm/cm^3 with seismic velocities ranging from 6 to 7.6 km/sec which correspond to the physical properties of basalts, gabbro and other basic rocks. This layer is, therefore, known as **basaltic layer**. The total thickness of the earth's crust beneath the continents ranges from 30 to 40 km, whereas under the mountain 'roots' the crustal thickness has a range of 50 to 80 km.

Structure of the earth's crust is very different beneath the oceanic areas. Total crustal thickness, predominantly made up of the basaltic layer, is greatly reduced to about 5 to 15 km only. Granitic layer beneath the oceanic areas are conspicuously missing. The upper layer of sedimentary rocks is reduced to an average thickness of 0.2 to 0.5 km. In some oceanic areas, the sedimentary rocks are as thick as 3 km, while in others they are completely missing. Structure of the earth's crust gradually changes from the oceanic to continental types through the areas marginal to continents and oceans characterised by the transitional type of the earth's crust.

TECTONIC ELEMENTS OF CONTINENTS

Continents are broadly divided into areas of **folded mountain belts** and **cratons** (Fig. 4.2). Folded mountain belts comprise thick sedimentary sequences of **geosynclinal** type which were deformed and uplifted during an orogenic phase of its evolution. Cratons are relatively stable areas of the earth's surface a major part of which, in general, has not undergone any violent (orogenic) type of tectonic activity since the end of the Middle Precambrian. The **Precambrian Basement** of the cratons, however, comprise the intensely deformed rock sequences. These rock sequences forming ancient mountain belts were violently twisted, folded and intensely metamorphosed. These mountains were peneplained prior to the deposition of the Middle and Upper Proterozoic rocks giving rise to a pronounced unconformity. The Precambrian Basements generally exposed in the central parts of the cratons are also known as **shields** (e.g., Indian Shield, Canadian

42 Elements of Historical Geology

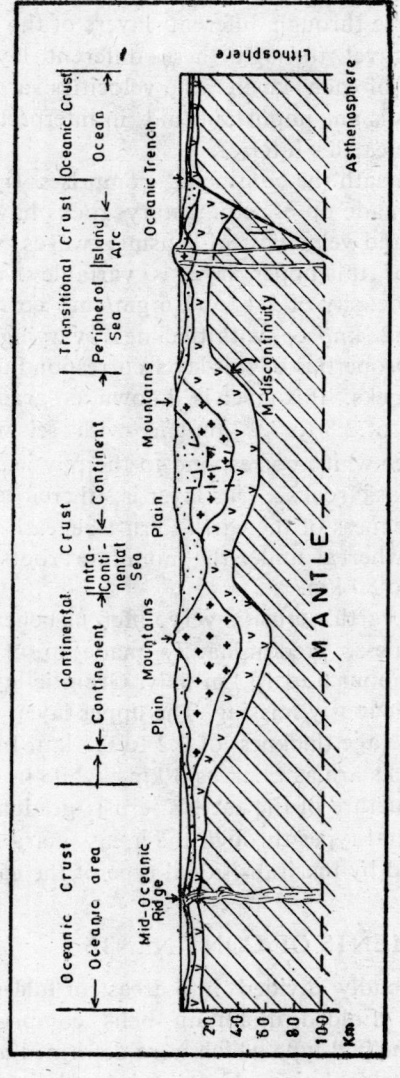

Fig. 4.1: Structure of the earth's crust beneath the continents and oceans.

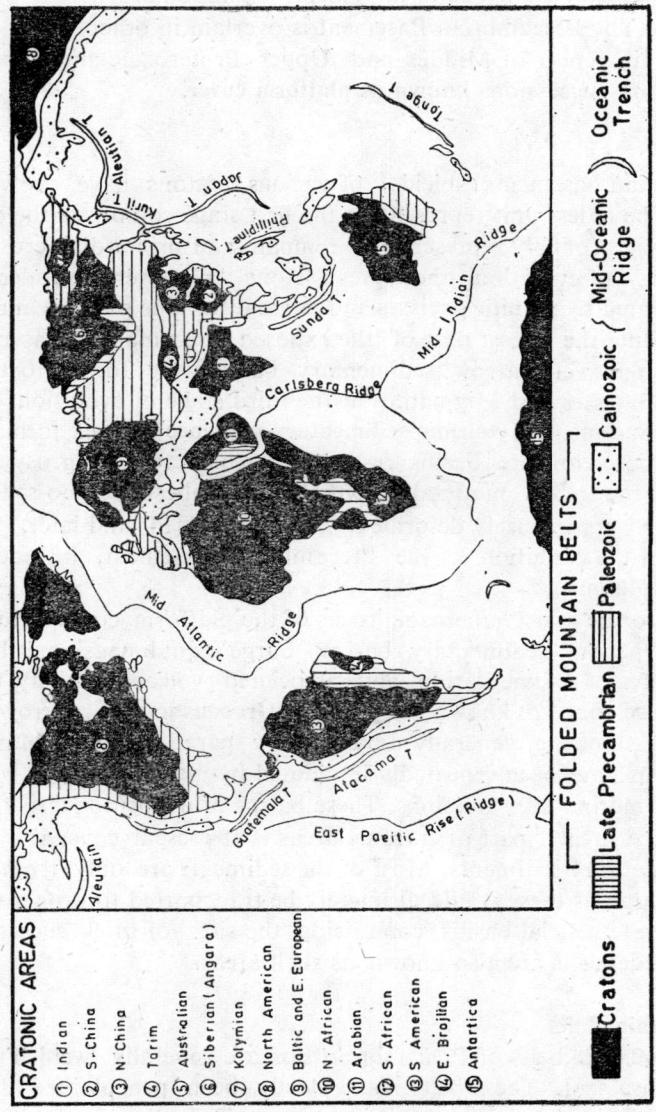

Fig. 4.2: Main tectonic divisions of the earth's surface.

Shield, Baltic Shield, etc.). The term 'shield' signifies hardened and stable rock complex. The Precambrian Basement is overlain in other parts of the craton by a thin veneer of Middle and Upper Proterozoic and Phanerozoic sedimentary successions known as **platform cover.**

Cratons
The Precambrian basements (shields) of various cratons have evolved in four stages. The oldest unit representing the first stage comprises the **greenstone belt** which probably represent the remnants of primordial crust. At the second stage of evolution, these green stone belts were intruded and largely assimilated by **granitic gneisses** and **migmatites**. The gneisses and migmatites constitute the largest part of the exposed Precambrian Basements. Elongate and relatively narrow sedimentary basins were formed over this basement of Gneisses and Migmatites at the third stage of evolution of the Precambrian Basement. A **volcano-sedimentary sequence** of rock formations deposited in these elongate basins resemble in character the geosynclinal successions of the folded mountain belts. These earliest volcano-sedimentary sequences were intensely deformed, metamorphosed and later, at the fourth stage of the evolution of the Precambrian Basement, intruded by large **granitic plutons**.

The Proterozoic and Phanerozoic rocks of the platform cover were laid in different types of sedimentary basins. Large equidimensional basins forming the sites of sedimentation and associated volcanic activity for a long duration of time are known as **syneclise. Grabens** having narrow and elongate depressions are generally bounded by parallel faults. Marginal areas of cratons have been repeatedly inundated by marine water in successive phases of marine transgressions. These basins of deposition are known as **shelf** areas. A greater part of cratonic areas is at present covered by thin veneer of Quaternary sediments. Most of the sediments are only temporary sites of deposition as these would ultimately be transported to proper tectonic basins. The terrestrial basins comprising the sites of thick successions of continental deposits are also known as shelf areas.

Folded Mountain Belts
The folded mountain belts of Precambrian Eon are generally 'welded' with the cratonic basement. These belts have mostly been peneplained due to continuous erosion since the Precambrian times. The Phanerozoic folded mountain belts, however, have still retained their oro-graphic characters for they stand out on the earth's relief. In Europe, three mountain chains, i.e., Caledonoid, Hercynoid and Alps, having their own distinct geological history, have been delineated. The Caledonoids were uplifted as mountain belts during the Early Palaeozoic time; the Hercynoids came into being towards the end of Palaeozoic Era while the Alps, considered as the youngest mountain chain, were formed during the Tertiary Period.

The Tertiary mountain chains constitute the loftiest mountains of the world. They are distributed on the earth's surface into two groups of mountain chains, viz., the **Circum-Pacific Chains** and the **Mediterranean Chains**. Both these groups of mountain chains meet in the East Indian Archipelagoes. The Circum-Pacific belt is bounded by continental plains on the one side and by the Pacific Ocean on the other. The Mediterranean belt, on the other hand, is bounded by the continental plains on both the sides. The Caledonoids, like the Tertiary Circum-Pacific belt, are circum continental belts around the Canadian, the Angara and the Australian shields. The Hercynoids are either circum-continental or intra-continental belts. The circum-continental Hercynoids envelop the Angara, the Baltic, the Canadian, the Brazilian and the Australian shields. The intra-continental Hercynoids include the mountain chains between the Gondwana Land and the Eurasia (Palaeozoic Tethys) on the one hand and between the Llandrian and the Canadian shield (Appalachians) on the other.

The folded mountain belts generally come into being at the culmination of three phased evolution of elongate and narrow basins of deposition known as **geosynclines**. The first phase, known as the **geosynclinal phase**, is characterized by a prolonged period of sedimentation. The rate of sedimentation is generally compensated by an equal amount of subsidence of the geosynclinal basins. During the second phase, known as **inversion phase**, the rock successions of the basin are subjected to intense deformation, metamorphism and plutonic magmatism. Tectonic movements are directed vertically downward at the first phase whereas at the second phase they are directed upwards leading to the uplift of the region. The inversion phase, also known as **orogenic phase**, is accomplished in a shorter duration of time (a few tens of million years) in comparison to the prolonged depositional cycle (a few hundreds of million years) of the geosynclinal phase. **Foredeeps** result from the uplift of the folded mountain belts in the frontal regions of the mountains. The sedimentary rocks of these basins comprising **post-orogenic succession** characterise the third, i.e., final phase of the development of folded mountain belts.

TECTONIC ELEMENTS OF OCEANS

The structure of the oceanic bottoms is distinctly different from that of the continents. Structural elements of the continents extend up to the continental shelves and then come to an abrupt end at the margins of the continental slopes. Studies made in marine geology during the last about three decades have revealed certain startling facts regarding the geological history of the ocean bottoms. These facts have been elucidated in the theory of **Plate Tectonics**.

The three main types of structural elements characterising the oceanic bottoms are the **mid-oceanic ridges**, the **oceanic trenches** and the **transform faults**. The four main mid-oceanic ridges are: Mid-Atlantic Ridge, East Pacific Ridge, the Carlsberg Ridge and the Indian Oceanic Ridge (Fig. 4.2).

These ridges comprise paired mountain chains of volcanic origin the median portion of which defines submarine rifts. The centre of the rifts is characterized by contemporary volcanism and shallow earthquakes. The volcanic rocks become older as one moves away from the central parts of the ridges. It has specially been demonstrated with the help of the palaeomagnetic data that the mid-oceanic ridges were formed by certain mechanism of **ocean floor spreading** resulting in the formation of rifts along the axis of mid-oceanic ridges. These rifts are filled by the introduction of magma from upper mantle leading to the formation of new crust.

The oceanic trenches are arcuate depressions on the ocean bottoms associated either with island arc system or young mountain belts on the continental side. A number of such trenches have been delineated on the ocean floors though only ten of the major ones have been shown on the tectonic map of the world (Fig. 4.2). The trenches are characterized by a zone of deep focus earthquakes shown to lie on a plane dipping at an angle of 45° towards the continental side. Magmatism in the island arcs is related with the generation of magma along this zone (Fig. 4.1).

The mid-oceanic ridges and the oceanic trenches are often dissected by oblique or transverse fracture planes known as transform faults, which resemble the normal transcurrent faults in as much as that movement is along the strike of the fault restricted only between the intersections of the displaced ridges or trenches.

According to the theory of Plate Tectonics, the earth's surface is divided into a number of 'plates' in relative motion. The plates regarded as rigid comprise 100 km thick **lithosphere** (i.e., earth's crust and the upper parts of the upper mantle). The plate motion is caused by 'convection currents' in the underlying 'plastic' layer of the upper mantle known as **asthenosphere**. At the mid-oceanic ridges, ocean floor spreading leads to the accretion of lithospheric plate by addition of volcanic rocks. Since the earth's circumference has presumably remained unchanged during the earth's later history, it is postulated that an equal amount of the surface area produced at the mid-oceanic ridges is consumed along the oceanic trenches by **subduction** of the lithospheric plate.

TECTONIC DIVISIONS OF INDIA

The broadest geomorphic divisions of India, viz., the Peninsular India, the Extra-peninsular India and the Indo-Gangetic Plain also correspond to the three broadest tectonic divisions. The peninsular India comprises the Indian Shield and its Proterozoic and Phanerozoic covers. The Extra-peninsula constitutes a part of the Alpine-Himalayan Tertiary mountain belt. The geo-morphic characters of these young mountains are in contrast to largely peneplained and plateau mountains of the Peninsular India. The Indo-Gangetic Plain extends from the mouth of Indus river draining into the Arabian Sea in the west through the northern plains of India to the great deltaic Sunderban where the Ganga and Bramhaputra river systems together drain into the Bay of Bengal. Quaternary alluvial sediments covering the Indo-Gangetic Plain were accumulated in a great trough between the Peninsular and Extra-peninsular regions.

The above mentioned three tectonic divisions are further subdivided into tectonic units of smaller order (Fig. 4.3). Each of these tectonic units is characterized by its own set of geological features. Areal extent of each tectonic unit has, at the same time, expanded and vaned during the course of its evolution. Thus, a tectonic unit often overlapped the neighbouring tectonic unit during a certain interval of geological time. Moreover, it is

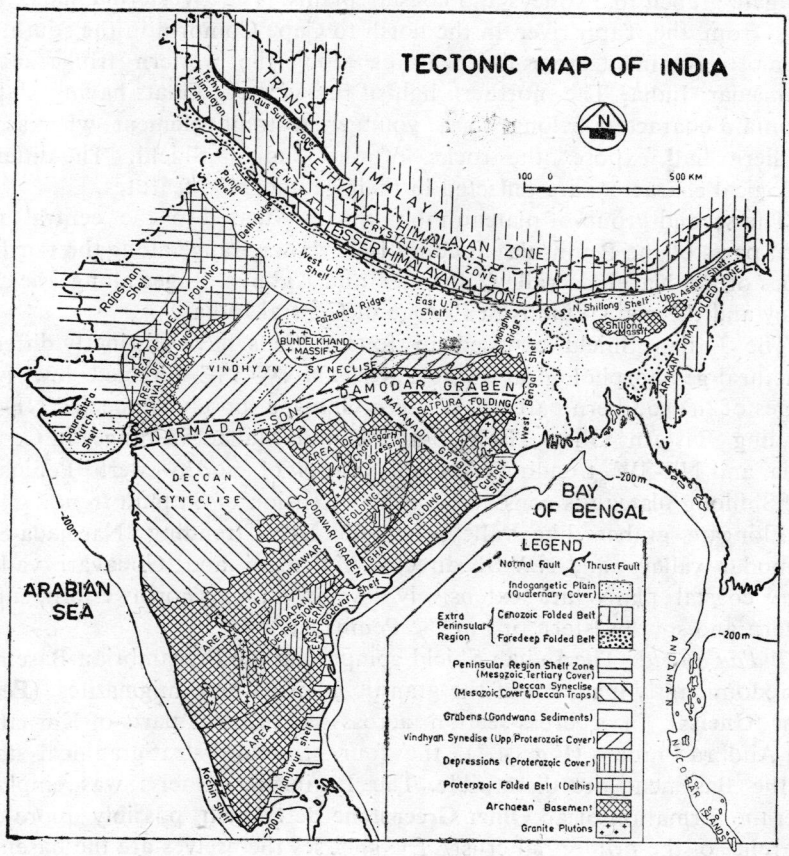

Fig. 4.3: Generalised tectonic map of India (compiled mainly from the tectonic maps of India by Geol. Surv. Ind., 1963 and Oil and Natural Gas Commission, 1968).

sometimes difficult to distinguish the basic criteria of tectonic zonation of the earth's surface as the natural boundaries are generally of gradational types. For example, certain orogenic movements recorded in the form of a rock succession may be exceptionally slow whereas certain epierogenic movements recorded in another succession of rocks of the neighbouring region may be relatively faster. At the boundaries of these two types of regions, i.e., orogenic and epierogenic, the rock characters may show extremely gradational and fused boundary.

Peninsular India

The Peninsular India is characterized by extremely varied physiography. It comprises a complex association of plateau mountains with widely distributed drainage, peneplained ancient folded mountains, massifs, elongate graben like valleys and coastal plains. The Western Ghats, running from the Tapti river in the north to Cape Comorin in the south, are the most prominent orographic features along the western fringe of the Peninsular India. The northern half of the Western Ghats having plateau mountain characters belong to a younger tectonic element whereas the southern half exposes the rocks of the Indian Shield. The differing geological characters are reflected in their geomorphic features.

The second group of plateau mountains is situated in the central northern parts of the Peninsula. It comprises Malwa Range along the northern slopes of the Narmada valley, Bhander and Kaimur ranges along the Son valley and the main Vindhyan Plateau of Central India.

The Indian Shield exposed in four regions have distinctly different structural-geomorphological trends, viz., NNW-SSE directed low lying ranges of the southern parts of Western Ghats, a loosely connected NE-SW trending Eastern Ghats, E-W Satpura and Mahadeva ranges in Central India, and NE-SW trending Aravalli ranges of northwestern Peninsula. The Shillong plateau is considered as an extension of Satpura trend.

Elongate graben like valleys include E-W trending Narmada-Son-Damodar valleys and NW-SE directed Mahanadi and Godavari valleys. Large coastal plains are extensively developed along northwestern, southeastern and southwestern tips of the Peninsula.

Indian Shield: The Indian Shield comprising the Precambrian Basement is predominantly made up of the granitic gneisses and migmatites (**Peninsular Gneiss**). In a cross-section across the central parts of Karnataka and Andhra Pradesh (Fig. 4.4), the four structural-stratigraphical stages of the Basement are discernible. The Peninsular Gneiss was emplaced after the formation of an **Older Greenstone Belt** which possibly represents the relic of the primordial crust. The gneisses themselves are the basement for the deposition of the first sedimentary-volcanic sequence which is now represented by a **Newer Greenstone Belt**. Accordingly, the Peninsular Gneiss is also known as **Fundamental Gneiss**, i.e. basement for all the succeeding rock formations. The fourth stage of evolution of the Indian

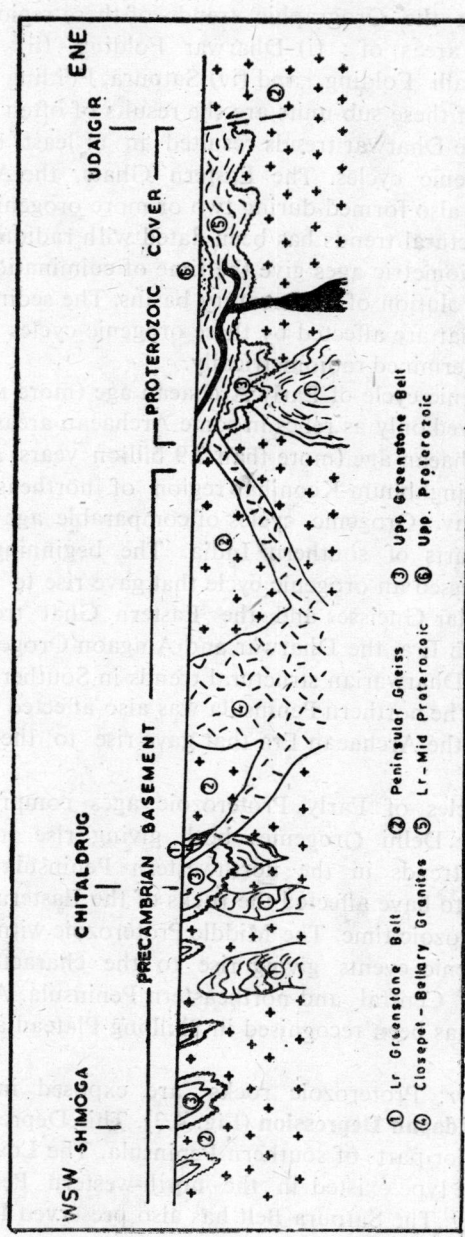

Fig. 4.4 : Main structural-stratigraphic stages in Precambrians of South India.

shield is characterised by the intrusion of large granitic plutons.

The Precambrian Basement of the Indian Craton has been delineated into four sub-units according to their characteristic structural trends that are well reflected in the Orographic trends of these regions. These sub-units are known as areas of : (i) Dharwar Folding, (ii) Eastern Ghats Folding, (iii) Aravalli Folding, and (iv) Satpura Folding (Fig. 4.3). The structural trends of these sub-units are the results of often more than one orogenic events. The Dharwar trends resulted in at least two or possibly three distinct orogenic cycles. The Eastern Ghats, the Aravalli and the Satpura trends were also formed during two or more orogenic cycles.

Each of the structural trends has been dated with radiometric methods (Table 4.1). The radiometric ages give the time of culmination of the orogenic cycle of the evolution of sedimentary basins. The sedimentary successions of the areas that are affected by these orogenic cycles are obviously older than those determined radiometrically.

The oldest orogenic cycle of Early Archaean age (more than 3.2 billion years ago) are observed only as relics in some Archaean areas. The orogenic cycle of Middle Archaean age (more than 2.9 billion years ago) has been well established in Singhbhum-Keonjhar region of northeastern Peninsula as Iron-Ore Orogeny. Orogenic cycles of comparable age have also been identified in some parts of southern India. The beginning of the Late Archaean time witnessed an orogenic cycle that gave rise to the formation of extensive Peninsular Gneisses and the Eastern Ghat trends. Towards the close of Archaean Era, the Dharwar and Amgaon Orogenies gave rise to the characteristic Dharwarian structural trends in Southern and Central Peninsular regions. The northern Peninsula was also affected by an orogenic cycle at the close of the Archaean Era that gave rise to the formation of **Bundelkhand Gneiss**.

The orogenic cycles of Early Proterozoic ages comprises an earlier Aravalli and a later Delhi Orogenies both giving rise to characteristic Aravalli structural trends in the northwestern Peninsula. Yet another orogenic cycle seem to have affected the rocks of the Eastern Ghats at the close of Early Proterozoic time. The Middle Proterozoic witnessed a successive phases of orogenic events giving rise to the characteristic Satpura structural trends in Central and northeastern Peninsula. A Late Proterozoic orogenic cycle has been recognised in Shillong Plateau and Travencore regions.

Proterozoic Cover: Proterozoic rocks are exposed in the southern Peninsula in the **Cuddapah Depression** (Fig. 4.3). This Depression seems to have covered a major part of southern Peninsula. The Lower Proterozoic basin of geosynclinal type existed in the north-western Peninsula in the *Area of Delhi Folding*. The Satpura Belt has also preserved Proterozoic formations of geosynclinal type.

Middle and Upper Proterozoic successions were deposited in a great

Table 4.1 : Radiometric Ages (in million years) of Precambrian Orogenic cycles in Indian Peninsula

Geological Age		Orogenic Cycles	Holmes 1955	Aswathanasayana 1964	Pitchamuthu 1971	Balasundaram and Balasubramanyam 1973	Sarkar 1972, 80
PROTEROZOIC	Late	Balagonda (Travencore & Ceylon)	c. 485				475-529 (Shillong)
	Middle	Satpura	c. 955	950±50 (Singhbhum)	950±50	c. 850 (Singhbhum & Erinpura)	c. 955 (Gaya) 850-1550 (Singhbhum)
	Early	Delhi/Eastern Ghat II	735±50 (Delhi) c. 1570 (Eastern Ghat)	735±50 (Delhi) 1600±80 (Eastern Ghat)	700±30 (Delhi) 1600±70 (Eastern Ghat)	c. 2200 (Eastern Ghat)	c. 1600 (Eastern Ghat II) c. 1660 (Delhi)
ARCHAEAN	Late	Aravalli Dharwar/Amgaon/ Bundelkhand Peninsular Gneiss II/Eastern Ghat I	c. 2300	2300±100	2400±100	c. 2600	c. 2000 c. 2450 c. 2600
	Middle	Iron Ore/Sargur/ Peninsular Gneiss I/Khondalites				c. 2000 (Iron-ore)	c. 2950
	Early	Older Metamorphics				c. 3000	c. 3200

Earth's Crustal Structure and Tectonic Divisions of India 51 basin demarcated on the tectonic map as **Vindhyan Syneclise** (Fig. 4.3). These rocks were laid directly over the Precambrian Basement emerging out of the sedimentary cover in the northern parts of the syneclise in the form of the **Bundelkhand Massif**. The Bundelkhand Massif comprises granitic gneisses and migmatites similar to the Peninsular Gneiss. The Middle and Upper Proterozoic basins, originally having a wider southerly extension, have left remnants in the form of **Chattisgarh** and **Bastar Depressions** of Central India (Fig. 4.3). The Vindhyan Syneclise is also considered to have extended northwards into the Lesser Himalayan region where the rocks of the basin have undergone a Tertiary orogenic deformation.

Upper Palaeozoic and Mesozoic Successions: The Lower Palaeozoic succession is conspicuously absent from the Indian Peninsula. Thick successions of the Upper Palaeozoic and Mesozoic rocks deposited in three great graben type basins are demarcated on the tectonic map of India as: **Narmada-Son-Damodar, Mahanadi** and **Godavari Grabens** (Fig. 4.3). Sedimentary rocks of these grabens are grouped into Gondwana Sequence known for its rich coal deposits. The rocks are almost entirely terrestrial with only marginal marine characters. In Narmada-Son-Damodar and Mahanadi Grabens, the Gondwana rocks rest directly over the Precambrian Basement. However, in Godavari Graben, an intervening Upper Proterozoic rock succession has also been preserved. The grabens were formed during the Late Palaeozoic times. The sedimentation phase continued till the close of Mesozoic Era.

The western and the central parts of the Narmada-Son-Damodar Graben are known to contain marine successions of Late Palaeozoic age. The marine basin is supposed to have extended over a vast area now entirely covered by the Deccan Trap. The basin is referred to as **Deccan Syneclise**. The syneclise achieved its maximum tectonic activation during the Late Mesozoic and Early Tertiary times when a thick succession of plateau basalts (Deccan Traps) together with Inter-trappean Beds were laid down. The Deccan Traps extending northwards covered a substantial part of the Vindhyan Syneclise. The western extension of the traps is observed in Saurashtra.

Cenozoic Cover: The greater part of north-western and south-eastern Peninsula was under marine transgressive basins during Mesozoic and Cenozoic Eras. A great thickness of marine rocks were deposited in these continental shelves. The **Rajasthan Shelf** is characterized by a succession of rocks beginning with a Proterozoic Basement (Delhis and Vindhyans) overlain by marine rocks of Palaeozoic, Mesozoic, Palaeogene and Neogene ages. The **Saurashtra-Kutch Shelf** comprises Aravalli-Delhi Basement, Mesozoic marine rocks, Deccan Traps and marine Palaeogene and Neogene successions.

The southeastern coastal region is demarcated into three shelves, i.e., **Thanjavur Shelf, Godavari Shelf** and **Cuttack Shelf**. Marine Mesozoic,

Palaeogene and Neogene rocks were deposited in these shelves over the Precambrian Basement of the area of Eastern Ghat Folding. The Indian Shield exposed as the **Shillong Massif** and the **Mikir Hill Massif** in the northeastern India is overlain by Palaeogene and Neogene successions in the **North Shillong Shelf** and **Upper Assam Shelf**. The shelves comprise a Precambrian Basement and an oil-producing Palaeogene-Neogene cover.

Extra-Peninsular India

The Extra-peninsular India is composed of the Himalayan mountain ranges in the north and the Arakan-Yoma ranges in the east. The latter is considered to extend up to Andaman Nicobar Islands. The ranges are made up of the Tertiary mountain belts and the frontal foredeep folded belts. The Himalayan belt extends for a total length of about 2400 km from Nanga Parbat in the west to Namcha Barwa in the east. At its western extremity, the orographic trend takes a sharp arcuate bent known as **syntaxial bend** and thence continues into the garland mountains of Baluchistan and Marakan coast in Pakistan. A similar arcuate bend is also observed at the eastern end of the Himalaya where the NE-SW trend seems to have ridden over the NNE-SSW trending Arakan-Yoma orogen. The northern limit of the Himalaya is marked by a lineament along the westerly flowing Upper Indus and easterly flowing Tsangpo (Upper Bramhaputra). The lineament is known as **Indus Suture**. The Trans-Himalayan mountain ranges lying to the north of the Indus Suture include the Ladakh and Karakoram ranges of Jammu & Kashmir. The Himalaya is further subdivided into the three longitudinal tectonic-geomorphic zones (Fig 4.5), viz., the Lesser Himalayan Zone, the Central Crystalline Zone of the Higher Himalaya and the Tethyan Himalayan Zone.

Foredeep Folded Belts: The foredeep folded belt known as the Siwalik Range (Outer Himalaya) has a maximum width of about 50 km in its western extremity in the neighbourhood of Jammu. The belt merges westwardly into the Potwar plateau of Pakistan. The width of the Outer Himalaya gradually decreases eastward to its complete elimination for a stretch of about 80 km but, further eastward, it reappears as a narrow strip in the southern Arunachal Pradesh.

Foredeep folded belt in the west of the Arakan-Yoma ranges comprises the low lying hills of Mizoram, Tripura and Manipur. The Siwalik foredeep folded belt and the foredeep in the west of Arakan-Yoma ranges are both made up of Neogene sediments. The boundary of the Siwalik belt with the Indo-Gangetic trough is poorly defined, however, with the Tertiary folded belts, it is marked by major faults known as **Main Boundary Thrusts**.

Lesser Himalayan Zone: Average altitude of the Lesser Himalayan Zone ranges between 2000 m. and 3000 m. The physiography of this zone is characterized by three main branches of mountain ranges obliquely

54 *Elements of Historical Geology*

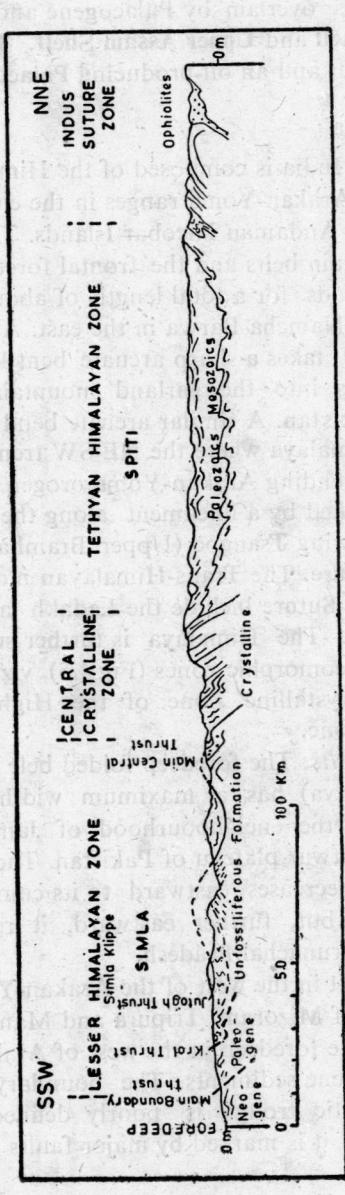

Fig. 4.5: Generalised geological cross-section across the Himalaya (based on Berthelsen, 1951).

emerging westwards from the Great Himalayan ranges. The Lesser Himalayan ranges are known as: (i) Nag-Tibba range (splaying off from Dhaulagiri mountains in Western Nepal), (ii) Dhauladhar range (emerging from the vicinity of Badrinath and (iii) Pir-Panjal range (arising from the vicinity of Kilar). The three main independent ranges occupying the southern parts of the Lesser Himalaya are known as Mahabharat Range (in Nepal), Mussoorie Range (between Ganga and Sutlej) and Ratanpir Range (of Jammu and Kashmir). The Lesser Himalayan ranges are drained by antecedent rivers. The rivers originating very often in the northern Tethyan Himalayan Zone cut across the Great Himalayan and Lesser Himalayan Ranges in a number of deep gorges and ultimately flow into the great alluvial flood plains of the Indo-Gangetic Plain.

The Lesser Himalayan Zone is characterized by a very complex structure comprising superimposed thrust sheets (Fig. 4.5). The highest thrust sheets consisting of Precambrian crystalline rocks were derived from the Central Crystalline Zone during the Tertiary deformation. The crystalline rocks tectonically rest over relatively less metamorphosed Lesser Himalayan formations. Largely unfossiliferous nature of the Lesser Himalayan formations makes their correlation with the standard stratigraphical scale highly controversial. A major part of the lower succession of these unfossiliferous rocks has been assigned a Precambrian age overlain by sediments roughly equivalent of the Gondwana Supergroup. The latter are, in turn, overlain by marine Palaeogene rocks in the southern parts of the Lesser Himalaya. A series of thrust faulting has brought about a general reversal in the stratigraphic succession. Thus, the youngest Palaeogene succession is overlain tectonically by Late Palaeozoic-Mesozoic succession of the **Krol Belt** which, in turn, is overlain by Precambrian rocks.

Central Crystalline and Tethyan Himalayan Zones: The two tectonic zones have been grouped into a single physiographic unit referred to as Higher (or Great) Himalaya with average altitude of more than 6,000 m. A greater part of this region is snowbound throughout the year. The Central Crystalline Zone comprises Precambrian basement intermixed with granitic plutons of Tertiary age. The Tethyan Himalayan Zone is composed of a fairly continuous marine succession of Palaeozoic and Mesozoic ages resting unconformably over the Precambrian basement of the Central Crystalline Zone.

The sedimentary record of the Tethyan Himalayan zone characterizes the geosynclinal stage of the evolution of the Himalayan Orogen. The Himalayan Geosyncline appeared over a Precambrian basement at the beginning of the Phanerozoic time and it was closed during the Tertiary orogenic phase. This geosynclinal history is marked by certain breaks in deposition, offlaps and overlaps.

The Phanerozoic successions of the Lesser Himalayan Zone are largely unfossiliferous whereas those of the Tethyan Himalayan Zone have yielded a rich record of fossil fauna. Some geologists regard this contrast in fossil record as an evidence of the existence of two parallel geosynclines, namely the Himalayan geosyncline in the south and the Tethyan geosyncline in the north. These two geosynclines were presumably separated by a geanticline (Central Crystalline Zone) which explains the differing facies of the two basins. Fossil fauna have recently been reported from a few sections of the Lesser Himalayan formations. These discoveries have upset the earlier established hypothesis of two parallel geosynclines in the Himalaya.

The northern parts of the Tethyan Himalayan Zone are overlapped by the Indus Suture Zone consisting of ophiolites (ultra-basic, basic and intermediate volcanic and hypabassal rocks) intermixed with Upper Cretaceous to Lower Tertiary sediments. The Indus Suture Zone is highly deformed and is characterized by the presence of nearly vertical thrust faults. The inversion stage of the Himalayan geosyncline is obviously related to the tectonics of the Indus Suture Zone. According to the theory of Plate-Tectonics, the Indus Suture Zone constitutes the subduction zone along which the "Indian Plate" collided with the "Tibetan Plate" giving rise to the formation of the Himalayan mountains.

Indo-Gangetic Plain

The Indo-Gangetic Plain is a deep crustal trough filled with Quaternary sediments. Its origin and structure are closely related with the rise of the Himalaya. Changes are still taking place at the bottom of this great trough giving rise to occasional earthquakes in the north Indian plains. The bottom of the basin seems to have assymetrical character. Northern fringe of the Indian Peninsula gently dips northwards beneath the plains. Maximum thickness of the Quaternary sediments is encountered along its northern fringes near the Foredeep Folded Belt of Siwalik hills.

The Indo-Gangetic Plain is divided into four shelf areas separated from one another by three transverse 'highs' (burried hills). The 'highs' are known, from west to east, as **Delhi-Haridwar Ridge, Faizabad Ridge** and **Monghyr-Saharsa Ridge** (Fig. 4.3). The Delhi-Haridwar Ridge is presumably the extension of the NE-SW trending Delhi orogen. The Faizabad Ridge is regarded as the extension of the Bundelkhand Massif. The Punjab Shelf lying to the west of the Delhi-Haridwar Ridge is composed of a Precambrian basement overlain by a succession of Palaeogene and Neogene sediments. East and West U.P. Shelves, lying to the east and west of the Faizabad Ridge, are made up of Precambrian basement overlain by Vindhyan rocks and Neogene sediments. The northern parts of these shelves are relatively deeper comprising the Mesozoic and Palaeogene rocks beneath the Quaternary sediments. The West Bengal Shelf lying to the east of the Monghyr-Saharsa Ridge contains in its succession the rocks of Gondwana age and elements of Rajmahal Traps.

Part II

STRATIGRAPHY OF INDIA

Chapter 5
Precambrian Basement of Indian Peninsula

The Precambrian time encompasses a vast expanse of time of about 3.5 billion years. It began at about 4 billion years ago when the earth was formed as planet of the solar system. The beginning of Precambrian phase of the geological history is lost in antiquity in view of the scanty geological records. The earth is considered to have originated by cooling of hot molten and gaseous substances. Thus, the earliest rocks of the earth's surface should be of igneous origin. With the formation of hydrosphere and atmosphere at a later date, the processes of erosion and sedimentation would have given rise to the deposition of the first sedimentary rocks. The earliest sedimentary rocks have been strongly deformed and metamorphosed during successive periods of orogeny leaving almost no trace of their sedimentary origin. Sedimentary rocks laid down during the Late Precambrian time were relatively less deformed often retaining their primary sedimentary characters.

Thus, the lower parts of the Precambrian succession are essentially composed of gneisses, high grade schists, quartzites, crystalline limestones, marbles, amphibolites and amphibolitic gneisses. The rocks have undergone an amphibolite to granulite facies of metamorphism. A considerable part of the older Precambrian succession has been affected by metasomatism whereby certain lighter chemical elements from deeper sources were added to give rise to the formation of granitic gneisses. The upper part of the Precambrian succession comprises relatively less metamorphosed sandstones, shales and limestones with intercalations of volcanic rocks.

J.D. Dana introduced the term 'Archaean' for the ancient rock succession of the Canadian Shield that had not afforded any fossil record. Later, this succession was divided into two units. The lower unit consisting of highly deformed and metamorphosed rocks was referred to as the **Archaean Group**. The upper unit made up of less deformed and metamorphosed rocks was named as the **Proterozoic Group**. The two groups of rocks are separated by a pronounced unconformity. The respective time units represented by the two groups of rocks were referred to as Archaean and Proterozoic eras. These terms were adopted by the International Geological Congress held in 1884.

The criteria of separating the two groups of the Precambrian rocks based on the degree of metamorphism and deformation were soon found to be invalid outside the Canadian Shield. Later classifications were based on three groups of data, namely, the geological, the palaeontological and the radiochronological. The geological data helped in identifying the major geological events such as plutonism, orogeny, magmatism, etc. The palaeontological data have been useful in classifying only the upper parts of the Precambrian succession. The radiochronology has been the most useful in classifying the Precambrian succession.

Based on primarily the radiochronological data, the Precambrian time has been divided into the Archaean and the Proterozoic eras separated by a time plane of 2.5 billion years (Sims, 1980). Each era has been further subdivided into three each sub-eras. The Archaean divisions are separated by the time planes of 3.2 and 2.9 billion years, and the Proterozoic divisions are separated by the time planes of 1.6 and 0.9 billion years (Sarkar, 1980). Some geologists have advocated for a fixed standard in rocks for the purpose of stratigraphic classification (Hedberg, 1974). Nisbet (1982) has, thus, suggested that the end of the Archean era be placed at the time of cooling of the Hartley Complex of the Great Dyke of Zimbabwe which has given an isotopic age of 2.5 billion years.

PRECAMBRIAN HISTORY OF THE EARTH

The earth was formed as a planet of the solar system some 5 to 7 billion years ago. According to an earlier view, the earth was formed from a gaseous liquid which gradually cooled down to form the first hard crust. Yet another view regards the solar system to have been formed by condensation of cold dust known as nebula. Continued condensation led to heating and melting of the planets. Subsequent cooling led to the formation of the earth's crust. The first stage of the earth's history involving about 2 billion years of time has left no geological record. The oldest rocks of the earth's surface isotopically dated as 3.7 to 4.0 billion years old are the first geological record of the earth's history.

During the first stage of the earth's history, earth's crust was very thin. The earth's surface was probably covered with several craters and seas of lava flows resembling the present lunar surface. The earth's first atmosphere had altogether different composition. It was primarily formed of gaseous volcanic products such as vapour, methane, carbondioxide, nitrogen, hydrogen, inert gases (Ar, Kr, Xe, He) and acidic fumes (HF, HCl, Boric acid, H_2S, etc.). The lighter gases (hydrogen and partly heleum) were latter dissipated into the universe. Free oxygen was first formed from the disintegration of H_2O and CO_2 under the influence of sun rays in the upper layers of the earth's atmosphere. The proportion of oxygen in the atmosphere increased gradually to its present level with the evolution of the life.

The temperature of the earth's surface gradually came down to less than 100°C which led to the condensation of vapour into water, filling the earliest basins of the earth's surface. The formation of the hydrosphere and the atmosphere led to intensive erosion and deposition of sedimentary rocks intercalated with volcanic rocks. Recorded geological history begins with the deposition of the sedimentary-volcanic succession at the beginning of the Archaean Era.

Hydrosphere of the Early Archaean Era contained dissolved gaseous products of volcanism such as HCl, HF, Boric acid, H_2S, CO_2, CH_4 and other hydro-carbons. Highly acidic water of the first formed hydrosphere (pH 1 to 2) kept silica in a dissolved state. The extreme acidic characters of the first hydrosphere, extensive volcanism and unusual composition of the atmosphere are reflected in the chemical composition of the Archaean rocks making them distinct from the younger rock formations.

The dissolved acids of the hydrosphere were gradually neutralised during the later part of the Archaean Era leading to the deposition of silica and carbonates of K, Na, Ca and Mg. Carbonates were formed on land as a product of chemical weathering under the influence of the atmosphere rich in carbonic acid. The atmosphere underwent a change with the release of more free nitrogen from ammonia and methane. The nitrogen gradually became the most predominant gas of the atmosphere.

Life in the form of anaeorobic bacteria seems to have evolved at the beginning of Late Archaean time. Simple forms of phytolites have been reported from the strata older than 2800 million years. The phytolites include stromatolites, oncoliths and catagraphics. These forms were formed by carbonate precipitation under the influence of organic activity. Stromatolites occurring in the upper part of the Precambrian succession have specially great stratigraphic significance (see Chapter 6).

PRECAMBRIAN BASEMENT

The cratonic areas of the world (Fig. 4.2) consists of a Precambrian Basement (also known as Shield) and a sedimentary cover. The Basement consists of 10 to 12 km thick succession of highly deformed and metamorphosed rocks. The rocks of the Canadian Shield were formed more than 2500 milllon years ago However, other shield areas contain some rock successions that were deposited during the Early Proterozoic time. The Indian Shield comprises the rocks that were formed more than about 2100 million years ago.

The Precambrian Basement evolved in four successive phases of sedimentation, magmatism, plutonism and orogeny. These phases are represented the four structural stages that are identifiable in many shield areas (Fig. 4.4). The lowest stage which has a strong influence of basic rocks represents the relicts of the premordial crust. These rocks are preserved as undigested masses or xenoliths in vast country of granitic gneisses that were formed at the second stage of the evolution of the Precambrian Basement. The granitic gneisses were formed either by the differentiation of the basaltic magma or by the metasomatic transformation of earlier existing rocks. At the third stage, the proto-geosynclines were formed over a basement of granitic rocks. A thick pile of volcano-sedimentary sequence was laid in these protogeosynclines. The sequence is known to contain some of the richest deposits of iron and manganese. The proto-geosynclines underwent an inversion phase at the close of the sedimentary history. The inversion phase is represented by the orogenic deformation and emplacement of granitic rocks that mark the fourth stage of the evolution of the Precambrian Basement.

The earliest phase of the earth's history representing the Archaean Era is preserved in the form of the rocks sequences of the Precambrian Basement. The Archaean rocks have undergone repeated phases of orogenic deformation, plutonism and high grade of metamorphism. The stratigraphic relationships in many cases have been rendered extremely complex. The stratigraphic classification and correlation of Archaean successions are based on a host of data obtained from petrographical, geochemical, structural and radiochronological investigations.

The Archaean successions of the Indian Shield are characterised by an extremely complex geological history. Successive orogenic cycles (Table 4.1) have imparted characteristic structural trends for various Archaean areas. At least five Archaean provinces having distinctly differing geological and structural history can be demarcated on the Peninsula (Table 5.1). The Dharwar Province of south India has a relatively continuous geological history for the most part of Archaean Era. The Eastern Ghats Province has preserved the rock records of Early, Middle and the beginning of the Late Archaean Periods. The Central Indian Province has exposed the rocks of Late Archaean Period only whereas the eastern Singhbhum-Orissa Province has preserved the rocks of Early and Middle Archaean Periods. In the Aravalli Bundelkhand Province of northern Peninsula, the Middle Archaean rocks are apparently missing.

Stabilisation of various provinces of the Indian Peninsula is marked by emplacement of granitic plutons. The five Archaean provinces achieved their stability at different times, either during the Archaean Era or at the beginning of the Proterozoic Era. The Singhbhum-Orissa, Aravalli-Bundelkhand and Eastern Ghats provinces were stabilised during the Archaean Era. The stabilisation of Dharwar and Central Indian provinces was formalised by the emplacement of Closepet Granite at the commencement of Proterozoic Era. Parts of these Archaean provinces were subsequently

Precambrian Basement of Indian Peninsula 63

remobilised during the younger orogenic cycles.

The stratigraphic boundary demarcating the Archaean rocks and Close pet Granite from the Cuddapah Supergroup of Proterozoic age is one of the best defined geological boundary (Fig. 4.4). This boundary was referred to as Eparchaean Unconformity implying that the rock formations above the unconformity are of Proterozoic age and those below are of Archaean age. Geochronological data for some of the rocks below the Unconformity has, however, given Early Proterozoic ages (Sarkar, 1980). These rocks have apparently closer geological affinity to the Archaean history and hence they are described in the present chapter.

Table 5.1. Generalised classification and correlation of Archaean and Lower Proterozoic formations of India (Chronometric data in million years, based on Sarkar, 1980; Sarkar et al 1981; Sarkar and Saha, 1982)

Dharwar (S. India)	Eastern Ghats	Central India	Singhbhum-Orissa	Aravalli-Bundelkhand
Closepet Granite (2000-2380)		Dongargarh Granite (c. 2200)	Mayurbhanj Granite (2000-2100)	
		Nandgaon Group (c. 2200)	Dhanjori Group (2100-2200)	
		Sakoli Sausar Groups	Singhbhum-Gangpur Groups	Aravalli Group
Dharwar Supergroup		Amgaon Group (2500)		Bundelkhand Gneiss (c. 2500) Banded Gneissic Complex
	Charnockite "Series" (c. 2600)			
Peninsular Gneissic Complex (2600-2950)			Singhbhum Granite (c. 2950)	
Sargur	Khondalite "Series" (c. 3000)		Iron Ore Group	
			Older Metamorphic Croup (c. 3200)	

Dharwar Province

Archaean rocks of the Dharwar Province has been studied since the second half of the last century. R. Bruce Foot (1888) introduced the term "Dharwar System" for the schistose rocks "buried in a number of synclinoria of large terrain of granitic gneissic rocks". The schistose rocks-occur in several NNW-SSE trending belts. Bruce Foot believed that the Dharwars were deposited over an eroded surface of gneisses. This view was contrary to the earlier held view of Newbold (1850) who regarded the granitic gneisses to have intruded the schistose rocks.

The granitic gneisses contain inclusions of amphibolites and schists which are regarded as relict of the host rocks in a large granitic intrusion. A greater part of the schistose formation was presumably absorbed by the invading magma. Intrusive nature of gneisses was also demonstrated by the evidence of forceful injection, cross cut relationships and contact effects typical of intrusions. Those favouring the view that the Dharwars were deposited over the gneissic basement argued the necessity of identifying the floor for the deposition of the Dharwar sediments. The other evidence favouring this view are: (*i*) relative abundance of gneisses over schist, (*ii*) normal order of superposition of schist over the gneisses and (*iii*) the presence of granitic pebbles in the basal conglomerates of the Dharwar succession. The problem: "whether the gneisses are basement for or intrusive into the Dharwar succession" is yet to be resolved to the satisfaction of all. The latest observations, however, appear to support the view first proposed by R. Bruce Foot.

Radhakrishnan (1964) observed that the schist inclusions within the gneisses represent basic igneous rocks and primitive sedimentary rocks older than the Dharwars in age. The older schists showing a relatively higher degree of metamorphism than the Dharwar schists, represent an **Older Green Stone Belt** known as **Sargur Schist Complex** (Radhakrishnan and Vasudev, 1977). The high grade schists were later intruded by the **Peninsular Gneissic Complex** and together they formed the basement for the deposition of the sedimentary-volcanic succession **(Newer Greenstone Belt)** known as **Dharwar Supergroup**. The youngest elements of the Precambrian Basement of the Dharwar region are the plutonic granitic rocks known as **Closepet Granite** (Fig. 5.1 & Table 5.2).

Sargur Schist Complex: Schistose inclusions in Peninsular gneisses known since long have recently been extensively exposed in exacavations of a long system of canals in southern Karnataka and specially around Sargur in the southeast of Mysore. The inclusions are largely composed of metamorphosed products of ultra-mafic rocks, quartzites, carbonates and argillaceous sediments. The rocks have undergone a high degree of metamorphism ranging from amphibolite to lower granulite facies. The schists contain abundant lenses of anorthosite and anorthositic gabbro. Quartzites are often fuchsitic and are characteristically bedded with sedimentary pyrite. The rocks have often been extensively migmatised and mobilised during the subsequent phase of emplacement of Peninsular Gneissic Complex. The schist belts of Kolar and Hutti containing gold deposits belong to this ancient schist complex.

Sargur Schist Complex can be correlated with the Keewatians of

Table 5.2: Precambrian Basement of Karnataka (Dharwar Province)

Approximate age in million years	Stratigraphic Units	Lithological Units	General Characters
2100-2380	Felsite and porphyry dikes **CLOSEPET GRANITE** **DHARWAR SUPERGROUP**		
	Rannibennur Group	Manganiferous phyllites, Greywackes, Chlorite phyllites	Least metamorphosed, gentle deformation, Mn & Fe ores
	Chitradurga Group	Agglomerates, tuffs, pillow lava, Ferruginous manganiferous chert, Dolomites and limestones, Phylites, Orthoquartzites, Conglomerates	Green-schist facies metamorphism, strong deformation, Placer gold
	Bababudan Group	Banded magnetite quartzite, Argillites, Mafic lavas, Orthoquartzites, Conglomerates	Green-schist to lower amphibolite facies of metamorphism, Fe & Au deposits
	Unconformity		
More than 2600	PENINSULAR GNEISSIC COMPLEX	Granites and Gneisses, Granites and Granodiorites	Migmatitic and amphibolite facies of metamorphism
More than 3200	SARGUR SCHIST	Magnetite quartzite, schist granulite, crystalline limestone and dolomite, mafic and ultra-mafic flows and anorthosite	Upper amphibolite to lower granulite facies of metamorphism, intense deformation, Au, Tungsten, Chromite, Ti-magnetite, Vanadium, barite

Canada and similar Green Stone Belts of other cratonic areas. K-Ar age of amphibolites of Hutti schist belt and lead age of Kolar amphibolites gave isochron dates as 3200 and 2900 million years respectively. Palynological microstructures from the schistose rocks of Sargur have been compared by Viswanathaiah et al. (1976) with similar forms from the oldest formations of South Africa which have been radiometrically dated as 3375 ± 100 million years old. Geochemical studies of mafic and ultramafic rocks of the Sargur Schist Complex have also shown their primitive character comparable in composition with those of the mantle (Radhakrishnan and Vasudev, 1977). The oldest lavas are exceptionally high in MgO and show higher concentration of Cr and Ni. The potash content is low as compared to the younger lava sequences.

Peninsular Gneissic Complex: A major part of the Indian craton exposes vast areas of gneissic and granitic rocks for which William Smeeth (1916) gave the name **Peninsular Gneiss**. The gneisses, also known as **Fundamental Gneiss**, constitute the basement for the deposition of the younger successions. They are typically migmatitic gneisses alternating with bands of amphibolites and tonalites. Variations shown by the Peninsular gneiss are attributed to intimate mixing and interaction of tonalitic magma with the pre-existing mafic and ultramafic rocks and associated sediments of the Sargur Schist Complex. Emplacement of this highly heterogeneous association of rocks took place in several tectonic impulses. Thus, it is more appropriate to refer to this association as "Peninsular Gneissic Complex". Radiometric age data from various sources point to an older event at about 3 billion years ago and a younger event at about 2.6 billion years ago giving a total spread of over 400 million years for the successive phases of emplacement of the gneissic rocks.

Dharwar Supergroup: The sediments of the Dharwar Super-group were laid in elongate proto-geosynclines over a basement of the Sargur Schist and Peninsular Gneissic Complex. The sedimentation was marked by intense magmatic activity during its earlier phase. A vast thickness of rocks were deposited in these basins. Kudermukh scarp of the Western Ghats alone exposes more than 2 kilometres thick gently dipping Dharwar beds resting over gneisses.

Smeeth (1916) proposed a two-fold division of the Dharwar succession, the lower referred to as "Hornblendic Division" and the upper as "Chloritic Division". B. Ramarao (1936) proposed a three-fold division: the lower division of Smeeth was named as Lower Dharwars, whereas the upper Cholritic division was divided into Middle and upper Dharwars. Each division is said to be characterised by a conglomeratic horizon at its base.

The two-fold division of Smeeth was based on the difference in chemical characters of the two units. The lower unit comprises mafic and ultramafic rocks of predominantly volcanic origin whereas the upper unit is

Precambrian Basement of Indian Peninsula 67

composed of geosynclinal sediments with relatively reduced volcanic contents. These two units were respectively named as **Bababudan Group** and **Chitradurga Group** (Radhakrishnan and Vasudev, 1977). A still younger group, called **Ranibennur Group** has also been recognised though its stratigraphic status has not yet been established.

Bababudan Group is well exposed in arc shaped mountains of that name in the north of Chikmaglur (Fig. 5.1). The succession begins with a basal orthoquartzite and conglomerate followed by thick flows of mafic and ultramafic lavas, massive beds of banded quartz-magnetite rocks and

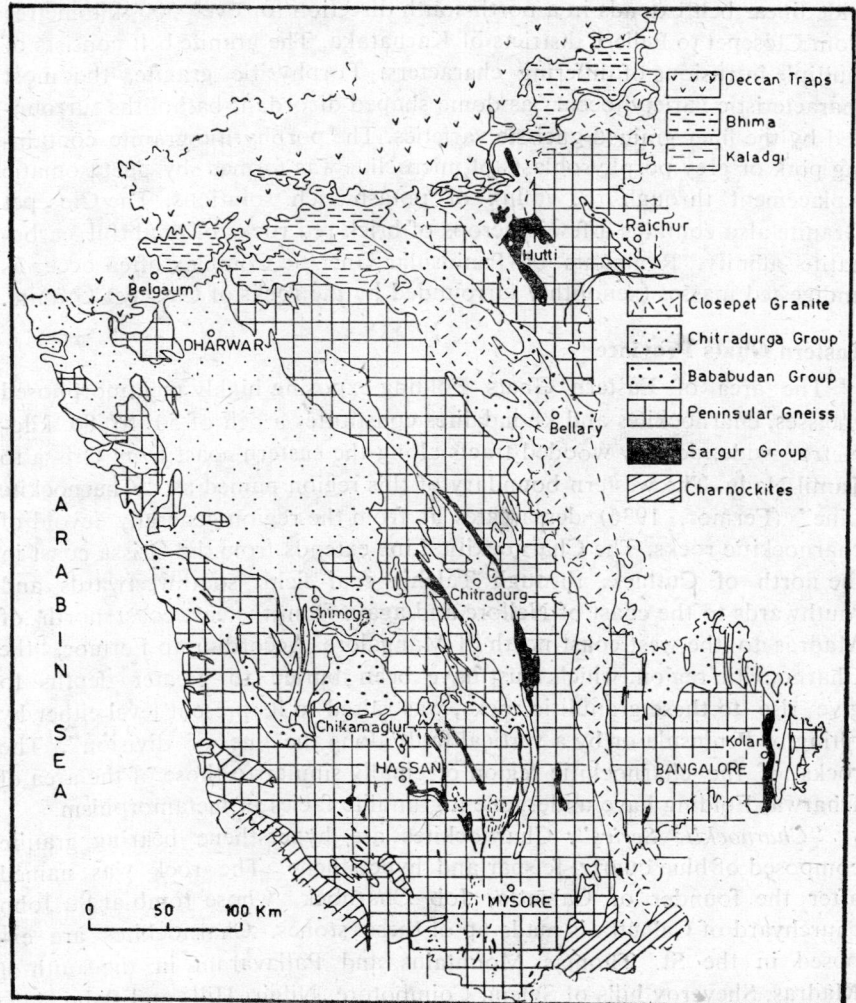

Fig. 5.1 : Precambrian Basement of Karnataka (based on Radhakrishnan and Vasudev, 1977).

argillites. Sedimentary structures, such as ripple marks and cross bedding in quartzites, suggest shallow water deposition. The overlying Chitrudurga Group covering a major portion of Dharwar outcrops has a conglomerate bed at the base of the succession comprising orthoquartzites, phyllites, limestones, dolomites, banded ferruginous manganeferous quartzites, cherts and greywackes. Contemporaneous volcanic products of basic and intermediate compositions are intercalated with this predominantly sedimentary succession. Lava flows often show pillow structures indicating submarine conditions of volcanic activity.

Closepet Granite: The Closepet Granite forming a 15 to 70 kilometre wide linear belt extends in a north-south direction for over 500 kilometres from Closepet to Bellery districts of Karnataka. The granite belt consists of multiple intrusions of differing characters. Porphyritic granite, the most characteristic variety, occurs as dome shaped dicordant batholiths surrounded by the finer grained gneissic varieties. The porphyritic granite containing pink or grey porphyroblasts of microcline was formed by metasomatic replacement through the influx of potash rich solutions. The Closepet Granite also contains a few outcrops of brick red rocks of doubtful carbonatite affinity. Remnants of Peninsular Gneissic Complex often occur as undigested masses (xenoliths) surrounded by the rocks of Closepet Granite.

Eastern Ghats Province

The area of Eastern Ghats Folding exposing highly metamorphosed gneisses, charnockites and khondalites constitutes a belt of 50 to 80 kilometres wide, densely wooded tracts along the eastern coast from Orissa to Tamil Nadu. The western boundary of this region named as "Charnockite Line" (Fermor, 1936) demarcates it from the region generally devoid of charnockitic rocks. The Charnockitic Line extends from the Orissa coast in the north of Cuttack, through Talchir coal field, southwestwards and southwards to the coast of Nellore and again from the east coast north of Madras to the west coast north of Mangalore. According to Fermor, "the Charnockitic region, which must have been buried to greater depths to give rise to these granulitic rocks, was raised to its present level either by tilting of Peninsula or by a vertical fault along the line of division". The rocks of the charnockitic region originally similar to those of the area of Dharwar Folding have undergone a granulite facies of metamorphism.

"Charnockite Series": Charnockites are hypersthene bearing granite composed of blue quartz, felspar and hypersthene. The rock was named after the founder of Calcutta, Job Charnock, whose tomb at St. John churchyard of Calcutta is made up of these stones. Charnockites are exposed in the St. Thomas Mountains and Pallavaram in the south of Madras, Sheveroy hills of Salem, Coimbotore, Nilgiri Hills and Palni hills of Madurai, Coorg, Travencore, Cape Comorin, South Arcot District, Ganjam, Vishakhapatnam, Kalahandi and other areas of the Eastern

Ghats. The charnockites are also known as Nilgiri Gneiss and Mountain Gneiss as they occupy the higher peaks of southern India.

Charnockites ranging in composition from basic to acidic varieties constitute a suite of rocks referred to as "Charnockite Series". Acidic varieties include alaskites, birkremites, enderbites and hypersthene-quartz syenites. Basic types are essentially pyroxene granulites. Intermediate varieties range from homogeneous hypersthene diorites to pyroxene granulites and granulitic migmatites.

Views differ regarding the origin of charnockites. Holland (1900) regarded these rocks to be of igneous origin. Ghosh (1941) considers them to have been derived from metamorphism of sediments rich in Fe, Mg and Ca. Charnockites of Karnataka have been regarded by Pitchamuthu (1965) as belonging to two types: the granulitic and the granitic. The granulitic type is considered to have been formed by high grade metamorphism whereas the granitic type is regarded to have been formed much later due to palingenetic fusion and metasomatism.

"*Khondalite Series*": Khondalites named after the Khonds tribes of Vishakhapatnam and Kalahandi of Andhra Pradesh are essentially grey and red foliated garnet-silliminite schists containing quartz, felspar and graphite. In hand specimens, the rocks exhibit numerous brownish red garnets. Other varieties of rocks associated with the "Khondalite Series" include garnetiferous quartzites and calciphyres. The rocks of the Series are also exposed in Guntur District in Andhra Pradesh, parts of Orissa, Bihar and Tamil Nadu. Highly acidic granulites (leptynites) resembling Khondalites are exposed in the Nilgiri, Annamallai, Sheveroy, Palni and Travencore areas.

Khondalites are 'dry' metamorphic rocks comprising minerals devoid of water and other volatile constituents. Physical conditions of their formation are comparable to those of Charnockites. According to Dunn (1942), the Charnockites were formed by the palingenesis of Khondalite under the plutonic conditions of metamorphism.

Khondalites often contain graphite deposits of commercial importance. The deposits occur in bands of 30 cm to 50 cm in thickness. Minute grains of diamond known from the stream sediments of the Khondalite terrains are regarded to be a form of graphite in Khondalites. Certain hydrocarbons similar to those of petroleum have also been reported to be associated with the graphite deposits.

"Khondalite Series" is often associated with manganese rich gneisses known as **Kodurites**. The kodurites are composed of potash felspar, manganese-garnet, manganese-pyroxene, rare biotite (spandite) and apatite. The rocks have yielded workable deposits of manganese ores. Chemical alteration of kodurites gives rise to the formation of litho-marge, chert and ochres. **Gondite** is yet another manganese bearing high grade metamorphic rock comprising quartz, spessartite, rhodonite and small percentage of apatite. Gondite is characteristically devoid of potash felspar. They were formed by the plutonic metamorphism of sediments consisting of sand, clay and manganese oxides. Gondites are better developed in the Balaghat-Chindwara region of the area of Satpura Folding.

The "Khondalite Series" is also known to be associated with igneous complex of alkaline gabbro, calc-silicate syenites and nephiline-syenites. Cordierite-silliminite gneisses are characteristically developed at the boundaries of the basic charnockites and khondalites. Saphirine (green spinel) bearing hypersthene rocks are known from Vishakhapatnam.

Central Indian Province

The Nagpur, Bhandara and Durg districts in Central India exposes a triangular belt of Precambrian rocks known as Sakoli Group. This belt is bordered on the north and north-west by ENE trending Sausar Group and on the southeast by NNE trending Dongargarh belt comprising a succession of Amgaon, Nandgaon and Khairagarh Groups. On structural considerations, the Sakoli Group is considered to be older than Sausar Group and younger than Amgaon Group (Sarkar and Saha, 1982; Sarkar, 1968). The Precambrian rocks of Sausar, Sakoli and Dongargarh belts that were considered to be stratigraphic equivalents of Dharwar succession in South India have given radiometric ages suggesting Early and Middle Proterozoic ages for the major part of the succession (Table 5.1).

Sausar Group : The type area of the Group is situated in the Sausar subdivision of the Chindwara district in Madhya Pradesh (Fermor, 1926). The Group consists of about 4.5 km thick succession of hornblende granulites, manganiferous marbles, gondites, calciphyres, quartzites and quartz schists. The rocks of the group largely of sedimentary origin were subsequently metamorphosed and invaded by acid and basic plutonic rocks.

The Sausar Group has been divided into eight units (Table 5.3). The rock types of the Group show variable characters in different areas that

Table 5.3 : Classification of Sausar Group (after Straczek et al. 1956)

Formations	Dominant Lithology
Bichua Formation	Dolomitic marbles and calc silicates
Junewani Formation	Muscovite-biotite schists and granulites
Chorbaoli Formation	Quartzites and schists
Mansar Formation	Muscovite schists and muscovite-biotite schists with Mn-ore zones
Lohangi Formation	Dolomitic marbles and calc silicates
Kadibikhera Formation	Quartz biotite granulites
Sitasaongi Formation	Felspathic schists and quartzites
Tirodi Biotite Gneiss	Biotite gneiss with amphibolites calc gneisses and quartz-biotite granulites

have been attributed to facies variation. The lowermost unit, **Tirodi Biotite Gneiss** was considered to have formed the basement for the deposition of the Sausar succession (Fermor, 1926). The formation of the gneisses could also be related to the migmatisation at a much later date after deposition of the Sausar Group.

The **Mansar Formation** contains some of the richest manganese deposits of India. The outcrops of the Mn-bearing rocks extend in the form of a 30 km wide and 200 km long arcuate belt from Chindwara to Balaghat districts. The Mn-horizons occur one each at the bottom (Lohangi Zone), middle (Middle zone) and top (Chorbaoli zone) of the Mansar Formation.

A part of the Sausar succession has been correlated with the **Chilpi Ghat Group** (Chilpi Ghat Beds; King, 1885) exposed along the northern margin of Chattisgarh Depression (Fig. 4.3). At Chilpi Ghat, the succession comprises a great thickness of phyllites, slates and mica schists with tuffaceous quartzites, coarse felspathic grit and conglomerates. This succession is also known to contain some manganese ores though none of them are of any economic importance.

The rocks of the Sausar Group have been intensely deformed (Vemban, 1961) giving rise to a southern belt of overturned isoclinal folds and a northern belt of recumbent folds and nappes (Fig. 5.2). The central part of the Sausar belt consists of gneissic rocks with linear outcrops of infolded schists. The K-Ar ages of muscovite and biotite from the schists suggest that the "Sausar Orogeny", regional metamorphism and granitisation had closed during the period 844-996 million years ago (Sarkar *et al.*, 1967).

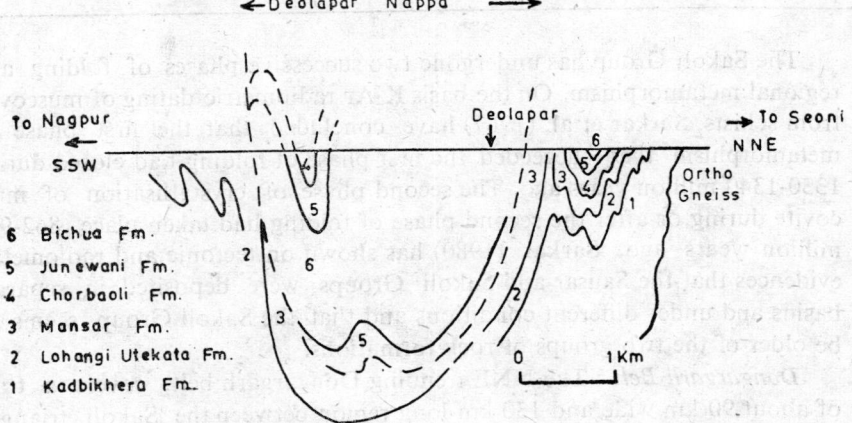

Fig. 5.2 : Geological cross-section across the Deolapar Nappe (based on West, 1935).

Sakoli Group : Rocks of Sakoli Group occur in a triangular tract (Sakoli triangle) of Nagpur, Bhandara and Chanda districts. The succession at the type area in Sakoli village of Bhandara district comprises schists, quartzites, slates and phyllites (Ball, 1877). The northeastern tip of the triangular outcrop (Fig. 5.3) exposes a closure of steeply south and southwesterly plunging synclinorium with sub-vertical limbs (Sarkar 1957-58). The stratigraphic succession of the two limbs are identical. The schists normally contain serecite and chlorite with occasional appearance of garnet. The succession is ramified with epidiorites and tourmaline-muscovite granite containing a host of minerals such as kyanite, tourmaline, rutile, topaz etc.

Table 5.4 : Stratigraphic Succession of the Sakoli Synclinorium (after Sarkar, 1968)

Northwestern limb of the Synclinorium	Eastern Limb of the Synclinorium
"Sakoli Orogeny" (metamorphism and granitisation)	
Phyllites with minor hematite quartzite	Phyllites with minor ferruginous quartzite
	Epidote amphibolite
Carbon phyllite, chlorite phyllite, chlorite-quartz schist	Carbonaceous phyllite interlayered with muscovite-biotite phyllite.
Banded hematite quartzite with chlorite phyllite, ferruginous phyllite	Hematite quartzite with impersistent felspathic serecite quartzite
Garnetiferous phyllite, chlorite-quartz schist, muscovite-quartz schist, ortho-hornblende schist chlorite-garnet-biotite-hornblende schist	Muscovite-biotite phyllite and higher grade equivalents, quartz schists and amphibolites
Muscovite-quartz schist amphibolite, garnet-biotite schist, staurolite quartz schist	Muscovite-quartz schist

The Sakoli Group has undergone two successive phases of folding and regional metamorphism. On the basis K-Ar radiometric dating of muscovite from schists, Sarkar et al. (1967) have concluded that the first phase of metamorphism that succeeded the first phase of folding had closed during 1330-1340 million years ago. The second phase of crystallisation of muscovite during or after the second phase of folding had taken place 862-952 million years ago. Sarkar (1980) has shown on tectonic and radiometric evidences that the Sausar and Sakoli Groups were deposited in separate basins and under different conditions and that the Sakoli Group seems to be older of the two groups of rock formations.

Dongargarh Belt : The NNE trending Dongargarh belt, covers a tract of about 90 km wide and 130 km long region between the 'Sakoli triangle' and the Chattisgarh Depression (Fig. 5.3). The belt consists of three groups of rock formations (Sarkar, 1957-58) each separated from the other by a pronounced unconformity (Table 5.5). The **Amgaon Group**, oldest of the three groups, consists of quartz schists and felspathic quartzites alternating with metabasic lava. The rocks of the Group have undergone orogenic deformation, amphibolite facies of metamorphism and granitisation marking the "Amgaon Orogeny".

Based on Rb-Sr and K-Ar isochron age data, Sarkar *et al.* (1981) have inferred that the "Amgaon Orogeny" closed earlier than about 2300 million years. The Amgaon Group formed the basement for the deposition of the Sakoli Group as well as the Nandgaon Group.

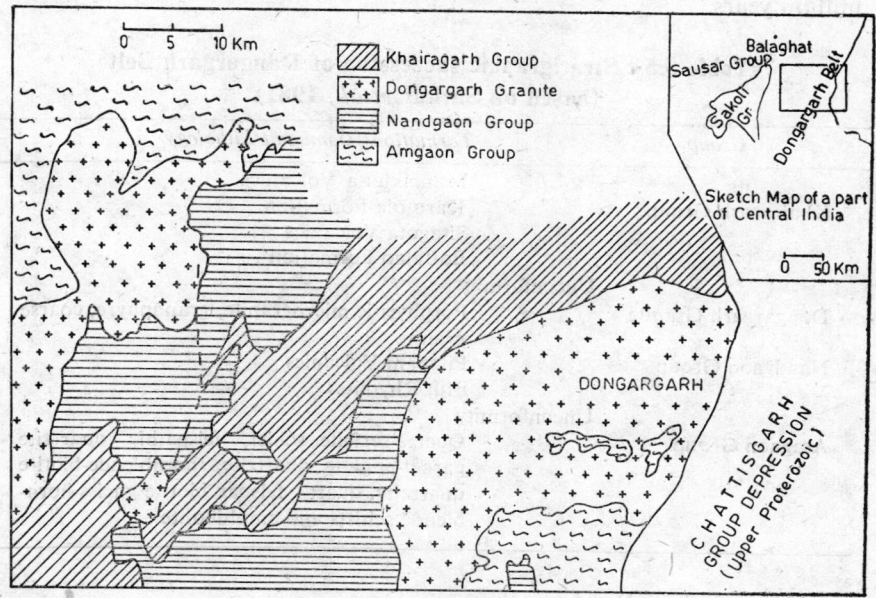

Fig. 5.3 : Generalised geological map of a part of Dongargarh Belt (based on Sarkar, 1957-58).

The **Nandgaon Group** comprising a lower Bijli Rhyolites and an upper Pitepani Volcanics marks a phase of extensive volcanism. The **Bijli Rhyolites** comprise about 4.5 km thick succession of rhyolites with intertrappean conglomerates, sandstones, siltstones, shales and tuffs. The rhyolites contain inclusions of amphibolites and quartzites of the Amgaon Group over which they unconformably overlie. The unconformity is well marked where the unmetamorphosed rhyolites occur in the neighbourhood of regionally metamorphosed rocks of the Amgaon Group. The rhyolites are also known to unconformably overlie steeply dipping sheared phyllites and banded hematite quartzites of presumably Sakoli Group (Sarkar *et al.*, 1981).

The volcanic phase is followed by the emplacement of porphyritic micro-granite, granophyre and coarse granite grouped as **Dongargarh Granite**. The Granite has developed a thermal metamorphic aureole in the adjoining Bijli Rhyolites and Pitepani Volcanics. The Bijli Rhyolites have given isochron age of 2180±25 million years and the time of emplacement for the Dongargarh Granite on the evidence of radiometric data has been placed at slightly younger than 2200 million years (Sarkar *et al.* 1981). Thus the emplacement of the Dongargarh Granite roughly corresponds to the time of emplacement of Closepet Granite in South India. The Khairagarh Group unconformably overlying the Dongargarh Granite has given the isochron age of sedimentation and volcanism younger than 1686 million years.

Table 5.5 : Stratigraphic succession of Dongargarh Belt (based on Sarkar *et al.*, 1981)

Group	Formations/Dominant Lithology
Khairagarh Group	Mangikhuta Volcanics Karutola Formation Sitagota Volcanics Bortalao Formation
	Unconformity
Dongargarh Granite	Porphyritic microgranite, granophyre, coarse granite
Nandgaon Group	Pitepani Volcanics Bijli Rhyolites
	Unconformity
Amgaon Group	Quartz-sericite schist, felspathic quartzite garnet-epidote quartzite, hornb'ende-biotite quartzite, quartz felspar biotite gneiss hornblende schists and amphibolite

Singhbhum-Orissa Province

The Singhbhum region of the southern Bihar and its contiguous regions of Mayurbhanj, Keonjhar and Bonai districts of Orissa are well known for their rich deposits of iron and copper. The region is traversed by **Singhbhum Shear Zone** (also known as Copper Belt Thrust) extending over a strike length of more than 160 km (Fig. 5.4). The shear zone trends eastwards from Parahat in western Singhbhum to Chakradharpur and then it takes a southwards swing in the neighbourhood of Jamshedpur. The Shear Zone separates a northern terrain of more highly metamorphosed rocks and a southern terrain of relatively less metamorphosed rocks. Sarkar and Saha (1977) have shown that this Shear Zone separates two Precambrian provinces of the Indian shield : an older province in the south which stabilised after the Iron-Ore Orogenic cycle closing about 2900 million years ago and a younger province in the north that underwent the Singhbhum Orogenic cycle closing at about 850 million years ago.

The rock succession of the tract in the south of the Shear Zone (Table 5.6) consists of a Lower Archaean Basement of Older Metamorphic Group invaded by the Biotitetonalite gneiss. The Iron Ore Group rocks were deposited over the eroded Lower Archaean Basement. These rocks were folded about NNE to NNW trending fold axes and low grade metamorphism culminating in the emplacement of the Singhbhum Granite (Iron Ore Orogeny). After a long period of erosion, rocks of Singhbhum and Gangpur Groups were laid down along the northern edge of the stabilised "Iron Ore Craton".

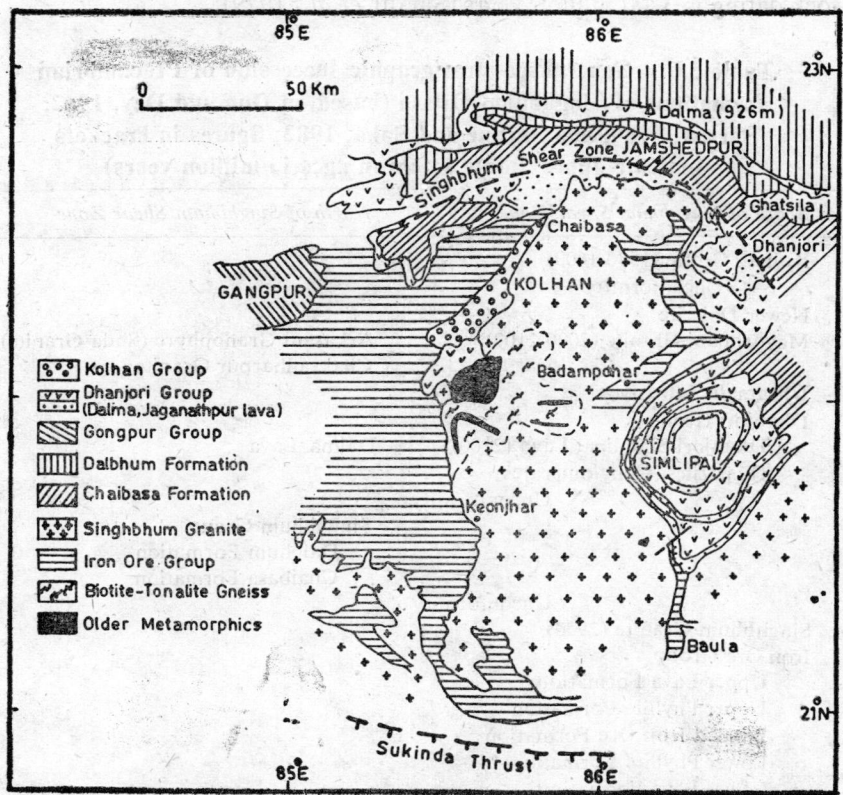

Fig. 5.4 : Genralised geological map of Singhbhum-Orissa Province (based on Sarkar and Saha, 1982 and Iyenger and Murty, 1982).

The rocks of the Singhbhum Group underwent a first generation of folding, uplift leading to retreat of ocean and subaerial erosion. A subsequent phase of regional tension led to the eruption of thoelitic Dalma and Dhanjori Lavas concurrent with the deposition of terrigenous sediments. A second phase of folding preceded by the emplacement of granitic rocks led to the development of Singhbhum Shear Zone that served as the favourable channel for the copper mineralization.

Older Metamorphic Group : The oldest rocks lying in the south of

Singhbhum Shear Zone were named by Dunn (1929) as the "Older Metamorphics". The rocks consist of predominantly hornblende schists with quartzites and quartz mica-schists. These rocks are known to occur as isolated exposures within the general mass of Singhbhum Granite. The "Older Metamorphics" are intruded and partly granitised by a suite of biotite tonalite gneiss. The closing date of metamorphism and emplacement of the tonalite gneiss as indicated by Rb-Sr and K-Ar mineral and whole rock dating is 3200 million years (Sarkar *et al.*, 1979).

Table 5.6 : Generalised stratigraphic succession of Precambrian formations of Singhbhum-Orissa (based on Dun and Dey, 1942; Sarkar, 1980; and Sarkar and Saha, 1982; figures in brackets indicate approximate isochron ages in million years)

South of Singhbhum Shear Zone	North of Singhbhum Shear Zone
Kolhan Group (1500-1600)	
Unconformity	
Newer Dolerite	
Mayurbhanj Granite (2000-2100)	Arkasani Granophyre (Soda Granite)
	Chakradharpur Granite
Ultramafic Intrusives	
Dhanjori Group	
Dhanjori-Simplipal Lava (2100)	Dalma Lava
Quartzites, Conglomerates	
Unconformity	
	Singhbhum Group
	Dalbhum Formation
	Chaibasa Formation
Unconformity	
Singhbhum Granite (2950)	
Iron Ore Group	
Upper Lava Formation	
Upper Phyllite Formation	
Banded Iron Ore Formation	
Lower Phyllite Formation	
Lower Lava Formation	
Sandstone and Conglomerate (local)	
Unconformity	
Biotite-tonalite Gneiss (3200)	
Older Metamorphic Group (3800)	

Iron Ore Group : The rock succession exposed in the southern Singhbhum and Keonjhar and lying unconformably over the "Older Metamorphics" was named as "Iron Ore Series" by H.C. Jones (1934). The order of superposition of the rock succession has been in dispute on account of their complex structure. Jones (1934) proposed a sequence and described the regional structure as a north plunging synclinorium with an overturned western limb. Dunn (1940) suggested three alternative stratigraphic successions. Sarkar and Saha (1962), Murty and Acharya (1975) and Sarangi and Acharya (1975) established the regional structure to be a low NNE plunging synclinorium overturned towards the southeast.

The stratigraphic succession of the Iron Ore Group begins with a locally developed basal sandstone and conglomerate. The basal beds are successively overlain by Lower Lava, Lower Phyllite, Banded Iron Ore, Upper Phyllite and Upper Lava Formations (Table 5.6). The Banded Iron Ore Formation crops up in ridges arranged in sharply bent horse-shoe patterns. The Formation has preserved primary depositional features such as current bedding ripple marks scour-and-fill structures etc., indicating a shallow water environment of deposition in a region proximal to the shore line (Rai et al., 1980). The Banded Iron Ore Formation has provided some of the richest iron-deposits of India.

Singhbhum Granite : The Singhbhum Granite is a great batholithic mass occupying an elongated tract of about 10,000 square kilometers in Singhbhum, Keonjhar and Mayurbhanj districts. The batholith consists of several domed up intrusions (Saha, 1975) varying in composition from biotite granodiorite to adamallite, biotite trondjhemite and leuco-granite. At margins of the batholith, chloritic or epidotic granodiorite and pyroxene diorite have developed. The main mass shows a distinct N-S or NNE-SSW foliation in parallelism with the foliation of the host rocks of the Iron-Ore Group.

Roof pendents of rocks of Iron Ore Group and patches of granitised amphibolites of the Older Metamorphic Group are known to occur within the Singhbhum Granite. The time of emplacement of the Singhbhum Granite is regarded as syn- to post-tectonic to the deformation of the rocks of the Iron Ore Group (Iron Ore Orogeny). Rb-Sr whole rock isochron date of around 2950 million years indicate at least three closely related phases of emplacement of granitic rocks (Sarkar and Saha, 1982).

Singhbhum Group : The rock succession lying to the north of the Singhbhum Shear Zone extends in a series of east-west folds for over 200 kilometres. Dunn and Dey (1942) correlated this succession with the succession of Iron Ore Group lying in the south of the Shear Zone on the basis of general lithological similarity and the presence of iron ore beds in the upper part of the succession. Sarkar and Saha (1963, 1977) have, however, renamed the succession lying to the north of the Singhbhum Shear Zone as Singhbhum Group. This succession has been divided into a lower **Chaibasa Formation** and an upper **Dalbhum Formation**. The lower formation is exposed in the cores of the anticlinal structures situated immediately to the north of the Singhbhum Shear Zone. The upper formation is preserved in synclinal troughs in the northern areas.

The name of the Chaibasa Formation is a misnomer as the nearest outcrops of this formation are exposed some 20 kilometres north of **Chaibasa.** The formation consists of high grade mica-schists, hornblende-schists and quartz granulites. The rocks were considered as more metamorphosed equivalents of rocks exposed at Chaibasa which was, however, later grouped into a much younger Kolhan Group of Middle to Late Proterozoic age. The term "Chaibasa Formation" has been retained in Indian Geology for its usage for a very long time and for want of any other proper stratigraphical name for the formation.

The Dalbhum Formation consists of phyllites and banded quartzites which show a lower grade of metamorphism than the underlying rocks of the Chaibasa Formation. Some of the phyllites are considered to represent tuffs and volcanic flows. The phyllites contain ferruginous beds and sometimes manganese minerals as well.

Dhanjori Group : The group consists of a basal conglomerate, arkose, quartzite and extensive lava flows. The succession was unconformably deposited over the Singhbhum Granite and the rocks of Iron Ore Group. Equivalent volcanic rocks exposed in the Dalma hills in the north of the Singhbhum Shear Zone are known as **'Dalma Lava'** (Fig. 5.4). Southern extension of these volcanic flows are observed in Simlipal of Keonjhar district and Bonai of Sundergarh district in Orissa. Basal sediments of the Dhanjori Group are regarded as non-marine facies of piedmont or alluvial fan types. The Dalma Lava, according to Naha and Ghosh (1960), represents an island arc system of the Archaean time, probably once covering an area of over 2000 square kilometres.

Gangpur Group : Archaean rocks of the Sundergarh district in Orissa were grouped into a "Gangpur Series" by Krishnan (1935). These rocks were regarded to be older than the rocks of the Iron Ore Group. The Gangpur rocks were shown to occupy the core of anticlinorium. Kanungo and Mahalik (1967) have, however, shown on the basis of detailed structural analysis that the anticlinorium of Krishnan is in fact a synclinorium comprising a younger succession of Gangpur Group. The Gangpur Group has been divided into five formations (Table 5.4). On the basis of their lithological similarities, the succession has been correlated with the Chilpi Ghat Group exposed at about 120 kilometres to the west of Gangpur area.

Table 5.4 : **Stratigraphic succession of Gangpur group**

Stratigraphic Units	Main Lithology
GANGPUR GROUP	
Ghorijhar Formation	Staurolite and garnet schists, calc schists, quartzites and conglomerates
Kumarmunda Formation	Carboniferous quartzites, slates and phyllites
Birmitrapur Formation	Limestone, dolomite quartzites and phyllites
Laingar Formation	Phyllites, carbonaceous slates and quartzites
Raghunathpalli Formation	Conglomerates, quartzites & slates
Unconformity	
IRON ORE GROUP	

Aravalli Bundelkhand Province

Aravalli mountains extending from Delhi in the northeast to the gulf of Cambay in the southwest constitute an important orographic feature of the northwestern Peninsula. Hilly tracts of the Aravalli range are composed of Proterozoic succession of the Delhi Supergroup. The Precambrian Basement of this Middle-Upper Proterozoic succession is exposed in the central plains between the Aravalli and Vindhyan ranges (Fig. 5.5). The Precambrian Basement of this region is made up of Banded Gneissic Complex and a sedimentary succession comprising the Aravalli Group. Bundelkhand Gneissic Complex exposed at the northeastern edge of the Vindhyan Syneclise as Bundelkhand Massif (Fig. 4.3) has been considered to be equivalent of the Banded Gneissic Complex. The gneissic rocks of Bundelkhand are succeeded by Gwalior Group.

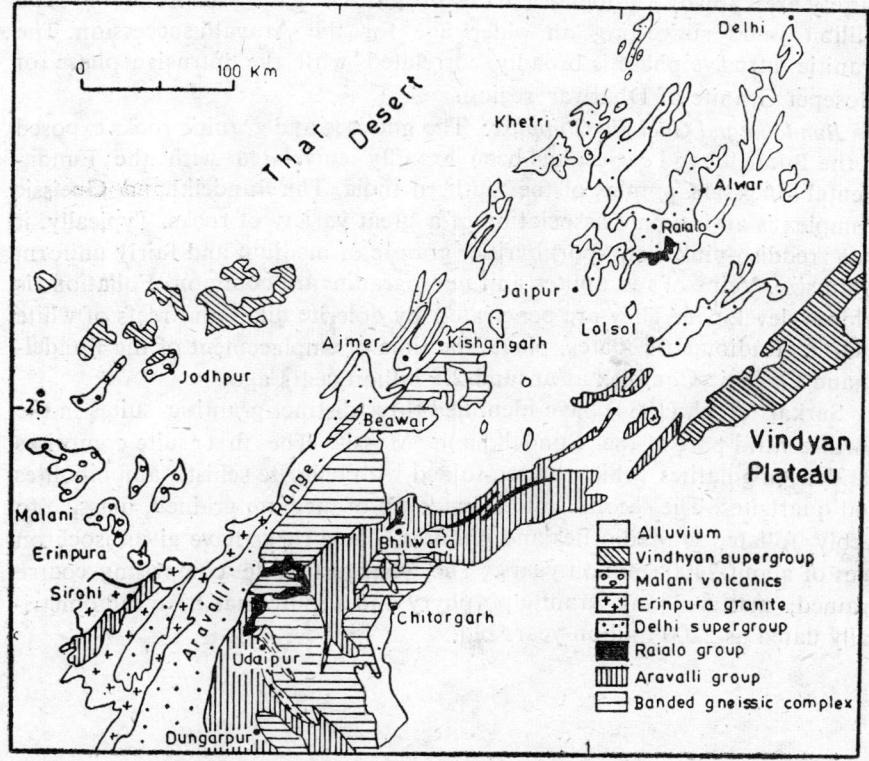

Fig. 5.5 : Geological map of the Aravalli mountains (simplified from Heron, 1935).

Aravalli Group: This group of rock formations comprises an immense thickness of predominantly argillaceous rocks which exhibit prograde metamorphosed characters towards the core of the Aravalli range grading from almost unaltered shales in the east through phyllites in the centre to high grade schists in the west. Impure argillaceous limestone occurs in subordinate amounts. The dolomitic limestone of Zawar area known for the only workable Pb-Zn deposits of India forms a part of the Aravalli succession. Ferruginous limestone of Bundi and dark limestone of Udaipur are characteristic calcareous facies of this succession. Quartzites occurring in the basal parts of the succession rests unconformably over the Banded Gneissic Complex. At places, the quartzites are succeeded by a great thickness of volcanic rocks. Arenaceous facies again appear towards the upper parts of the succession. Unmetamorphosed rocks of argillaceous composition of Chittor known as Binota Shales are olive to brown variegated shales with purple sandy, micaceous and often ferruginous beds.

Aravalli succession is intruded by granites and ultra-basic rocks. The latter rocks are represented by talc-serpentine-chlorite schists. Granites of two types, viz., a fine grained variety and a younger coarse-porphyritic variety are known. Radiometric dates of granite gave isochrons of 1900 million years suggesting an older age for the Aravalli succession. The granitic intrusive phase is broadly correlated with the intrusive phase of Closepet Granite of Dharwar region.

Bundelkhand Gneissic Complex: The gneisses and granitic rocks exposed in the Bundelkhand region has been broadly correlated with the Fundamental Gneissic Complex of the southern India. The Bundelkhand Gneissic Complex is an intimate association of a great variety of rocks. Typically, it is a reddish-pink, non-porphyritic granite of medium and fairly uniform grain size. Veins of red aplites and microgranite are common. Foliation is seldom developed. They are penetrated by dolerite dikes and reefs of white quartz. Radiometric dates place the time of emplacement of the Bundelkhand Gneissic Complex at around 2.5 billion years ago.

Sarkar et al. (1984) have identified three distinct granitic suites in the north-central part of the Bundelkhand Massif. The first suite comprises banded migmatites which are co-folded with chlorite schists, amphibolites and quartzites. The second suite is made up of medium grained, massive to feebly foliated granodiorites and granites. These rocks have given isochron ages of about 2400 million years. The youngest suite comprising coarse grained, massive pink granite porphyry and aplites has been radiometrically dated as 2200 million years old.

Chapter 6

Proterozoic Formations of Indian Peninsula

The Proterozoic formations were first described from the Canadian Craton. These rocks were deposited over a basement of Archaean age. A pronounced unconformity known as the Main Unconformity separates the Proterozoic and Archaean successions. The Proterozoic succession is unconformably overlain by the rocks of the Cambrian System. The gaps in deposition represented by the two surfaces of unconformity appearing at the base and at the top of the Proterozoic succession of Canadian Craton are represented by the rock successions deposited in other parts of the world.

Beginning of the Proterozoic Era is placed at about 2500 million years ago and the Era came to an end at about 570 million years ago. The Proterozoic succession has been often divided into three units. The Lower and Middle units representing the time duration from 2500 to 1600 million years ago (Fig. 2.5) are identified on the basis of mainly radiochronological data. The Upper Proterozoic has been further subdivided in to **Riphean** and **Vendian** units on the basis of characteristic forms of stromatolites occurring in these successions (Table 6.1). The Riphean comprising the major part of the Upper Proterozoic succession has been further subdivided into three sub-units. The Vendian, also known as **Eocambrian**, represents the transitional phase from Precambrian to Phanerozoic eons of the earth's history.

Table 6.1: Classification of Upper Proterozoic

Divisions of Upper Proterozoic		Characteristic forms of Stromatolites	Age in Million years
		Cambrian System	570
Vendian		Boxonia, Colleniella, Linella	680-700
Riphean	Upper	Gymnosolen, Inseria, Tungussia	700-1000
	Middle	Baicalia, Tungussia	1000-1350
	Lower	Kussiella, Gongilina, Nucleela	1350-1600
		Lower and Middle Proterozoic	

The Proterozoic succession of United States of America has been subdivided into three sub-units referred to as Precambrians X, Y and Z (James, 1972). In a similar scheme of classification, Sarkar (1980) has identified three subdivisions in the Proterozoic formations of India. The subdivisions have been named as Precambrians IV, V and VI representing the time durations of 2500 to 1600 m.y., 1600 to 900 m.y., and 900 to 570 m.y. respectively.

PROTEROZOIC HISTORY

The sedimentary basins of the first phase of the earth's history (Archaean Era) had uniform characters resembling those of the geosynclines. The Proterozoic Era began with a tectonic zonation of the earth's surface into sedimentary basins of platform and geosynclinal types. The geosynclinal basins are represented by a vast thickness of sedimentary and volcanic rocks that were deposited in shallow marine conditions. The marine basins of the platform type are characterised by the deposition of a relatively smaller thickness of rock formations during the same duration of time. Such a tectonic zonation was accomplished at somewhat differing time over different continents. In some cases, this tectonic zonation marks the Archaean-Proterozoic boundary. In other areas, the tectonic zonation took place during the Early Proterozoic time. These two primary types of sedimentary basins have since existed over the earth's surface for the remaining part of the earth's history.

The composition of the hydrosphere and atmosphere underwent very little change during the transition period of Archaean and Proterozoic eras. The carbon-dioxide content of the atmosphere gradually declined throughout the Proterozoic Era. The atmosphere attained its predominant nitrogen-oxygen character at the close of the Era. The carbon-dioxide content of the hydrosphere also declined during the Era. The volcanic products such as sulphur and hydrogen sulphide of the hydrosphere were converted into sulphates in the presence of oxygen which were deposited in the sedimentary succession of the time. The rock succession of the Era is also characterised by extensive deposition of carbonate rocks such as limestones and dolomites. Phosphorites known from the upper parts of these successions were the first rocks of evaporite facies deposited in a regressing marine basin.

An increase in oxygen and a corresponding decrease in carbon-dioxide in the atmosphere led to favourable conditions for the growth of life. With the removal of toxic contents of marine waters such as sulphur, hydrogen sulphide and carbon-dioxide, algae and other primitive forms of life thrived in the Proterozoic sedimentary basins. However, the early forms devoid of hard parts could not be preserved as fossils. Imprints of algal life preserved in the form of stromatolites are commonly observed in the Proterozoic successions. Other forms such as worms *Sabellidites*, jellyfish

Beltanella and primitive corals *Rangea* are often preserved as trace fossils.

BASEMENT COVER TRANSITION

The tectonic cycles in the history of the earth caused radical changes in the palaeogeography of the geological past. These changes had also caused significant changes in the composition of hydrosphere and atmosphere that is often well recorded in the form of composition and textures of rocks deposited during that period. The organic life also felt the transformations in environment. Radical changes are observed in the form and morphology of animals and plants that lived during the successive stages of earth's evolution. The biostratigraphic boundaries in the history of the earth followed immediately after the major tectonic cycles which might have been caused by an increased radiation due to introduction of radiogenic acidic magma in the upper parts of the earth's crust.

The major orogenic cycles normally envolve a long duration of time sometimes stretching in several tens of million years. The synchronity of such tectonic cycles in all parts of the earth's crust has also not been established. It is more likely that they were diachronous. However, the orogenic cycles marked the closing phases of successive stages of earth's evolution and they are also significant in stratigraphic reconstructions.

The Precambrian Basement of the Canadian Shield is separated from the Proterozoic Cover by the **Main (Eparchaean) Unconformity** that represents a major tectonic cycle. Similar pronounced unconformities separating the strongly deformed and metamorphosed rocks of the Precambrian Basement from the less deformed and less metamorphosed rocks of the sedimentary cover have also been identified over the Indian Peninsula (Fig. 4.4). The isotopic data from the Basement rocks of the Indian Peninsula, however, indicate the presence of some Lower Proterozoic elements in the Precambrian Basement. These Lower Proterozoic rocks were formed at the culmination of earliest phase of the evolution of the Indian Craton.

The most parts of the Indian shield had stablised at about 2100 million years ago. They were partly submerged beneath the shelf seas giving rise to the platform type of sediments. The Aravalli Province, however, continued to remain under the geosynclinal influence during the Early and Middle Proterozoic time. The Aravalli Group of rocks that forms a part of the Basement is overlain by the **Raialo Group** that is considered as a transitional stratigraphic unit represented elsewhere on the Peninsula by a gap in deposition.

A predominantly carbonate succession exposed near Raialo village in Alwar district of Rajasthan (Fig. 5.5) was named as "Raialo Series" by Hacket (1877). The succession, comprising mainly limestone with thin beds of quartzites and occasional conglomerates, attains an average thickness of over 700 metres. Frequently the basal arenaceous beds are missing and the carbonate rocks rest with distinct and sharp boundaries over the argillaceous rich meta-sediments of the Aravalli Group or over the Banded Gneissic Complex. The higher units of the Raialo Group consist of thin beds of quartzites and quartz-mica schist having resemblance with similar arenaceous rocks of the basal Delhi Supergroup.

The basin of the deposition of the Raialo Group extended in two ENE-WSW trending belts along the western and eastern flanks of the "Delhi Synclinorium". The Raialo Group gradually becomes thinner to its complete elimination in the southern Aravalli where the Delhi succession rests directly over the Archaean rocks. The western outcrops occur in two NNE-SSW trending discontinuous belts around Makrana and Ras in the southwest of Ajmer. The Makrana Marble of the succession provided excellent building stones used in the construction of Taj Mahal at Agra.

PROTEROZOIC SUCCESSION

The Proterozoic rocks are extensively exposed both in the northern as well as the southern parts of the Indian Peninsula (Table 6.2). The rock formations of the Cuddapah Supergroup and its equivalent were earlier referred to as Purana formations. In some later work, the Purana time refers to the time of deposition of the rock formations of both, the Cuddapah and Vindhyan Supergroups. The Cuddapah Supergroup and its equivalent have been assigned to Precambrian V (1600 to 900 Ma) and the Vindhyan Supergroup and equivalent formations were included in Precambrian VI (900 to 570 Ma) by Sarkar (1980). The Delhi Supergroup which has been generally correlated with the Cuddapah Supergroup has been included by Sarkar (1980) in the Precambrian IV (2500 to 1600 Ma).

Table 6.2: Proterozoic formations of the Indian Peninsula comprising the sedimentary cover

Main Divisions	Southern Peninsula	Northern Peninsula
Upper Purana	Kurnool Group, Bhima Group, Sullavai Group, Indravati Group, Chattisgarh Group	Upper Vindhyan Group Malani Volcanics
	Unconformity	
Lower Purana	Cuddapah Supergroup, Kaladgi Group, Pakhal Group	Lower Vindhyan Group Gwalior Group Bijawar Group Kolhan Group Delhi Supergroup

From the radiometric data (Sarkar, 1980) it appears that the Lower Purana rocks were deposited during the Middle Proterozoic to Middle Riphean time whereas the Upper Purana rocks were deposited during the

Upper Riphean to Vendian time. Upper parts of the Upper Purana succession in the northern Peninsula presumably transgressed in time to Early Cambrian (see Chapter 7). The stratigraphic boundaries in the Proterozoic rocks of the Indian Peninsula do not correspond to those identified in the International Stratigraphic Scale. The Indian Proterozoic rocks have been described in the sequel under the two headings, namely the Lower and the Upper Purana Successions. The Lower Vindhyans belonging to the Lower Purana Succession have been taken along with the Upper Vindhyans in a unified description of the Vindhyan Supergroup.

LOWER PURANA SUCCESSION

The Lower Purana rocks of the southern Indian Peninsula were deposited in the platform type basins. These rocks are preserved in the Cuddapah Depression, the Kaladgi Basin and the Godavari Graben (Fig. 6.4). The sedimentary successions were presumably deposited in a single sedimentary

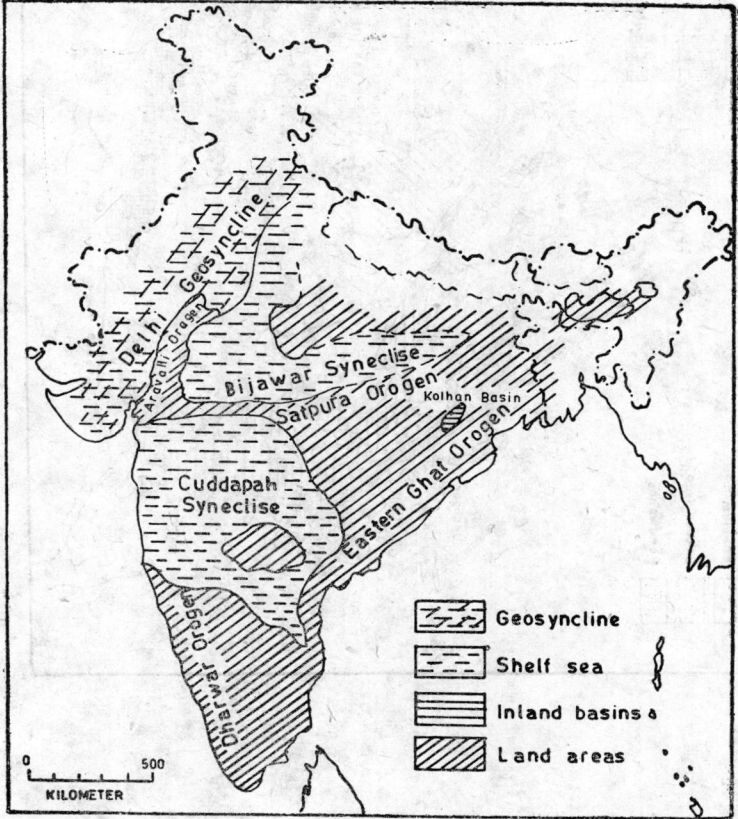

Fig. 6.1: Palaeogeographic map of the Indian Peninsula for the Early Purana time.

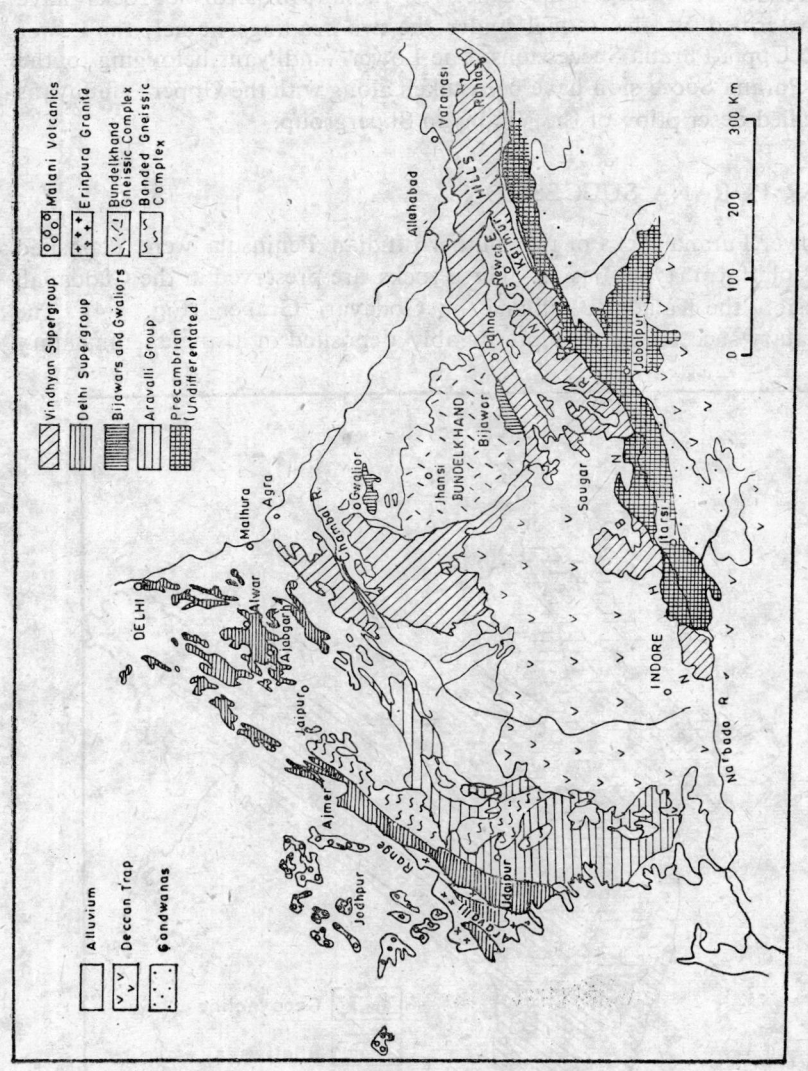

Fig. 6.2: Distribution of the Proterozoic formations of the northern Indian Peninsula (based on various sources).

basin covering the southern central parts of the Indian Peninsula. The Middle Proterozoic succession of the northern India exposed in the "Delhi Synclinorium" of the Aravalli range (Fig. 6.2) is represented by the geosynclinal sequence of the Delhi Supergroup. The Middle Proterozoic rocks known from the southeastern and northwestern edges of the Bundelkhand massif have been grouped into Bijawar and Gwalior Groups. These rock groups show platformal characters similar to Lower Purana succession of the southern India. The northern and the southern palaeo-seas of the Early Purana time were presumably separated by a land barrier constituting the Satpura Orogen (Fig. 6.1).

Delhi Supergroup

The rocks of the Delhi Supergroup are exposed in the main Aravalli mountain chain extending from Delhi in a southwestward direction through parts of Haryana and Rajasthan to the parts of Gujarat. The Aravalli mountains in its northern parts seem to have almost completely drowned leaving a few remnants in the form of isolated ridges of Delhi rocks protruding out of the great sand desert in the south of Delhi and in the northeastern parts of Rajasthan.

The Delhi Supergroup estimated to be about 7 kilometres thick comprises a quartzite-slate succession. The lower part of the succession known as 'Alwar Group' consists of predominantly arenaceous facies (Fig. 6.3). The upper part of the succession known as 'Ajabgarh Group' is composed of argillaceous-calcareous slates and schists. In Alwar area, the two groups of rock formations are separated by an intervening calcareous succession known as 'Kushalgarh Limestone'. The rock succession of the Delhi Supergroup has been intruded by Erinpura Granite exposed in the south-central parts of the Aravalli mountains.

Alwar Group: The rocks of the Alwar Group are distributed, though unevenly, throughout the northern parts of the Aravalli range. Southwards, it disappears in the central parts of the range and reappears in the southern parts. In the northern areas, the Alwar Group rests over the Raialo Group with an unconformable contact. In the southern areas, this succession rests directly over the rocks of the Aravalli Group. Basal arkose and quartzites of the Alwar Group contain fragments of granitic rocks presumably derived from the underlying Banded Gneissic Complex and other granitic rocks of the Archaean Basement. In the type area in the northern Rajasthan, the Alwar succession estimated to be more than 3 kilometres in thickness is almost entirely made up of compact quartzites, grits and conglomerates with a few argillaceous beds and impure limestones. The succession contains several sill like bodies of epidiorites showing both the intrusive and extrusive characters.

Kushalgarh Limestone: The Kushalgarh Limestone usually banded with dark grey and black layers has an impure character containing small amounts of quartz, mica and some form of carbon. The limestone is coarse crystalline when it is highly metamorphosed. The succession attaining a maximum thickness of about 700 metres contains a few horizons of brecciated rocks known as **Hornstone Breccia**. Typically, the breccia consists of angular pieces of quartz embedded in a fine grained matrix of ferruginous and siliceous matter.

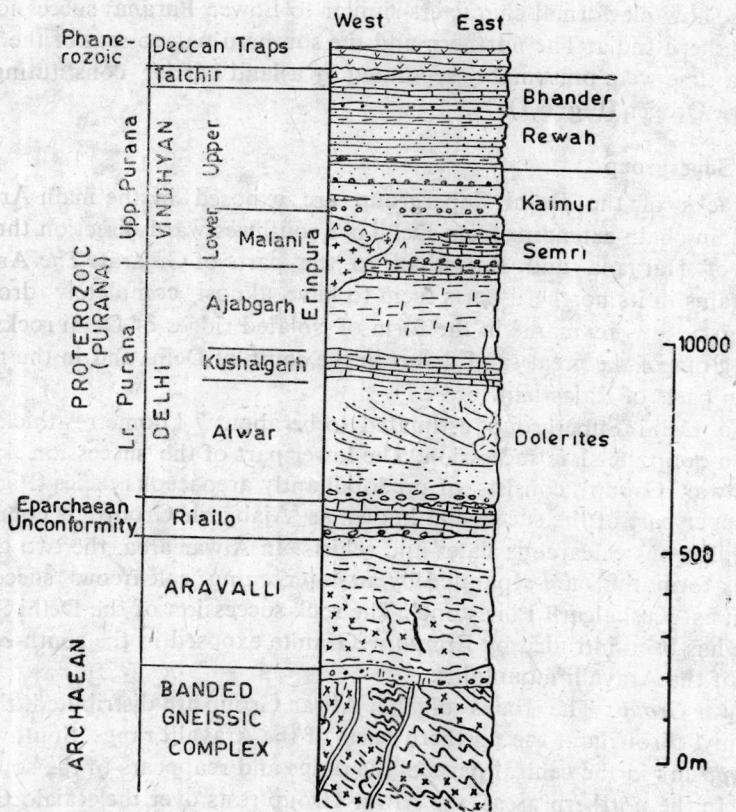

Fig. 6.3: Stratigraphic classification of the rocks of the Aravalli mountains (based on Gansser, 1964).

Ajabgarh Group: The rocks of the Ajabgarh Group representing a deep water argillaceous-calcareous facies vary in mineral composition depending upon the degree of metamorphism. The commonest rock types are splintery slates and sheeny phyllites. Upper contacts of the Ajabgarh Group is nowhere exposed. The Aravalli region has remained a positive relief area of the Indian Peninsula since the end of the time of the deposition of the Ajabgarh Group. The minimum thickness of the Ajabgarh Group has been estimated to be of the same order as that of the Alwar Group.

Erinpura Granite: Granitic intrusions in the succession of the Delhi Supergroup observed at many places along the Aravalli range include the intrusions of Palanpur, Idar, Sirohi, Erinpura, Beawar, Jaipur and Alwar. They are together known as 'Erinpura Granite' as it was first studied in some detail in this type area. The granite is essentially a coarse-grained non-foliated rock. Foliation has developed at only a few places. It is primarily a biotite granite but the associated pegmatites contain some muscovite and tourmaline. Emplacement of the Erinpura Granite is related with the phase of Delhi Orogeny which led to the deformation and metamorphism of the rocks of the Delhi Supergroup. The Bairath Granite representing the closing phase of the Delhi Orogeny has given an isochron age of 1660 million years (Sarkar, 1980).

Bijawar and Gwalior Groups

The Middle Proterozic formations deposited over the Bundelkhand massif along its southeastern and northwestern margins are known as Bijawar Group and Gwalior Group respectively (Fig. 6.7). The Bijawar Group comprises a succession of a basal conglomerate and quartzite overlain by hornstone breccia, limestone, phyllitic shales, red jaspars and dioritic traps. These rocks are again well exposed along the southeastern edges of the Vindhyan Syneclise where they dip under the Semri Group of the Vindhyan Supergroup.

The rocks of the Gwalior Group exposed in the hills surrounding the Gwalior city of Madhya Pradesh are overlain by the rocks of the Upper Vindhyan succession. The Gwalior Group consists of a lower arenaceous facies unit known as **Par Formation** and an upper unit of varied facies known as **Morar Formation**. The lithology of the succession has much similarity to that of the Bijawar Group. However, the Gwalior rocks appear to be relatively less deformed as compared to the Bijawars.

Kolhan Group

The Kolhan Group comprising basal purple sandstones and conglomerates overlain by limestones and slates is exposed in Singhbhum region (Fig. 5.3) in the form of an elliptical outcrop stretching in SSW direction from Chaibasa to the north of Noamandi. The rocks of the Kolhan Group having gentle dips along the southeastern margin of the outcrop rest unconformably over the Singhbhum Granite. The rocks of the group become increasingly more deformed along the northeastern margins where they come in contact with the rocks of the Iron Ore Group. Pebbles in the conglomerate derived from the Singhbhum Granite and the Banded Hematite Jaspar of the Iron Ore Group indicate that the Kolhan Group was deposited over a basement comprising the Iron Ore Group and the Singhbhum Granite.

Cuddapah Supergroup

The crescent shaped basin of the Cuddapah Depression is concave towards the east in parallelism with the general configuration of the Coromondal coast line (Fig. 6.4). Physiographically, the region, about 300 kilometres along its length and 140 kilometres across, is characterised by a few north-south parallel ridges. The Velikonda range along the eastern flank of the basin overlooks the gneissic coastal plains of Nellore and Guntur. The Nallamalai range along the middle of the crescent consists of high hills of folded Cuddapah rocks. The ridges at the western flank extending from Triputi in the south to Kurnool in the north overlook the gneissic uplands of Mysore and Bellary.

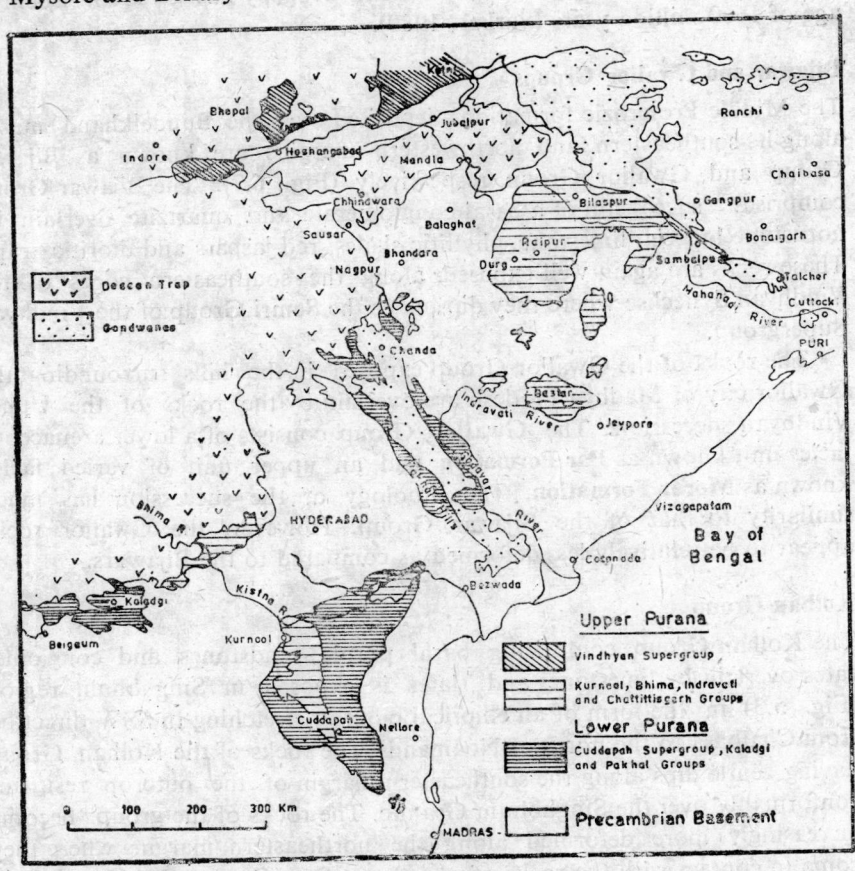

Fig. 6.4: Distribution of the Proterozoic formations of the southern Indian Peninsula (based on various sources).

The Cuddapah Supergroup comprises over 6 kilometres thick succession of essentially quartzites, quartzitic sandstones and slates or shales with subordinate limestones. The lower parts of the succession are intermixed with a great thickness of volcanic rocks. Rich iron deposits occur at various horizons of the succession. The rocks show, in general, a higher degree of metamorphism and deformation along the eastern flank of the crescent as compared to the rocks of the western flank. The succession has been divided into four groups of rock formations each separated from one another by marked unconformities (Table 6.3).

Table 6.3: Classification of the Cuddapah Supergroup (after King, 1872)

Groups	Formations	Thickness in metres
	KURNOOL GROUP	
	Unconformity	
CUDDAPAH SUPERGROUP		
Kistna Group	Srisailam Quartzite	600
	Kolamnala Shales	
	Irlakonda Quartzite	
	Unconformity	
Nallamalai Group	Cumbum Shales	1000
	Bairakonda Quartzite	
	Unconformity	
Cheyair Group	Tadpatri (Pullampset) Shales	3300
	Pulivendla (Nagri) Quartzites	
	Unconformity	
Papaghani Group	Vempalle Shales & Limestones	1400
	Gulcheru Quartzite	
	Eparchean Unconformity	
	ARCHAEAN BASEMENT	

The Papaghani and Cheyair Groups are exposed along the southwestern margin and upper central parts (Eshwarakupam hill) of the Cuddapah crescent (Fig. 6.5). The Nallamalai Group is exposed in the range of that name along the middle part of the crescent. The rocks of the Kistna Group are confined along the eastern margin and the northwestern areas of the crescent. The rocks of the Cuddapah Supergroup are presumably concealed beneath the younger Kurnool succession in the western and northern parts of the crescent. The four divisions of the Cuddapah succession have overlapping characters over one another indicating a shifting nature of the basin during the deposition of the succession. Each division of the succession begins with coarse sediments (quartzites and conglomerates) succeeded by the finer sediments (shales and slates) and carbonate rocks indicating a cycle of deposition. The four major cycles of deposition represented by the four groups of the rock formations were intermittently broken by phases of upheaval and erosion.

Papaghani Group: The 1400 m thick succession of the Papaghani Group best exposed in the Papaghani river valley has been further subdivided into two formations. The lower formation, named as **Gulcheru**

Quartzite, rests over the undulating denuded surface of the Archaean gneisses. The quartzites rise along the southwestern margin of the crescent in the form of steep cliffs. This formation often contains beds of conglomerates whose pebbles are derived from the Archaean Basement. The upper formation, known as **Vempelle Shales** and **Limestones** is three times thicker than the lower arenaceous formation. In the northwestern outcrops, the upper unit has overlapped the Gulcheru Quartzite bringing it to rest

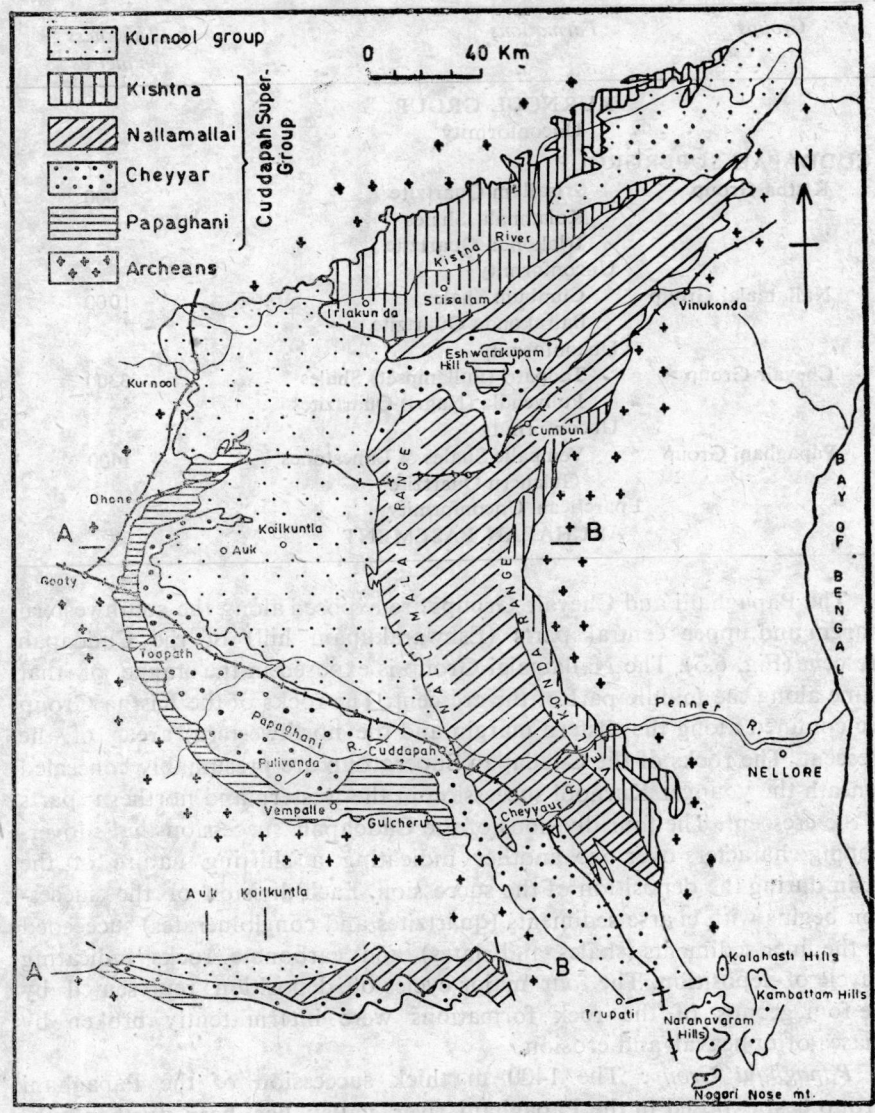

Fig. 6.5: Geological map of the Cuddapah Depression (based on Narayanaswamy, 1966).

directly over the Archaean Basement. Stromatolites (*Collenia vajurkari*) have been reported from the limestone of the formation (Prasad and Verma, 1967).

Cheyair Group: The Cheyair Group well exposed in the Cheyair river and in the upstream areas of the Penner river around Tadpatri comprises about 3300 m thick succession of quartzites and shales. The **Pulivendla Quartzites** constituting the lower formation of the group contains pebbles derived from the underlying Papaghani Group and the Archaean Basement. The southeastern extension of Pulivendla Quartzite is known as **Nagri Quartzite**. The upper formation known as **Tadpatri Shales** comprises a thick succession of shales with thin beds of siliceous limestones, basic rocks, ferruginous chert and jaspar. Discoidal markings of doubtful organic origin have been observed from the limestone of the upper formation.

Two main phases of emplacement of volcanic rocks are characteristically well marked in the Cheyair succession. Basic sills having intruded the limestones of the underlying Papaghani Group contain at their contacts with the host rocks some economic deposits of asbestos, serpentine and steatite. A volcanic focii, which was presumably the source of eruption represented by the volcanic rocks of the Cuddapah Supergroup, is located at about 40 kilometres from the western edge of the Cuddapah crescent at Wajrakarur village in the Anantpur district of Andhra Pradesh. The bluish grey rock of the volcanic neck resembles the diamond bearing kimberlites of Africa. Specks of diamond were reported to occur in the stream sediments of the region. The diamond occurrence in the conglomerates of the overlying younger succession of the Kurnool Group is also presumably derived from this source.

Nallamalai Group: The succession best exposed in the Nallamalai range has been divided into two formations. The lower formation known as **Bairakonda Quartzite** is best exposed in the hills of that name. The upper formation known as **Cumbum Shales** occupies the synclinal troughs of the folded Cuddapah rocks. The shales are often intercalated with beds of limestones and quartzites. Lead and copper mineralisation has been reported from the rocks of the Nallamalai Group.

Kistna Group: The rocks of the group are well exposed along the Krishna (Kistna) river cutting a deep gorge in the Krishna-Nallamalai plateau in the northwestern parts of the crescent. The succession has been divided into three well differentiated formations. The lowermost **Irlakonda Quartzite** forming a plateau of that name consists of about 400 metres thick massive bedded quartzites. The quartzites are overlain by about 200 metre thick **Kolamnala Shales** named after a tributary of Krishna river. The shales often contain thick beds of ferruginous quartzites. The **Srisailam Quartzite** overlying the Kolmanala Shales form the higher plateau region of the well known shrine of Srisailam overlooking the Krishna river.

Kaladgi and Pakhal Groups

The **Kaladgi Group** named after the Kaladgi town in Bijapur District of Maharashtra rests unconformably over the Peninsular Gneissic Complex of Archaean age and it is unconformably overlain by the Deccan Traps of Cretaceous-Tertiary age. A major portion of the Kaladgi succession is presumably concealed beneath the volcanic flows of the Deccan Traps. A few isolated outcrops of rock succession similar to the Kaladgi Group are known from deeper valleys of the Deccan country lying in the north of the Kaladgi basin.

The Kaladgi Group comprises about 3 to 5 kilometres thick succession of arenaceous, argillaceous, and calcareous rocks. The sediments show a general decrease in coarseness from bottom to top representing at least two major cycles of deposition corresponding to the lower and the upper formations of the Kaladgi Group. The lower formation begins with arenaceous coarse sediments (quartzites and conglomerates) followed by siliceous limestones, sandstones and shales. The upper formation comprises a basal coarse arenaceoes unit followed by ferruginous limestones and shales. Stromatolites such as *Collenia compacta, C. columnaris, Cryptozoan proliferum, Cryptozoan* sp. and a rich assemblage of acritarchs have been reported from the succession of the Kaladgi Group.

The Proterozoic rocks exposed on the southwestern and northeastern Pakhal hills of the Godavari valley have been grouped as **Pakhal Group**. The succession divisible into a 2800 m thick lower formation and a 3500 m thick upper formation comprises coarse pebbly sandstones, conglomerates, siliceous limestones, shales and slates. The two subdivisions represent two major cycles of deposition similar to those of the Kaladgi Group. Presence of glauconite, bedded cherts and high MgO contents in carbonates, moderate to well sorted nature of sediments and secondary overgrowths of felspars indicate the marine epineritic environment of deposition.

UPPER PURANA SUCCESSION

Paleogeography of the Upper Purana time had undergone a marked change from that of its precursor. The shelf sea of the southern central Peninsula had slightly shifted northward and, at the same time, it had enlarged northeastward to include the Chattisgarh and Bastar Depressions (Fig. 6.6). In the northern part of the Peninsula, the Delhi geosyncline has closed with the formation of the Delhi Orogen. The Bijawar shelf sea was replaced by the Vindhyan Synclise over the northern edge of the Peninsula. The Vindhyan Sea presumably extended northward into the Lesser Himalayan region. To the west of the Delhi Orogen, a series of mainly acidic volcanic rocks known as '**Malani Volcanics**, were laid during the time of deposition of the lower parts of the Upper Vindhyans. The volcanics are overlain by arenaceous formation which resemble the uppermost succession of the Vindhyan Supergroup deposited in the Vindhyan Syncline during the Late Proterozoic time. The uppermost parts of the Vindhyan succession presumably extended in time to the lowermost Palaeozoic Era.

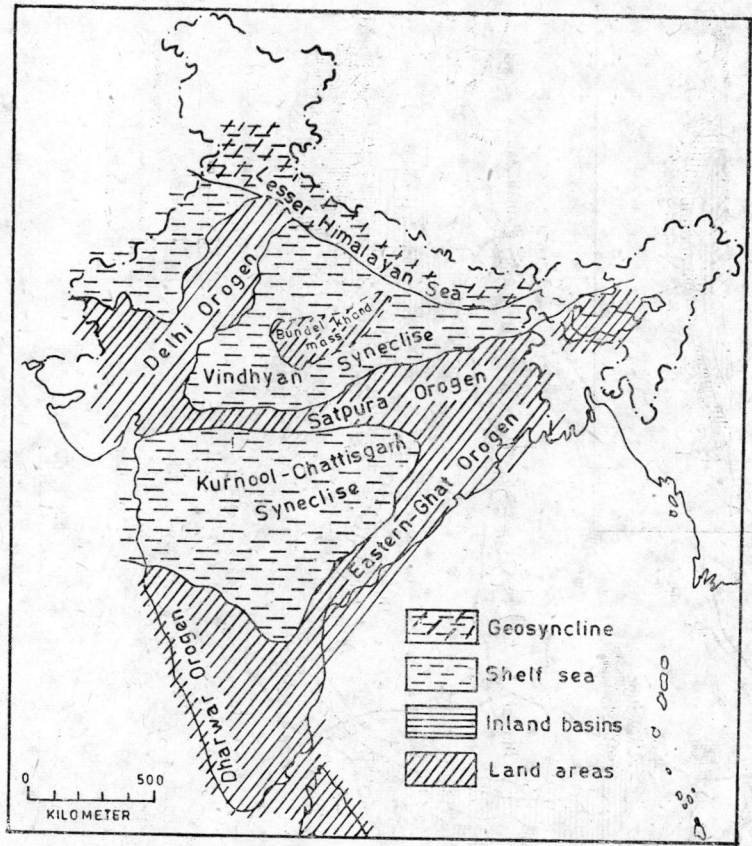

Fig. 6.6: Palaeogeographic map of the Indian Peninsula for the Late Purana time.

Vindhyan Supergroup

The Vindhyan Supergroup named after the scarped plateau mountains on the north side of the Narmada valley is exposed in an area of about 104,000 square kilometres (Fig 6.7). An area of about 78,000 square kilometres of the Vindhyan rocks are concealed under the Deccan Traps of the Malwa plateau and the Alluvium of the Indo-Gangetic Plain. The term "Vindhyans" was first used for a thick succession of sandstone formation of Bundelkhand and Malwa regions where this formation rests directly over the rocks of Gwalior Group. The sandstone formation is underlain in southeastern and southwestern parts of the Vindhyan Syneclise by a calcareous-argillaceous succession named earlier as "Semri Series" that was separated from the overlying "Vindhyans". The contact between them is marked by a well defined unconformity. Later, the "Semri Series" was

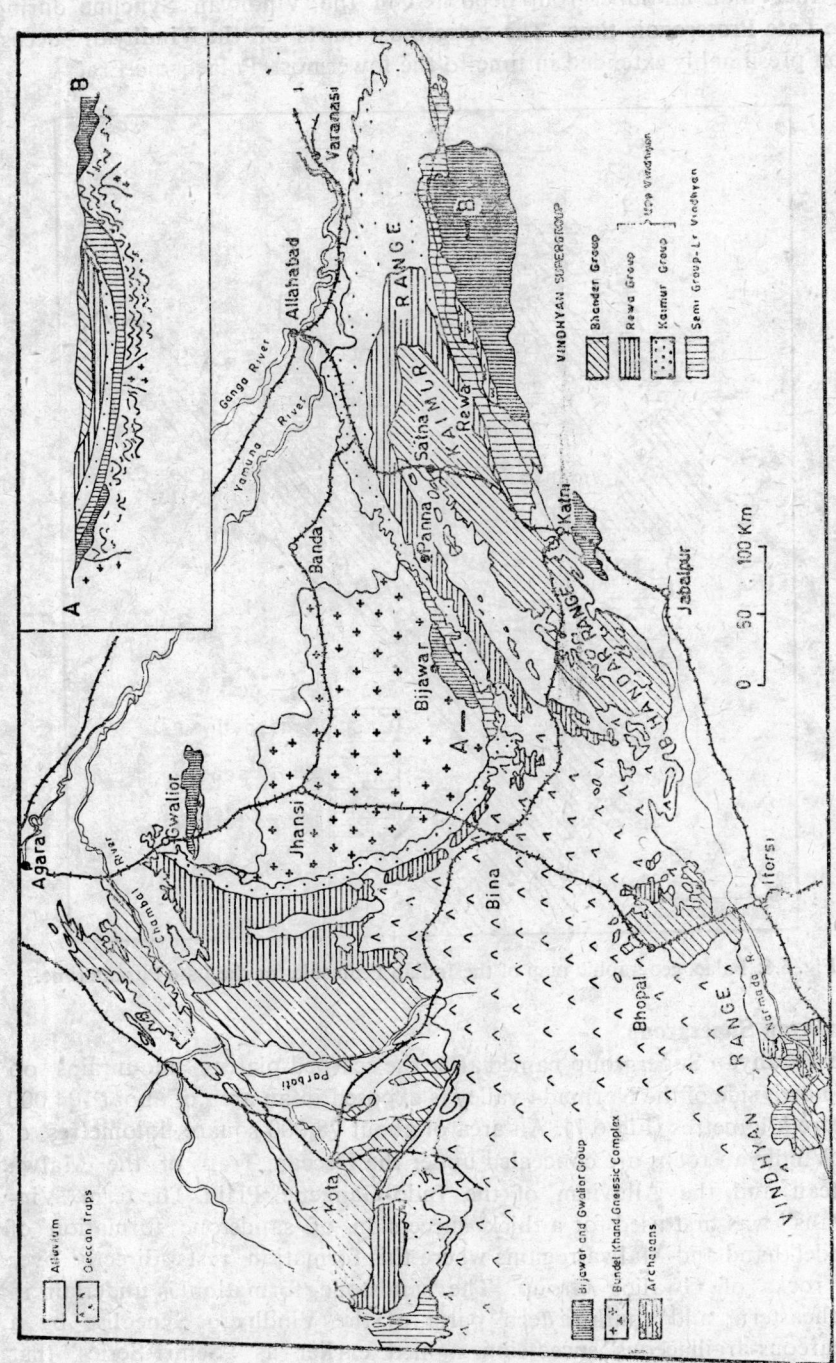

Fig. 6.7: Geological map of the Vindhyan Syneclise (based on Auden, 1933).

included within the Vindhyans sequence constituting the Lower Vindhyans. The Upper Vindhyans consisting of a predominantly sandstone succession are three times larger in thickness than the Lower Vindhyans. The Upper Vindhyan Succession has been divided into three groups of rock formation (Table 6.4).

Table 6.4: Classification of the Vindhyan Supergroup

	Groups	Formations	Thickness in metres
Upper Vindhyans	Bhander Group	Upper Bhander Sandstone Sirbu Shales Lower Bhander Sandstone Bhander Limestone Ganurgarh Shales	1000
	Rewah Group	Upper Rewah Sandstone Jhiri Shales Lower Rewah Sandstone Panna Shales	2000
	Kaimur Group	Upper Kaimur Sandstone Bijagarh Shales Lower Kaimur Sandstone	400
		Unconformity	

Lower Vindhyans	Semri Group	Rohtas Formation:
		  Suket Shales
		  Nimbahara Limestone and Shale
		  Conglomerate and Sandstone
		Kheinjua Formation:
		  Glauconitic Member
		  Fawn Limestone
		  Olive Shales
		Porcellinite Formation:
		  Porcellinic Shales
		  Trappoid Beds
		  Porcellinic Shales
		Basal Formation:
		  Kajrahat Limestone
		  Bleaching Shales
		  Basal Quartzites and Conglomerates
			1300

Semri Group (*Lower Vindhyans*): The rocks of the Semri Group are best exposed in the Son valley, Dhar Forest. along the Kaimur Scarp on the north side of the Vindhyan Plateau and in the neighbourhood of Chittor in Rajasthan. The group comprising about 1300 metre thick shale and limestone succession with a basal conglomerate horizon has been further subdivided into four formations (Table 6.4). The basal rocks rest unconformably over the Bijawar Group along the Son valley where the dips of the Semri rocks as well as the Bijawar succession have become steep. The Porcellinite Formation contains volcanic products which were presumably contemporaneous with the Malani Volcanies of the Western Rajasthan.

The Fawn Limestone of Kheinjua Formation contains well preserved stromatolites identified as *Conophyton cylindricus*, *Collenia columnaris* and *Collenia* sp. The forms have been correlated with those found in the Lower Ripheans of U.S.S.R. regarded to be older in age than 1260 million years. Glauconitic beds from the Kheinjua Formation have given radiometric dates as 1100 ± 70 million years (Vinogradov and Tugarinov, 1964).

The Suket Shales from the outcrops of the Semri Group of Rajasthan have afforded disc shaped remains of doubtful organic origin (Jones, 1909). A great controversy ensued regarding the significance of these disc shaped objects. Some regarded them to be of organic origin representing the genus *Obolella* described from the Precambrian rocks of Arizona (U.S.A.). Others considered them to be remains of primitive brachiopods similar to *Acrothele* known from the Cambrian of Salt Range (Pakistan). The disc shaped forms were regarded by yet another group of experts to represent the genera *Protobolella* and *Fermoria*, the primitive organic forms having characters common to plant and animal kingdoms. Yet others consider them to be of algal origin. Some experts have questioned the organic origin of the disc shaped objects found in the Suket Shales and consider them to have been formed by colloidal precipitation of mineral matter.

Upper Vindhyans: The Upper Vindhyan succession rests over the Semri Group with a well marked unconformity in the Son Valley and Dhar Forest areas. In other places, the Lower Vindhyans are missing and the Upper Vindhyans rest over the Bundelkhand Gneisses, Gwalior Group and Bijawar Group. The Upper Vindhyans having a fairly uniform character consist of a predominantly arenaceous succession with subordinate shales and limestones. Shales often contain carbonaceous matter and microscopic plant remains. The sandstones constituting large open plateau frequently show shallow water structures characteristic of flood plains and deltaic deposits. Presence of coarser sediments in the south indicates that the (Vindhyan) sea was open towards the north (Fig. 6.6). Grain textures and red colours of the sandstones suggest eolian origin of the sediments under dry climatic conditions. The Upper Vindhyan succession has been divided into three groups of rock formations (Table 6.4).

The **Kaimur Group** begins with a thin conglomerate horizon containing pebbles of red jaspars. The jaspars were presumably derived from the Bijawar Group and Gwalior Group. The 400 m thick succession that follows the conglomeratic horizon has been further subdivided into Lower Kaimur Sandstone, Bijagarh Shales and Upper Kaimur Sandstone. The Lower and the Upper Kaimur Sandstone formations comprise flagstones, siltstones and quartzites. All the major sandstone horizons form scarps while the shale horizons constitute gentle slopes of the region.

The **Rewa Group** comprising a 2000 m thick succession of red shale and sandstone beds is separated from the underlying succession of the Kaimur Group by a diamond bearing horizon of conglomerate. The conglomerate bed about 70 cm in thickness consists of pebbles of vein quartz, jaspar and green quartzites in a matrix of sand and clay. Diamond occurs as placer deposit in the form of granules and grains in the matrix. The Rewa succession having a less consistent lithological characters as compared to the other two groups have been further subdivided into alternating two shale and two sandstone formations (Table 6.4). The upper limit of the Rewa Group is again marked by a diamond bearing conglomerate horizon. The famous Panna diamond field of India is situated in the country of Rewa rocks.

The **Bhander Group** comprising the most widely distributed group of the Upper Vindhyans consists of about a 1000 m thick succession of shales, limestones, dolomites and sandstones. The succession has been divided into five formations (Table 6.4). The Upper Bhander Sandstone having a uniformly fine grained character has been the most popular and versatile building stone of northern India. It has a lithological resemblance with the Purple Sandstone of Salt Range that has been assigned an Early Cambrian age.

A succession of predominantly arenaceous rocks with subordinate limestone resembling the Upper Vindhyans is known from the western parts of the Aravalli mountains (Fig. 5.5). The sandstone formation extending from Jodhpur to Pokharan in Rajasthan has been informally grouped as **Jodhpur Sandstones**.

Malani Volcanics: The Jodhpur Sandstone is unconformably underlain by acidic volcanic rocks known as Malani Volcanics. The Volcanics exposed in the west of the Aravalli mountains (Fig. 5.5) are associated with igneous rocks of intrusive phase. The extrusive and the intrusive phases together are grouped into the **Malani Igneous Suite**. The extrusive phase includes a cyclic sequence of rhyolites and tuffs. The intrusive phase is made up of co-magmatic Siwana and Jalor granites, and dikes of felsic to mafic composition. The volcanic rocks are potassic in chemical characters whereas the dikes show sodic characters (Pareek, 1981).

The Malani Igneous Suite represents one of the world's largest example of felsic magmatism. The lava flows are solely acidic in composition which prohibited their mobility to extend for longer distances. A total of 52 flows aggregating to a total thickness of about 3500 metres have been observed in the northern parts of the Malani Igneous Province. These rocks were formed after the deposition of the rocks of the Delhi Group and before the onset of the sedimentation of the Jodhpur Sandstone (Table 6.5).

Table 6.5: Precambrian Geochronology of the North Indian Peninsula (Based on Kochhar, 1982)

Stratigraphic Units	Radiometric ages in million years
Jodhpur Sandstone	
unconformity	
Malani Igneous Suite	
Rajasthan	745±10
Kirana Hills (Pakistan)	870±40
Tosham Hills (Haryana)	940±20
unconformity	
Delhi Supergroup	1650
Raialo Group	
Aravalli Group	2000 to 2500 (?)
Banded Gneissic Complex	more than 2500

Stratigraphy after Heron, 1953; isotopic ages from Crawford, 1970; Crawford and Compston, 1970; and Davies and Crawford, 1971.

Kurnool Group

The Upper Proterozoic succession of the Cuddapah Depression (Fig. 6.5) known as Kurnool Group rests over the rocks of the Cuddapah Supergroup with a marked unconformity. The Kurnool Group, unlike the Cuddapah succession, is almost completely devoid of extrusive or intrusive phase of igneous activity.

The Kurnool Group named after the Kurnool district in southwestern Andhra Pradesh comprises alternating sequence of quartzites, shales and flaggy and massive limestones. The rocks of the Kurnool Group are exposed over a greater part of the western central and the northeastern parts of the Cuddapah crescent. The succession shows frequent lateral thinning and thickening indicative of frequent oscillation of the sea level during the deposition of the Kurnool succession. Each cycle of deposition culminated in the deposition of red beds, salt pseudomorphs and gypsum at the time of the withdrawal of marine water when the dry arid conditions were set in for some time before the beginning of the next cycle of sedimentation.

The Kurnool succession has been divided into four formations (Table 6.6) representing two major cycles of deposition. Each cycle begins with an arenaceous and rudaceous formation followed by a thick succession of argillaceous and carbonate rocks. The rocks of the Kurnool Group show gentle dips and broad north-south trending folds. Micro-planktons of Late Precambrian to Cambrian age such as *Archaeofavosia, Granomarginata, Priscogalea,* etc., have been reported from various horizons of the Kurnool succession.

Table 6.6: Stratigraphic subdivisions of Kurnool Group

Formation	Members	Thickness in metres
Kundair Formation	Nandayal Shale Member Koilakundla Limestone Member	200
Paniam Quartzites	Pinnacled Quartzite Member Plateau Quartzite Member	50
Jamalamadugu Formation	Auk Shale Member Narji Limestone Member	200
Banganpalle Formation	Quartzites, sandstones & basal diamond bearing conglomerates	10
Unconformity		
Cuddapah Supergroup		

Northern Extension of Kurnool Group

The Kurnool basin presumably covered almost the entire central Peninsula (Fig. 6.6). The rock successions considered equivalents of the Kurnool Group are exposed in the Bhima Basin, Godavari Graben and Bastar and Chattisgarh Depressions (Fig. 6.4). A part of the Upper Proterozoic succession is presumably buried under the Deccan Traps of the western India. The northernmost outcrops of the Upper Proterozoic succession of the southern India are barely 200 kilometres south of the Vindhyan succession of the northern India. The Vindhyan and Kurnool basins of the Late Proterozoic age were separated from each other by the Satpura orogen.

Bhima Group: The Bhima Group exposed in the Bhima river at the southern edge of the Deccan Syneclise rests unconformably over the Precambrian rocks of the Dharwar region. The succession has been divided into three formations. The lower formation consists of about 110 metre thick succession of a basal conglomerate followed by sandstone, green and purple shales. The middle formation having a gradational contact with the lower unit consists of about 165 metre thick flaggy limestones. The upper formation comprises a 100 metre thick succession of buff and purple shales, sandstones and flaggy limestone. Lithological characters of the three successive formations of the Bhima Group indicate a gradual deepening of the marine basin associated with a marine transgression during the deposition of the lower and middle formations and a gradual withdrawal of the marine water during the deposition of the upper formation.

Sullavai Group: The Sullavai Group exposed in Sullavai and Dewalmari hills of the Godavari valley is considered to be an equivalent of the Kurnool Group. The Sullavai Group comprising a 400 metre thick succession of conglomerates, sandstones, quartzites and slates rests unconformably over the rocks of the Pakhal Group of Middle Proterozoic age. The quartzites of the Sullavai Group have a great lithological similarity with the Pinnacled Quartzites of the Kurnool Group.

Indravati Group: The Indravati Group considered as equivalent of the Kurnool Group succession is exposed in the Indravati river traversing westward across the Bastar Depression (Fig. 6.4). The succession having a thickness of about 400 metres consists of basal quartzites and conglomerates followed by quartzites, shales, dolomites and limestones.

Chattisgarh Group: The Chattisgarh Depression covering an area of about 30,000 sq. km extends over a large portion of Durg, Raipur and Bilaspur districts of Madhya Pradesh and parts of Sambalpur district of Orissa (Fig. 6.4). The depression consists of about 1600 metre thick succession of sandstones, shales and limestones known as Chattisgarh Group. The Chattisgarh Group resting unconformably over the Lower Precambrian rocks has been divided into two formations. The lower formation, known as **Chandrapur Sandstone**, comprises about 500 metre thick quartzite succession. The upper formation known as **Raipur Formation** disconformably overlies the lower formation. The Raipur Formation consists of about 1000 metre thick sediments of variable characters. The succession of the upper formation was deposited in five cyclic phases of sedimentation. Each cyclic phase begins with dark limestone or light sandstone and extends upwards in the succession into red beds comprising clay-stones and red shales. The carbonate rocks contain certain well preserved stromatolites.

Badami Group: A part of the Kaladgi succession has been separated and grouped into Badami Group (Viswanathiah, 1968) on the basis of a revised geological mapping of the Kaladgi Basin. The Badami Group has been shown to unconformably overlie the rocks of the Kaladgi Group. The Badami Group consists of buff white sandstones with purplish pebbly sandstones at its lower horizons. The sedimentary structures such as current bedding and convolute bedding of shallow water origin are commonly observed in the succession. The Badami Group has been correlated with the Bhima and Kurnool groups on the basis of the similarity in stratigraphic position and micro-floral record. Chandrashekhar Gowda (1981-82) has, however, doubted the presence of an independent group within the Kaladgi succession.

Chapter 7
Precambrians of Extra-Peninsula

Extra-peninsular formations of Precambrian age exposed in various sections of the Lesser, Higher and Tethyan Himalayan Zones are generally identified on the basis of the absence of fossil record, a relatively higher grade of metamorphism and their relative stratigraphic positions below the known Phanerozoic successions. The absence of the fossil records characteristic for certain Phanerozoic sections is attributed to either unfavourable conditions of fossilizations or lack of detailed search for fossils. Criterion of a relatively higher grade of metamorphism is also not entirely reliable for demarcation of Precambrian successions in the Himalaya in view of the fact that at least one major event of metamorphism of Tertiary age has affected the Himalayan rocks ranging in age from Precambrian to Early Tertiary. The only reliable criterion for the identification of the Precambrian successions in the Himalaya pertains to their relative stratigraphic positions below the known Phanerozoic successions. However, the order of superpositions of the rock formations is often extremely difficult to decipher in view of a highly complex fold and thrust structures of the Himalaya.

The Precambrian succession of the Indian Peninsula has been grouped into Precambrian Basement and Cover (Proterozoic) sequences separated from each other by a profound unconformity. Such an unconformity has not yet been identified in the known Precambrian rock successions of the Himalaya. Correlation of certain high grade schists and gneisses of the Himalaya with the Archaeans of the Peninsula on the basis of their lithological similarity is doubtful. The rocks of Archaean age are presumably absent in the Himalaya. The Precambrian rocks showing a relatively higher grade of metamorphism and normally constituting the basement for the Tethyan Phanerozoic succession may be the equivalents of the Middle Proterozoic successions of the Peninsula. The Precambrian succession of the Lesser Himalaya has been generally regarded as the northern facies of the Vindhyan Supergroup.

PRECAMBRIANS OF THE TETHYAN BASEMENT

The Precambrian successions constituting the basement for the Phanerozoic succession of the Tethyan Himalaya comprise the Salkhala Group in the northwestern Himalaya (Jammu & Kashmir), the Vaikrita Group in the Spiti region of Himachal Pradesh and the Bhimphedi Group in the Central Nepal. The groups are overlain in a normal stratigraphic order by an unfossiliferous sequence of Late Precambrian to Early Palaeozoic age. The latter, in turn, conformably grades into a Lower Palaeozoic fossiliferous succession. The Jutogh Group and Daling Group occurring as highest structural nappes in the Lesser Himalayan regions have been presumably derived from the basement succession of the Tethyan Himalaya.

Salkhala Group

The oldest rock succession of the northwest Himalaya was named by D.N. Wadia (1934) as "Salkhala Series" after a village of that name on the left bank of Kishanganga river in the northwestern Kashmir (Fig. 7.1). The Salkhala outcrops have been traced in the form of a hair-pin bend around the north-western end of the Kashmir valley. In the Pir Panjal Range, the Salkhala rocks constitute the base of the Kashmir Nappe. In the northern parts, the Salkhala beds have been profusely injected with veins, sills and bosses of granitic gneisses and granites. The Salkhala Group comprises a succession of carbonaceous slates, grahitic phyllites, schists, carbonaceous limestones, dolomites, calcareous slates, marbles, mica schists and flaggy quartzites. The carbon rich sediments are usually associated with pyrite indicating an euxinic (anaerobic) facies for a part of the Salkhala succession. The basal parts of the Salkhala succession are frequently associated with bands of migmatitic and augen gneisses. The total succession of the Salkhala Group has been highly deformed into several isoclinal folds. The base of the succession is not exposed and, therefore, it is difficult to estimate the total thickness of the group. A minimum thickness is estimated to be of the order of 2 to 3 kilometres.

The Salkhala Group is stratigraphically overlain by Dogra Slates which conformably grades into the Lower Palaeozoic succession yielding trilobites and *Orthis* and *Obolus* like brachiopods. The contact between the Salkhala and Dogra successions is tectonised at most of the places. Only at a few places, the normal stratigraphic contacts have been observed. Even in these sections, the contacts are uncertain and at best they are regarded to represent a gradual passage from Salkhalas to Dogras.

The gneissic and granitic rocks of the Higher Himalaya known as **Central Gneisses** constitute the core of the central range of the highest peaks of the Himalaya. The Central Gneisses comprise a highly variable group of granitic rocks which probably range in age from Precambrian to Tertiary. The Nanga Parbat massif of the northern Kashmir constitutes one

Precambrians of Extra-Peninsula 105

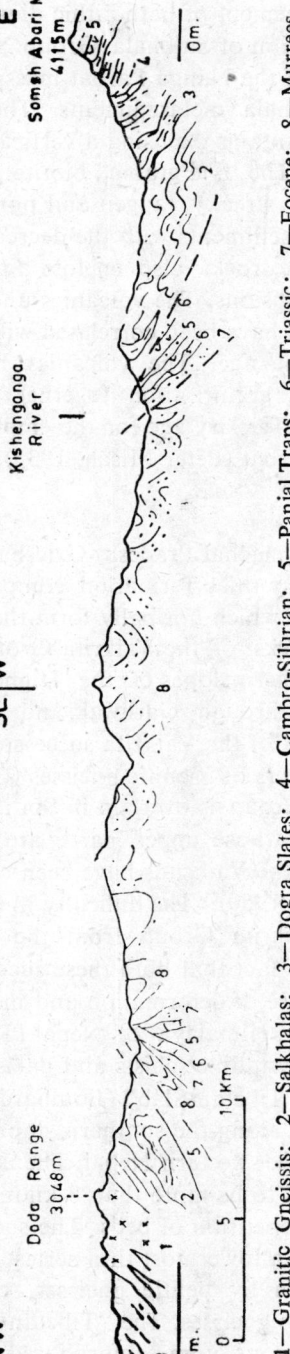

Fig. 7.1: Geological cross-section across the Western Himalaya (based on Wadia, 1934)

1—Granitic Gneissis; 2—Salkhalas; 3—Dogra Slates; 4—Cambro-Silurian; 5—Panjal Traps; 6—Triassic; 7—Eocene; 8—Murrees

such granitic emplacement of batholithic dimension of Tertiary age in the Precambrian succession of Salkhala Group.

Gneissic rocks of the Nanga Parbat massif are the products of granitisation of the Salkhala metasediments. The massif rises abruptly from a relatively lower mountains exposing a vertical section of about 4 kilometres of cliff face. The fine grained biotite gneisses form the predominant variety among the streaky, augen and porphyritic gneisses. The gneisses grade into the metasediments with the decrease in the grade of metamorphism. The gneissic rocks also enclose a thick succession of basic lavas, tuffs and noritic intrusions. The volcanics unconformably resting over the Salkhala succession have been correlated with the Dras Volcanics of Late Cretaceous to Eocene age. The volcanics have undergone deformation, metamorphism and granitisation together with the rocks of the Salkhala Group suggesting a Tertiary age for the emplacement of the biotite granites of the Nanga Parbat (Peter Misch, 1935).

Vaikrita Group

In Spiti area of Himachal Pradesh, Griesbach (1891) introduced the term "Vaikrita" for a very thick formation comprising largely mica and talc schists and phyllites which normally form the basement for the Palaeozoic succession. The rocks of the Vaikrita Group are observed in several sections along the northern slopes of the Higher Himalaya extending from Spiti in the west through Garhwal and Kumaun regions of U.P. into Nepal. A major part of the Vaikrita succession is almost always emplaced with concordant layers of granitic gneisses (Central Gneisses).

The Vaikrita Group is overlain in Spiti region by a succession known as **Haimanta Group** whose upper parts are regarded of Cambrian age. Rocks resembling the Vaikritas have been observed to grade laterally into Haimantas (Pascoe, 1950). The difficulty in differentiating the older succession of the Vaikrita Group from the younger Haimanta succession arises in view of the fact that both these successions have together undergone a Tertiary phase of deformation and metamorphism.

In the Tethyan Himalaya of Nepal (Thakhola region), the basement rocks comprising crystalline schists and gneisses have been grouped under an informal term "Tibetan Slab" (Lombard, 1953). The schists and gneisses are well exposed along the southern slopes of Dhauladhar and Annapurna ranges of the western Nepal. The unusually large thickness of the succession estimated to be more than 8 kilometres is probably due to intense folding and repetition of beds. The succession consists of four units, viz., beginning with a lowermost thin series of quartz, two mica schist, successively overlain by pelitic gneisses, calcareous gneisses and a thick pile of bedded augen gneisses. The "Tibetan Slab" is thrust in the south over the unfossiliferous unmetamorphosed sediments of the Lesser Himalaya. In the north, they are overlain by the Palaeozoic succession which have yielded from about 2 kilometres up in the succession the Ordovician fossils. The lower parts of the Palaeozoic succession have also been metamorphosed often up to amphibolite facies.

The contact between the Precambrian rocks comprising the "Tibetan Slab" and Palaeozoic succession has been obscured due to the high grade of metamorphism and intense deformation. However, this contact has been envisaged to be an unconformity representing a considerable hiatus in deposition. Such an unconformity has been observed in the case of the Precambrian-Palaeozoic contact in the succession of the Kathmandu Nappe of Central Nepal.

Bhimphedi Group

The Tethyan element, comprising a Precambrian basement and a Lower Palaeozoic cover, has come down in the Mahabharat range of the Nepalese Lesser Himalaya as Kathmandu Nappe (Fig. 7.2). The Precambrian succession of this nappe has been named as Bhimphedi Group after a village of that name in the south of Kathmandu (Stocklin, 1980). The succession has been divided into six formations (Fig. 7.3). The base of the succession is marked by a thrust named as Kathmandu or Mahabharat Thrust. The upper limit of the Bhimphedi Group is marked by an angular unconformity (Kumar et al., 1978) separating it from the overlying Lower Palaeozoic rocks. The angular unconformity was inferred on the basis of angular discordance of the older and younger beds at the outcrops, a sharp metamorphic break, an additional phase of deformation exhibited only by the underlying Precambrian rocks and a sudden change in sedimentary facies from the Precambrian succession to the Lower Palaeozoic succession.

Radua Formation: The lowermost stratigraphic unit of the Bhimphedi Group known as the Radua Formation consists of about a kilometre thick succession of garnet—two mica schist with occasional bands of amphibolites. Locally lenses and tongues of felspathic augen gneisses appear concordantly within the schists. The lithology is comparable with those of the lowermost unit of the "Tibetan Slab". Towards the base, the mica-schist changes into chloritic schists. The upper parts of the succession containing calcareous bands grade into the overlying carbonate formation.

Bhainsedoban Marble: The formation comprises about 800 m thick succession of coarse crystalline, well bedded to massive marbles containing mica in fine dispersion. In the basal and the upper parts of the succession, the mica partings and intercalations become frequent which characterises the gradual passage both from the underlying and into the overlying formations.

Kalitar Formation: This formation consists of about 2 kilometres thick succession of predominantly mica schists with occasional bands of micaceous quartzites. Garnet and amphibole minerals are common in lower parts but disappear higher in the section. Towards the top, a conglomeratic member appears locally between this formation and the overlying forma-

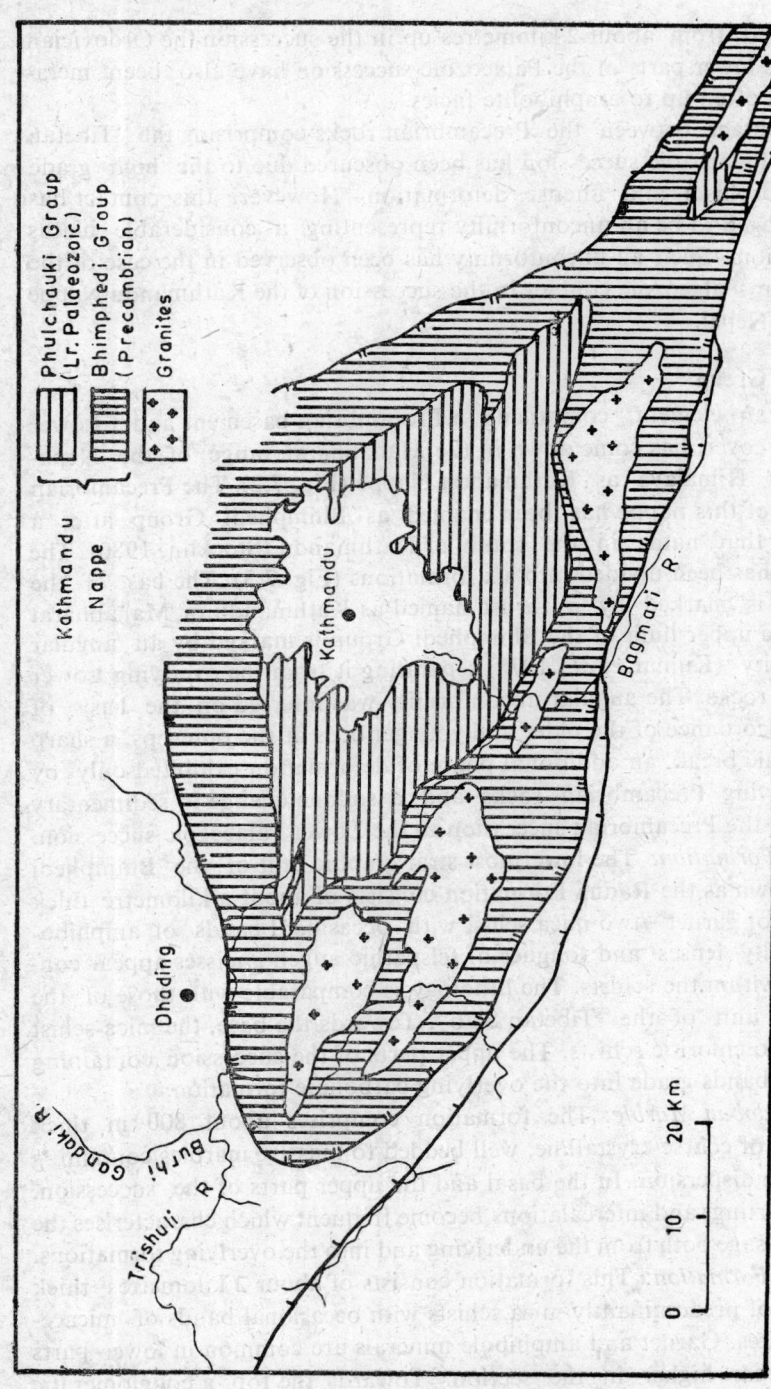

Fig. 7.2: Geological map of the Kathmandu Nappe (simplified from Kumar, 1980)

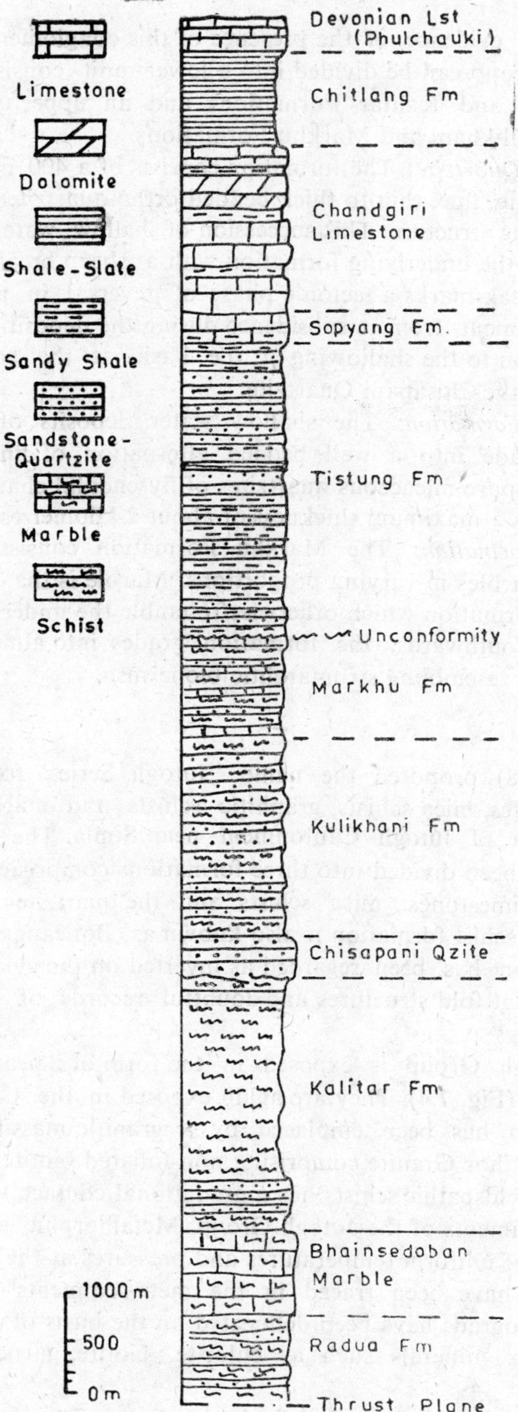

Precambrians of Extra-Peninsula

Fig. 7.3: Stratigraphical column showing the lithostratigraphic units of the Kathmandu Nappe (based on Kumar, 1980)

tion. On the evidence of the presence of this conglomerate member, the Bhimphedi Group can be divided into a lower unit consisting of Radua, Bhainsedoban and Kalitar Formations and an upper unit consisting of Chisapani, Kulikhani and Markhu Formations.

Chisapani Quartzite: The formation consists of a 400 metre thick succession of white, fine, thin to thick bedded ortho-quartzites showing strong current bedding structure. This succession of shallow water coastal deposits rests over the underlying formation with a sharp break in sedimentary facies. The break marks a tectonic phase of reversal in the direction of vertical movement from a subsidence during the deposition of the underlying formation to the shallowing of the basin at the beginning of the deposition of the Chisapani Quartzite.

Kulikhani Formation: The shallow water deposits of the Chisapani Quartzites grade into a well bedded alternation of fine grained biotite schists and impure micaceous quartzites of flyschoidal character. The formation attains a maximum thickness of about 2 kilometres.

Markhu Formation: The Markhu Formation consists of quartzites, schists and marbles in varying proportions. Marble is the distinctive lithology of this formation which otherwise resemble the underlying Kulikhani Formation. Southwards, the formation grades into almost pure massive marble bodies resembling stromatolitic bioherms.

Jutogh Group

Oldham (1888) proposed the name "Jutogh Series" for a succession of flaggy quartzites, mica schists, graphitic schists and marbles exposed in the type area of Jutogh Cantonment near Simla. The succession in the type area has been divided into three formations comprising carbonaceous schists and limestones, mica schists and the quartzites and schists. The quartzite and schist formation is also known as Boileaugunge Formation. The succession has been regarded as inverted on the evidence of inferred large recumbent fold structures and doubtful records of the sedimentary structures.

The Jutogh Group is exposed in the form of a pear shaped klippen around Simla (Fig. 7.4). They are again exposed in the Chor area where the succession has been emplaced by a granitic massif known as **Chor Granite**. The Chor Granite comprising non-foliated biotite-granite, granitic gneisses and feldspathic schists has a gradational contact with the surrounding meta-sediments of the Jutogh Group. Metamorphic isograds representing surfaces of uniform temperatures and pressures at the time of granitic emplacement have been traced in the meta-sediments surrounding the massif. The isograds have been delineated on the basis of the first appearance of index minerals such as chlorite, biotite, garnet and staurolite-kyanite.

At least three phases of deformation have affected the Jutogh meta-

Precambrians of Extra-Peninsula 111

sediments. The metamorphic isograds cut across the recumbent and isoclinal folds of first deformation indicating that the emplacement of the Chor Granite took place after the Jutogh metasediments were deformed into isoclinal and recumbent type folds. According to Pilgrim and West (1928), the recumbent folding, regional metamorphism and emplacement of Chor Granite took place prior to the deposition of the "Chail Series" of upper Proterozoic age. Thus, the Jutogh Group and the Chor Granite were assigned Archaean ages. However, some recent views on the age of the Chor Granite favour a Tertiary age in conformity with the age of the Nanga Parbat massif of northern Kashmir.

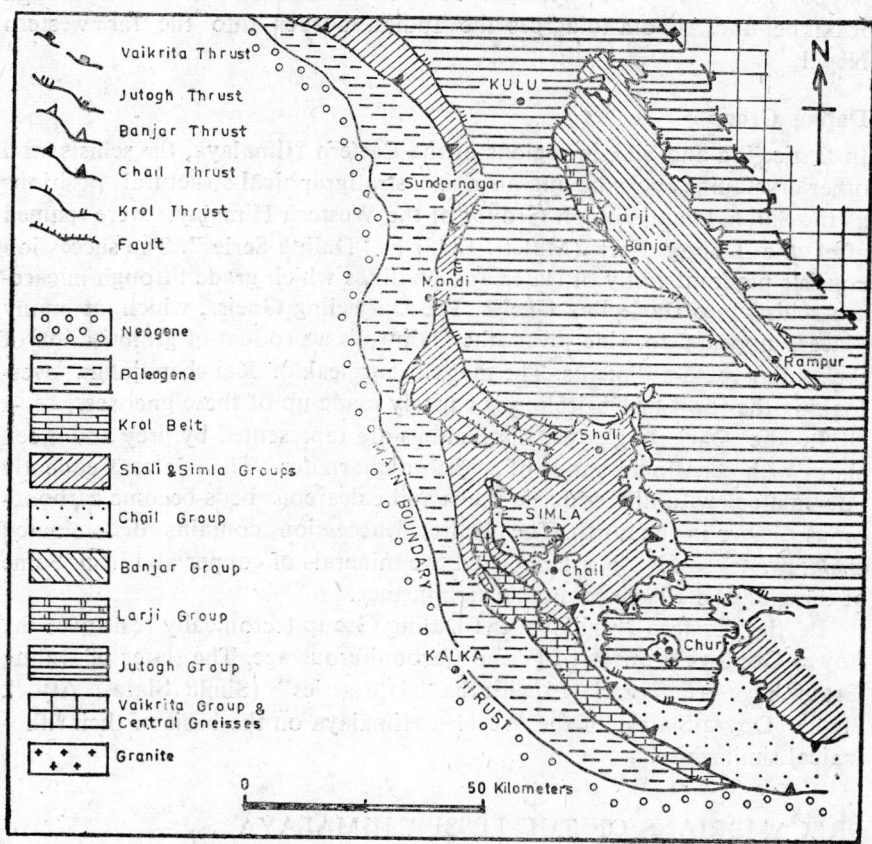

Fig. 7.4: Generalised geological map of the Simla Himalaya (based on Sharma, 1976 and author's own observations)

The rocks of the Jutogh Group have been traced as a huge nappe sheet (Jutogh Nappe) over an extensive hill tract of the northern Lesser Himalaya of Himachal Pradesh (Fig. 7.4). The Jutogh outcrops are composed of two broad litho-units viz., (i) the para-gneisses and (ii) the schists, phyllites and schistose quartzites. The para-gneisses commonly occur at the base of the Jutogh Nappe. Granitic massifs having characters similar to the Chor Granite include the Dhauladhar Gneisses, the Mandi Granite and the Karsog Granite. The schists and phyllites comprise the dominant lithology of the Jutogh Nappe and are frequently intercalated with amphibolites.

The Jutogh Nappe when traced eastward into the eastern Kumaun Himalaya gradually closes towards the northern Higher Himalayan ranges. An isolated klippen of the rocks considered equivalent of the Jutogh Group occurs around Lansdown in the northeast of Rishikesh. A larger nappe belt known as "Almora Nappe" extends from Dudatoli through Ranikhet and Almora to across the Indian border into the far western Nepal.

Daling Group

In Darjeeling and Sikkim regions of the Eastern Himalaya, the schists and other crystalline rocks having a similar stratigraphical-structural positions as the Salkhala and Jutogh Groups of the Western Himalaya were named after the Daling Fort (Malet, 1875) as "Daling Series". The succession consists predominantly of slates and phyllites which grade through micaceous schists into **Darjeeling Gneiss**. The Darjeeling Gneiss, which, at many places, appears to ride over the Dalings is a product of granitisation of the Daling meta-sediments. The magnificent peak of Kanchendjonga overlooking the town of Darjeeling is entirely made up of these gneisses.

In the Darjeeling area, the Dalings are represented by grey and green slates with occasional bands of schistose quartzites. The slates frequently intercalated with altered basic rocks and calcareous beds become carbonaceous further in the north. The Daling succession contains disseminated chalcopyrites and other associated ore minerals of copper and iron. Some of these deposits have economic significance.

In the Eastern Himalaya, the Daling Group tectonically rests over the Lower Gondwana rocks of Permo-Carboniferous age. The slates of Daling Group have been correlated with the "slate series" (Simla Slates, Attock Slates, Dogra Slates) of the Western Himalaya on the basis of their lithological similarities.

PRECAMBRIANS OF THE LESSER HIMALAYA

Precambrian formations of the Lesser Himalaya have been demarcated on the basis of their largely unfossiliferous nature and relative stratigraphic position below the known Phanerozoic rocks. The Precambrian formations comprising a complex association of clastic and carbonate sediments have remained almost completely unaffected by metamorphism. Unfossiliferous nature of the Phanerozoic formations makes it difficult to distinguish them from the Precambrian rocks of the Lesser Himalaya.

Structure of the Lesser Himalaya comprises a number of superimposed thrust sheets. The highest structural nappe consists of metamorphic rocks of the Jutogh Group. The lower structural levels comprise the rock formations occurring in a complex association of overlapping thrust sheets. The stratigraphic succession of the thrust sheets has generally retained its normal order of superposition. However, correlation of the rock formations of these slices is an enormous zig-saw puzzle. In certain sections, the Precambrian succession has been well studied while in others the data are as yet insufficient for a detailed classification.

Western Sector

Jammu Limestone: The Jammu Limestone, also known as Great Limestone or Riasi Limestone, occurs in a chain of inliers in the Tertiary expanse of the Sub-Himalaya of Jammu. The succession consists of a maximum of about 3 kilometres thick carbonate beds topped by about 200 m thick ortho-quartzite. The base of the limestone formation is not exposed anywhere in the area. The formation is unconformably overlain by Subathu Formation of Palaeocene-Eocene age.

The Precambrian age for the Jammu Limestone has been inferred on the basis of the occurrence of stromatolites and Pb-isotopic age of galena (967 million years) from the topmost orthoquartzite member. Three assemblage zones characterised by typical stromatolitic forms have been recognised in the upper parts of the carbonate succession (Fig. 7.5). The stromatolitic forms have been correlated with the Lower and Middle Ripheans of U.S.S.R. (Raha and Sastry, 1973).

Petrographic analysis of the carbonate rocks of Jammu Limestone has shown that the rocks were deposited in palaeo-environments which fluctuated from subtidal shelf through intertidal zones to supratidal mudflats (Raha, 1980). The generalised stratigraphic column (Fig. 7.5) shows four phases of marine environments intermittently broken by three phases of marine regression and a final phase of closure of the basin due to the complete withdrawal of marine water during the Late Precambrian. North-easterly palaeo-currents inferred on the basis of bent columnal stromatolites suggest a southern shoreline of the Lesser Himalayan Sea in consistence with that of the Vindhyan Sea during the Late Precambrian (Chapter 6).

Shali-Simla Groups: The Shali Group exposed in a belt extending from Shali type area in the northeast of Simla (Fig. 7.4) to Mandi in the northwest consists of over 3500 metre thick sequence of a predominantly carbonate succession. The succession begins with red shale, siltstone and dolomite with local development of salt, grit and marly litho-complex ("Lokhan") overlain by purple and white quartzite (Khaira Member) and alternation of massive dolomite, limestone, shales and siltstones and cherty dolomite.

114 Stratigraphy of India

Carbonate rocks of the Shali Group often contain well preserved stromatolites such as *Collenia collumnaris, C. baicalica* and *C. symmetrica* suggesting a Middle Riphean age for the succession (Valdiya, 1967). The Shali succession has been correlated with the Jammu Limestone. The presence of the salt beds in the lower part of the Shali succession indicates an arid climate during their deposition. Similar salt beds are known from the

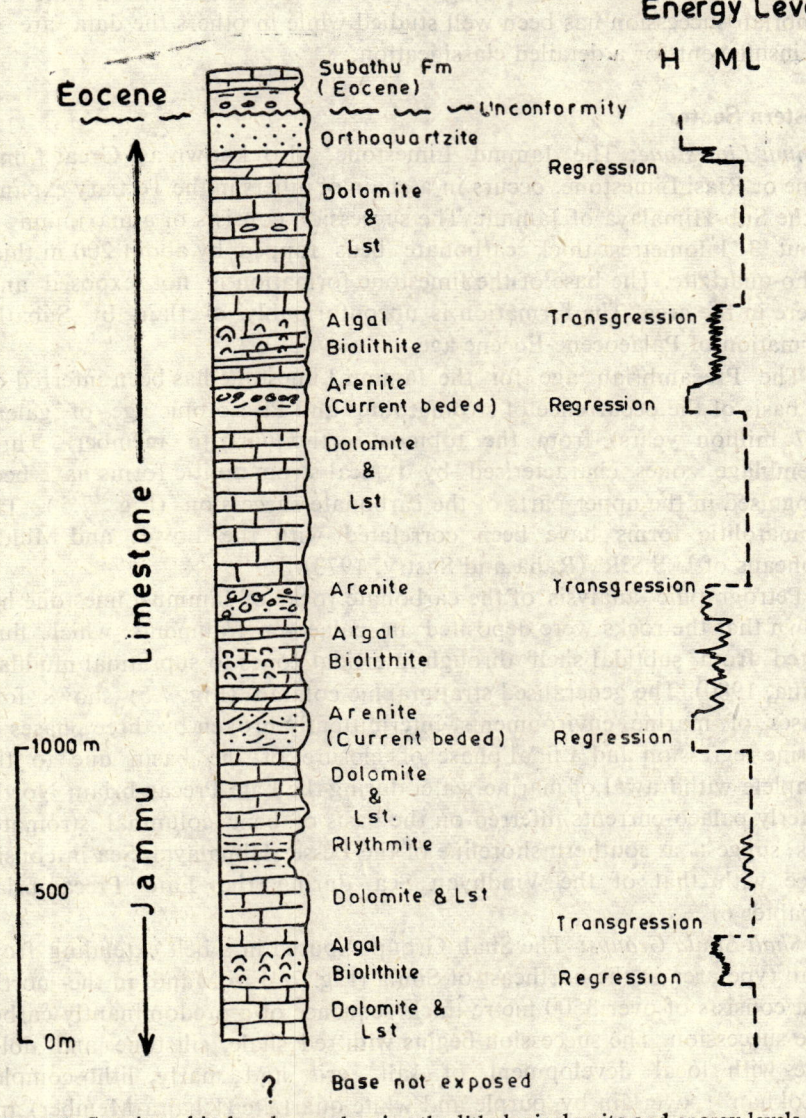

Fig. 7.5: Stratigraphical column showing the lithological units and energy levels in Jammu Limestone (based on Raha, 1980)
H—high, M—medium, L—low energy levels

"Saline Series" of Salt Range in Pakistan that is overlain by fossiliferous Cambrian beds.

A number of detached inliers of carbonate rocks and phyllites locally known as **Tunda Pathar Limestone** are exposed along the main boundary fault in the southeast and northwest of Kalka. The carbonate beds of the Shali Group, the Tundapathar Limestone and the Jammu Limestone were deposited on a broad shallow marginal shelf sea on the northwestern fringe of the Indian Peninsula during the Early and Middle Riphean time.

The Shali succession in the Simla area is disconformably overlain by a rock succession consisting predominantly of shales, siltstones, sandstones, greywackes and orthoquartzites named as "Simla Slates" by Medlicott (1864). The succession renamed as Simla Group on the basis of its highly varied lithology has been divided into four formations (Table 7.1). The lowermost Basantpur Formation contains a significant proportion of carbonate rocks. Lithofacies of Kunihar, Chaosa and Sanjauli Formations suggest a gradual regression of the marine conditions in the Lesser Himalaya of Himachal Pradesh.

Table 7.1: Lithostratigraphic classification of the Simla Group (after Srikantia and Sharma, 1971)

Formations	Lithology	Thickness in metres
Sanjauli Formation	Upper : Conglomerate, arkoses sandstone, shale Lower : Greywacke, shale & siltstone, orthoquartzite	1600
Chaosa Formation	Shale, siltstone, greywacke, orthoquartzite	1300
Kunihar Formation	Shale, siltstone with limestone interbeds	450
Basantpur Formation	Bedded & massive limestone, shale and siltstone and a basal greyish white quartzite and conglomerate	1000
Disconformity SHALI GROUP		

Age of the Simla Group is as yet uncertain. The succession is unconformably overlain by Upper Palaeozoic formations which places the upper age limit of the succession. The Simla Group is underlain by the Shali Group of Early to Middle Riphean age. The carbonate rocks of the Basantpur Formation have yielded stromatolites of Jurusania Group representing an Upper Riphean to Vendian age (Sinha, 1977). The upper formations of the Simla Group probably extend in time to Lower Palaeozoic. However, there is no fossil record to support this contention.

Central Sector

Precambrian rocks in the Kumaun and Garhwal regions of the U.P. Himalaya are exposed in the "Inner Belt" of autochthonous rocks extending from Deoban-Chakrata in the west through Tehri and Rudraprayag to Tejam and Pithoragarh in the east. The eastern part of the belt consists of predominantly carbonate facies of the Tejam Group. The Tejam Group is underlain by a sedimentary succession of varied lithology known as Damatha Group (Valdiya, 1980).

Damatha Group: The group named after the village Damatha on the bank of Yamuna river in the north of Mussoorie consists of a lower **Chakrata Formation** and an upper **Rautgarha Formation**. The Chakrata Formation consists of variegated coloured greywackes, siltstones and slates showing rhythmic alternations and turbidite features such as graded bedding, flute casts, load casts, etc. The Rautgarha Formation consists of fine to medium grained impure quartzites and purple slates. The succession contains abundant basic sills, dikes and lava flows. The type area of the formation is situated in Rautgarha village on the bank of Thuli Gad in eastern Pithoragarh district.

The Damatha Group occupies an older stratigraphic position than the Shali Group of Himachal Pradesh. The latter is underlain by Sundernagar Formation having a lithology similar to the Damatha Group. Radiometric dating of the basic volcanics exposed at Darlaghat constituting a part of the Sundernagar Formation gives K/Ar ages between 410 ± 10 and 1190 ± 35 million years (Sinha, 1977) of which the higher figure probably represents the true age of the volcanics.

Tejam Group : The Tejam Group conformably rests over the Rautgarha Formation of the Damatha Group with a rather abrupt change in facies from a flyschoidal succession to a predominantly shelf carbonate facies. At a few places, for example, in northern Chakrata and upper Tons valley, unconformable contact between the older flyschoidal and younger carbonate successions has been observed. The Tejam Group has been subdivided into Deoban and Mandhali Formations (Table 7.2). The Mandhali Formation occurring at the base of the Krol Nappe has been conformably overlain by the Chandpur Formation.

The **Deoban Formation** named after a peak of that name in the north of Chakrata comprises a succession of about 1200 metre thick stromatolite bearing cherty dolomites and dolomitic limestones with bands and intercalations of blue limestones and grey slates. Bioherms of branching columnar stromatolite belonging predominantly to the *Baicalica* group constitute a very characteristic feature of the peletal calcarenites. The stromatolitic limestone is poorly phosphatic. Lentiform magnesite deposits are known from various parts of Kumaun. Other stromatolitic forms include *Masloviella, Kussiella* and *Minjaria* of Middle Riphean age. The Deoban Formation has been correlated with similar carbonate formations

Table 7.2: Lithological succession of the inner Autochthonous Zone of the Kumaun Himalaya (after Valdiya, 1980)

Group	Formation	Lithology	Thickness in metres
	Sirmur Group (Palaeogene)		
	Unconformity		
Tejam Group	Mandhali Formation	Carbonaceous, pyritic phyllites, limestones and conglomerate	More than 2500
	Deoban Formation	Stromatolitic cherty limestones & slates	1200
	Sharp break in facies		
Damatha Group	Rautgraha Formation	Quartzites, slates with volcanic flows	2000
	Chakrata Formation	Greywackes, siltstones and shales	More than 2000
	Base not exposed		

of the Western Himalaya described as Jammu Limestone and Shali Group.

The **Mandhali Formation** named after a village of that name in the northern Chakrata consists of more than 2500 metre thick succession of greyish green and black carbonaceous phyllites (pyritic) or slates interbedded with blue limestones. The succession contains lentiform para-conglomerates at various horizons. The basal conglomerate is made up of clasts derived from the Deoban and Rautgarha Formations. The Mandhali Formation has been correlated with the Basantpur Formation, the lowermost unit of the Simla Group.

Chandpur Formation: The Chandpur Formation first described by Auden (1934) from the Chandpur peak in the southeastern Himachal Pradesh consists of meta-greywackes, meta-siltstones, slates, phyllites and locally carbonaceous greyish phyllites. The upper part of the succession contains certain well preserved load casts, graded bedding and other structures typical of flyschoidal succession. An angular unconformity marking the upper limit of the formation separates the formation from the overlying Nagthat Formation of Palaeozoic age (Kumar and Kapila, 1980). The rocks of the Chandpur Formation show a higher grade of metamorphism and emplacement of the "Ramgarh Quartz Porphyry" in the southeastern Kumaun Himalaya.

The lower succession of the Krol Nappe comprising the Mandhali, Chandpur and Nagthat Formations has been grouped into the Jaunsar Group ("Jaunsar Series", Auden, 1934). The Jaunsar succession has been regarded as a facies variant of the Simla Group of Himachal Pradesh (Fig 7.6, Bhargava, 1972). The latter is tectonically overlain by a succession of rocks known as **Chail Group** ("Chail Series", Pilgrim & West, 1928). The Chail Group has been reportedly traced to merge in its eastern extension with the Jaunsar Group of Garhwal and Kumaun.

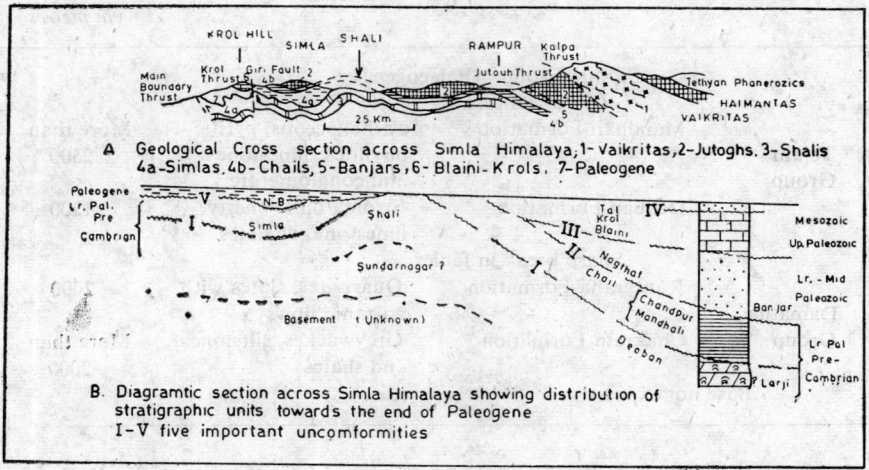

Fig. 7.6: Geological cross-section and section showing the palaeogeographic reconstruction of the Simla Himalaya

Nepal Himalaya

The Lesser Himalayan Succession of Precambrian age of Nepal has been divided into Lower, Middle and Upper Nawakot Groups (Table 7.3). The succession unconformably overlain by Upper Palaeozoic and Triassic rocks has yielded Riphean stromatolites.

Table 7.3: **Stratigraphic classification of autochthonous Precambrian succession of Lesser Himalaya in Central Nepal** (after Kumar, 1980)

Group	Formation	Approximate thickness in metres
Upper Nawakot Group		
	Dhading Dolomite	More than 1000
	Nourpur Formation	800
	Unconformity	
Middle Nawakot Group	Malekhu Limestone	800
	Benighat Slates	1500
	Fagfog Quartzite	350
	Disconformity	
Lower Nawakot Group		
	Kuncha Formation	3500
	Metasandstone Formation	More than 2500
Basement not exposed		

Metasandstone Formation: This formation described from the Nawakot-Burhi Gandaki area (Arita *et al.*, 1973) comprises greywacke type metasandstone containing a large amount of ovoidal quartz grains. A conglomerate horizon is known to occur in the middle of the formation. The formation estimated to be more than 2500 m thick has a gradational contact with the overlying Kuncha Formation.

Kuncha Formation: The Kuncha Formation described by Bordet (1961) as "an enormous complex of a kind of flysch" is exposed in the Kali Gandaki region and in the north of the Trishuli Ganga in Central Nepal. The succession comprises more than 3500 m thick succession of phyllites, phyllitic quartzites and phyllitic gritstones resembling greywackes. Radiometric dating of the rocks of the Kuncha Formation has given isochrons ages as 872 m.y. and 1150 m.y.

Fagfog Quartzite: The Fagfog Quartzite rests over the Kuncha Formation with a sharp facies change from a flysch sequence to a 350 metre thick fine to coarse orthoquartzites of shallow water basin. The quartzites showing frequent ripple marks and current bedding structures are locally interbedded with phyllites. The Fagfog Quartzite formation is overlain by the Benighat Slates through a rapid zone of alternation of slates and quartzites.

Benighat Slates: The Benighat Formation exposed around Benighat at the confluence of Burhi Gandaki and Trishuli Ganga consists of about 1500 m thick succession of dark slates and phyllites with interbeds of fine quartzites. The slates are rich in graphite with occasional disseminated pyrite. Locally, the slates are associated with about 100 metre thick carbonate horizon in the middle part of the formation. The formation is overlain by Malekhu Limestone with a zone of alternation of slates, thin bands of limestones and impure quartzites.

Malekhu Limestone: The formation named after Malekhu village on the bank of Trishuli Ganga consists of about 800 m thick succession of thin platy dolomitic and siliceous limestone of light-yellow colour with partings and thin intercalations of green sericite-chlorite phyllite.

Nourpur Formation: The Nourpur Formation consists of about 800 m thick succession of predominantly mudstones and impure quartzitic sandstones with a few carbonate bands. The rocks show frequent mud-cracks, ripple marks and current bedding structures. Red and purple colours characteristic for arid environments predominate. A basal quartzite member of the formation rests directly over the Benighat Slates. The formation has been overlapped by the Dhading Dolomite in the vicinity of the Trishuli Ganga valley where the Dhading Dolomite comes in direct contact with the Benighat Slates. Radiometric dating of the rocks of Nourpur Formation has given an isochron age of 600 million years for the formation.

Dhading Dolomite: The cliff forming Dhading Dolomite is generally massive and thickly bedded with the development of stromatolite throughout the formation. The stromatolites of the formation resemble Conophyton of Jammu Limestone of the Western Himalaya. The stromatolites become profused in the type area where they have been deformed into isoclinal looking folds in the rocks with broad open folds. The Dhading Dolomite extending in a linear belt along the middle course of the Trishuli Ganga attains a maximum thickness of about 100 metres.

Eastern Himalaya

The autochthonous Precambrian rocks of the Eastern Himalaya comprise a succession of quartzites, shales and dolomites named as **Buxa Group** ("Buxa Series", Mallet, 1875) after the old Buxa Fort in Western Duars. The succession is tectonically overlain in the north by the rocks of Daling Group. In the south, the Buxa succession is thrust over the Lower Gondwana rocks. In less disturbed areas, the Buxa Group rests with normal stratigraphic contacts over the older Precambrian Daling Group and is unconformably overlain by Lower Gondwana (Upper Palaeozoic) rocks.

The Buxa succession in the southern foothills of Bhutan rests over a sequence of crumpled slates and schists resembling the Daling Group. The Buxa succession of Bhutan having a thickness of about 1800 metres has been divided into three formations. The lowermost formation known as **Sinchula Formation** consists of massive and bedded quartzites in the lower parts and argillaceous rocks in the upper parts. The succession often contains sills and dikes of amphibolitic epidiorites. This formation is overlain by **Jainti Quartzite** consisting of pink quartzites and minor slates. The quartzite in upper parts of the succession contains jasperic pebbles, hematitic beds, pyrites and copper ores. They are overlain by a carbonate succession known as **Buxa Dolomite**. The Buxa Dolomite consists of several massive beds of dolomitic limestones and thickly bedded dolomitic layers with some beds of orthoquartzite. Stromatolitic forms observed in the dolomites are comparable with those observed in Dhading Dolomite, Deoban Formation, Shali Group and Jammu Limestone.

Chapter 8
Palaeozoic History

The Palaeozoic Era began at about 570 million years ago and came to an end at about 225 million years ago. The Era has been subdivided into Cambrian, Ordovician, Silurian, Devonian, Carboniferous and Permian Periods. The marine life that was greatly diversified and abundant during the Cambrian and early Ordovician gradually declined through the Late Ordovician and Silurian time. Devonian Period is characterised with a fresh period of evolution and diversification of life that faced a great crisis at the close of the Permian Period. The Palaeozoic Era has thus been often subdivided into an Early and a Late Palaeozoic sub-eras. The six periods of the Era have been further subdivided into epochs and ages (Table 8.1) each characterised by specific assemblages of fossil fauna.

The **Cambrian System** comprising the rock formations laid during the Cambrian Period was established in 1839 by Adam Sedgewick (1785-1873), a Professor of Geology at the Cambridge University. The type section of the System is located in North Wales for which the old Latin name is *Cambria*. The **Silurian System** established in the same year in Central Wales by Ruderich Murchison (1792-1871) was named after the *Silurs*, the ancient tribe of Wales. The upper parts of the Cambrian System when traced to Central Wales were regarded by Murchison as the lower parts of his Silurian System. Thus ensued a stiff controversy as regards to the boundary between the Cambrian and Silurian Systems. The transitional beds are known to contain fossil fauna having similarity to both the primitive Cambrian forms and the relatively advanced Silurian forms. The controversy was resolved in 1879 by Charles Lapworth (1842-1920) by separating the transitional beds and giving them the status of an independent system. The transitional beds were grouped into the **Ordovician System** named after yet another aboriginal tribe of Wales.

The **Devonian System** derives its name from the county of Devonshire in southwest England. Equivalent rock formations are exposed in northeastern France, western Germany and southern Belgium. Devonian succession of Europe has been found more useful as standard for International Stratigraphic Scale as the rocks of the succession are relatively less deformed as compared to the Devonian succession of England.

Table 8.1: Subdivisions of the Palaeozoic Era

System/Period	Series/Epochs	Stage/Age
Permian	Upper	Tatarian / Kazanian
	Middle	Kungurian
	Lower	Artinskian / Sakmarian
Carboniferous	Upper (Pennsylvanian)	Stephanian (Uralian) / Westphalian (Moscovian) / Namurian
	Lower (Mississipian)	Visean / Tournaisian
Devonian	Upper	Famennian / Frasnian
	Middle	Givetian / Couvinian
	Lower	Emsian / Siegenian / Gedinnian
Silurian (Gothlandian)		Downtonian / Ludlovian / Wenlockian / Llandoverian
Ordovician	Upper	Ashgillian / Caradocian
	Lower	Llandeillian / Llanvirnian / Arenigian / Tremadocian
Cambrian	Upper	(Potsdamian)
	Middle	(Acadian)
	Lower	(Georgian)

The **Carboniferous System** derives its name from the coal bearing rock formations overlying the Devonian System of England. In North America, the Carboniferous System has been divided into Mississipian and Pennsylvanian successions having independent status of system. **Permian System** was first described from the Perm in northeastern Russia. Permian successions exposed in western Texas and New Mexico of United States of America are rich in fossil records useful for subdividing the System into Series and Stages.

TECTONIC HISTORY

Stratigraphic records indicate a typical character for the tectonic movement during the early Palaeozoic time. The Cambrian and the early parts of the Ordovician periods experienced a general subsidence which led to the deposition of thick sedimentary sequences. This trend was reversed during the later parts of the Ordovician Period and during the Silurian Period leading to a world-wide marine regression and emergence of several mountain chains. This tectonic phase of mountain building is known as **Caledonian Orogeny**.

The Caledonian mountains of Europe were prominent relief on the surface of the earth at the beginning of the Devonian Period. These mountains were rapidly eroded leading to the deposition of a thick succession of the continental deposits during the Devonian Period known as **Old Red Sandstone**. The marine sediments of the Devonian Period show an influence of shallowing of the marine basins which gradually became deeper during the Carboniferous Period. The sedimentation phase was followed by yet another phase of mountain building processes known as **Hercynian** or **Variscan Orogeny** at the close of the Palaeozoic Era.

The palaeosea (Vindhyan Sea) that covered the northern part of the Peninsula during the Late Precambrian time gradually receded at the dawn of the Palaeozoic Era. A fairly continuous phase of sedimentation is, however, recorded in the Tethyan Himalayan Zone for almost entire Palaeozoic Era. The Tethyan marine basin of the Himalaya remained practically unaffected by Caledonian and Hercynian orogenic phases. The Caledonian tectonic phase is felt in the Himalaya only in the form of a relative shallowing of the basin and, at places by a break in deposition. The Hercynian tectonic phase is represented in the northwestern and eastern Himalayas by an episode of submarine and coastal volcanism.

The Upper Carboniferous Epoch marks the beginning of a major cycle of continental sedimentation over the Indian Peninsula. The sedimentation began with a continental glaciation that has left its record in the form of tillite beds. The glacial deposits are overlain by a vast thickness of rock succession of continental facies known as Gondwana Sequence (Chapter 10) ranging in age from Permian to Early Cretaceous.

Palaeoclimatic and **palaeomagnetic** studies indicate that during the

Palaeozoic Era, India was joined with the southern continents forming a supercontinent known as Gondwanaland (Fig 8.1). This supercontinent was encircled by a series of geosynclines. The Gondwanaland drifted over the earth's surface as a single 'plate' which led to the wandering of South Pole positions with respect to the supercontinent. The South Pole was located during the Cambrian Period, at the present northwestern tip of Africa. The pole position gradually shifted through Ordovician, Silurian and Devonian Periods across the Africa to Antarctica during the Carboniferous and Permian Periods.

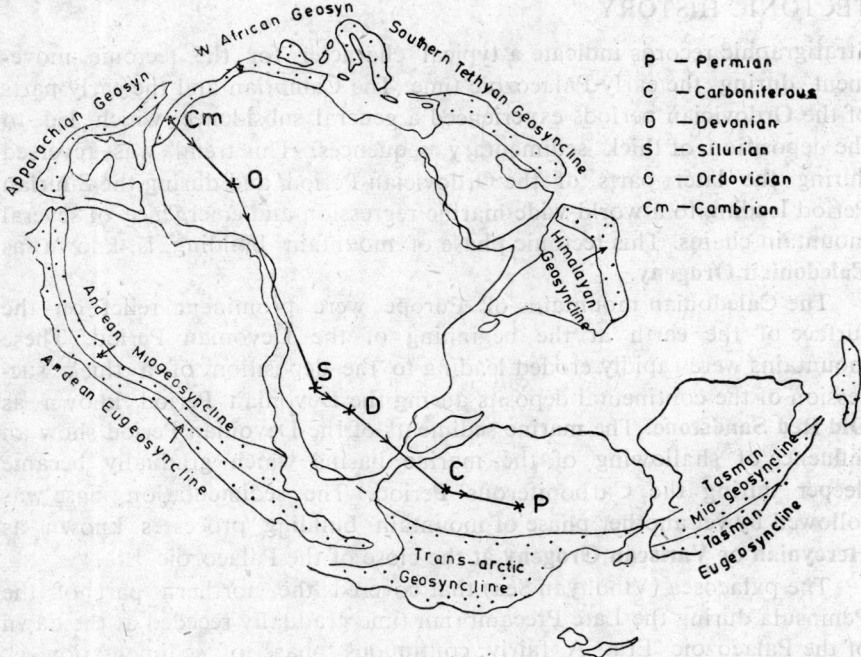

Fig. 8.1: Tectonic elements and polar wandering curves of the Gondwanaland during the Palaeozoic Era

PALAEOZOIC LIFE

All the phyla of plants and animals are represented in the fossil record of the Palaeozoic Era. Early Palaeozoic life lived primarily in sea. Invertebrates were the most predominant forms of life. Various diverse plants were also common in the marine waters. Only a few primitive vertebrates are known from the records of the Early Palaeozoic time. Unicellular plants and bacteria normally not preserved as fossils thrived on land at the beginning of the Era. Land animals and higher plants made their first appearance towards the end of the Silurian Period.

Life of the Cambrian Period is characterised by the abundance of invertebrate and algae. Trilobites, the most dominant forms of the Cambrian sea, made their first appearance at the dawn of the Palaeozoic Era. In a very short time, they reached the zenith of their evolution. Their history after Cambrian was one of steady decline and final extinction at the close of the Palaeozoic Era. Archaeacyathids, the reef building calcareous invertebrates, were abundant during the Early Cambrian Epoch and became extinct by the end of the Middle Cambrian. The Cambrian Period is also characterised by the first appearance of foraminifers and land plants.

Primitive jawless vertebrates, ostracoderms, evolved to replace the dominance of trilobites in the Ordovician sea. Graptolites, the extinct colonial invertebrates, were also dominant in the marine life of the Ordovician Period. Brachiopods were greatly diversified and increased in number. Tetracorals, small agglutinated foraminifers, nautiloids, cystoids, conodonts, sea weeds and algae are the other forms of life represented in the fossil record of the Ordovician Period.

Eurypterids, the great marine scorpions, were the dominant sea animals of the Silurian Period. Ostracoderms still survived and became numerous in shallow waters of the Silurian seas and estuaries. Tabulate corals and tetracorals were also abundant. Lycopsid, the land plants, made their first appearance during the Silurian Period.

Life in sea became more diversified during the Devonian Period. The sea bottoms were inhabited by a great number of brachiopods, corals, echinoderms, sponges and mulluscs. Jawless ostracoderms, which developed armours of diverse types during the Silurian Period, had their maximum development during this period. Placoderms and acanthodians reached the zenith of their evolution. The Devonian Period is regarded as a very crucial period in the evolution of vertebrates. The bony fishes, crossopterygians, which evolved during this period, led to the emergence of land living tetrapods.

The first indication of life on the land comes from the rocks of the Upper Silurian Series that contains primitive land plants and primitive arthropods. The Devonian successions of the continental facies show presence of green marshes and forests on land during the Devonian Period. The flora comprised primarily of spore bearing thallophytes, lycopods, scouring rushes and primitive ferns. During this period, amphibians, the first back-boned animals, migrated out of streams and ponds on to the marshy land. The primitive amphibians, ichthyostegids had evolved from the fishes of the Silurian Period.

During the Carboniferous Period, tropical jungles enveloped the continents from Equator to high latitudes. The forests contained the luxurious growth of ferns, giant lycopods, tree ferns, calamites, lepidodendrons, sigillarians and other forms of vegetation. Amphibians inhabiting the forests were rapidly diversified. Arthropods, such as dragon flies, cockroaches and insects also shared the land with the amphibians. Insects, which could fly and had no enemy in the air, grew to maximum sizes that could be attained by such animals. The Carboniferous forests were also inhabited by land living invertebrates such as centipeds and scorpions, ancestors of the spiders and air breathing gastropods.

The marine life of the Carboniferous Period consisted of coral reefs, echinoderms, blastoids, crinoids and protozoans. Other Devonian forms continued to evolve and became more diversified during the Carboniferous Period. Ostracoderms and placoderms became extinct thus marking a change in the fish fauna during the transition from the Devonian to Carboniferous Periods. The acanthodians persisted through the Carboniferous and into the beginning of the Permian Period.

Disappearance of the low marshy forests of the Carboniferous Period and the appearance of the new land areas and mountain chains due to the Hercynian Orogeny brought about a tremendous change in the evolutionary history of the life both on land and in marine areas. Many organisms of the marine life became extinct towards the close of the Permian Period. The extinction was so profound that it marks the boundary of the Palaeozoic and Mesozoic Eras.

PRECAMBRIAN CAMBRIAN BOUNDARY

Palaeozoic rock formations contain abundant fossil records which are in contrast to largely unfossiliferous nature of Precambrian rocks. Appearance of abundant and complex forms of organisms at the beginning of the Palaeozoic Era is rather puzzling. Many suggestions have been put forward to explain the paucity and primitive characters of fossils records in Precambrian rocks. It seems that there has been a sudden increase in the rate of evolution from primitive Late Precambrian forms to complex forms of Palaeozoic Era. It is also probable that the Precambrian forms were devoid of hard parts and, therefore, they could not be preserved as fossils. Or, perhaps the Precambrian forms inhabited deeper marine waters where sedimentation could not take place and, therefore, these forms were not preserved as fossils. None of these explanations, however, can satisfactorily explain the absence of requisite fossils in rocks older than Cambrian.

The Cambrian System of the North Wales consists of a sequence of sandstones and shales. The rocks are known to contain trilobites, brachiopods and graptolites. The organisms differ from bed to bed and they are distributed in a distinctive order of succession. Characteristic trilobites, *Paradoxides, Olenus, Orusia, Peltura* and *Niobe* have been reported to occur in the succession in that order. *Paradoxides* appear in the type section at about 1400 metres above the base of the exposed Cambrian succession. Cambrian succession having a similar order of succession of fossil genera is also known from the southern parts of Sweden where the oldest rocks of the Cambrian succession have yielded yet another trilobite genus, *Holmia*. These oldest Cambrian rocks rest unconformably over the Precambrian gneisses and schists.

Some excellent sections of Lower Cambrian rocks are exposed over the Siberian Platform (Mathews and Missarzhevsky, 1975). The Lower Cambrian succession that rests conformably over Upper Precambrian rocks have yielded a variety of fossils such as archaeocyathids, gastropods, hyolithids, hyolithelminthids, poryferids and tommotiids. The Lower Cambrian has been subdivided on the basis of the fossil fauna into **Tommotian, Atdabanian** and **Lenian** stages.

The working group on Precambrian-Cambrian boundary has suggested that the boundary should be placed close to the base of the stratigraphic unit yielding Tommotian fossil assemblage (Cowie, 1978). Typical Tommotian fossil assemblage of the Siberian Platform include the species of Torellella, Spinulitheca and Bemella. Besides the Tommotian fossil assemblage occurring above the Precambrian-Cambrian boundary, other features characterising the boundary seem to be the occurrence of soft bodied metazoa and the most ancient tubular organisms in the uppermost Precambrian rocks. In some parts of the world, the uppermost Precambrian succession is also characterised by the tillite horizon suggesting a phase of world wide phase of glaciation at the close of the Precambrian time.

Part of the Bhander Group (Chapter 6) of the Indian Peninsula were presumably deposited during the Precambrian-Cambrian transition. However, these rocks have not yielded the characteristic fossils demarcating the boundary. The equivalent formation exposed in Salt range that is succeeded by fossiliferous Lower Cambrian beds elucidates the problem inherent in demarcating the Precambrian-Cambrian boundary in conformable sequences.

In Salt Range (Fig. 8.2), the oldest rocks exposed at the base of the southern cliffs are known as **"Saline Series"** consisting of gypsum, marls and rock salt. This succession is conformably overlain by **Purple Sandstone, Neobolus Shales, Magnesian Sandstone,** and **Salt Pseudomorph Beds** in that order of succession. Lower Cambrian trilobite fauna (*Redlichia*) and well preserved trails and burrows are known from *Neobolus* Shales. The Magnesian Sandstone, a succession of dolo-arenites, contains a horizon of shales in the middle of the succession which has yielded fossil trails and burrows resembling those of the *Neobolus* Shales.

The nature of contact between the Lower Cambrian fossiliferous succession and the underlying "Saline Series" of Salt Range has been one of the most debated problems of the Indian Geology. This contact is very often characterised by tectonic disturbance that was explained in terms of disharmonic folding of competent rocks of Purple Sandstone resting over highly incompetent salt and gypsum beds of the "Saline Series" of Precambrian age. The other view considers a Tertiary age for the "Saline Series" that has yielded certain Tertiary plant and insect remains of microscopic sizes. The Lower Cambrian succession was thus thrust over the Tertiary rocks of the "Saline Series". Those, who opposed the Tertiary age for the

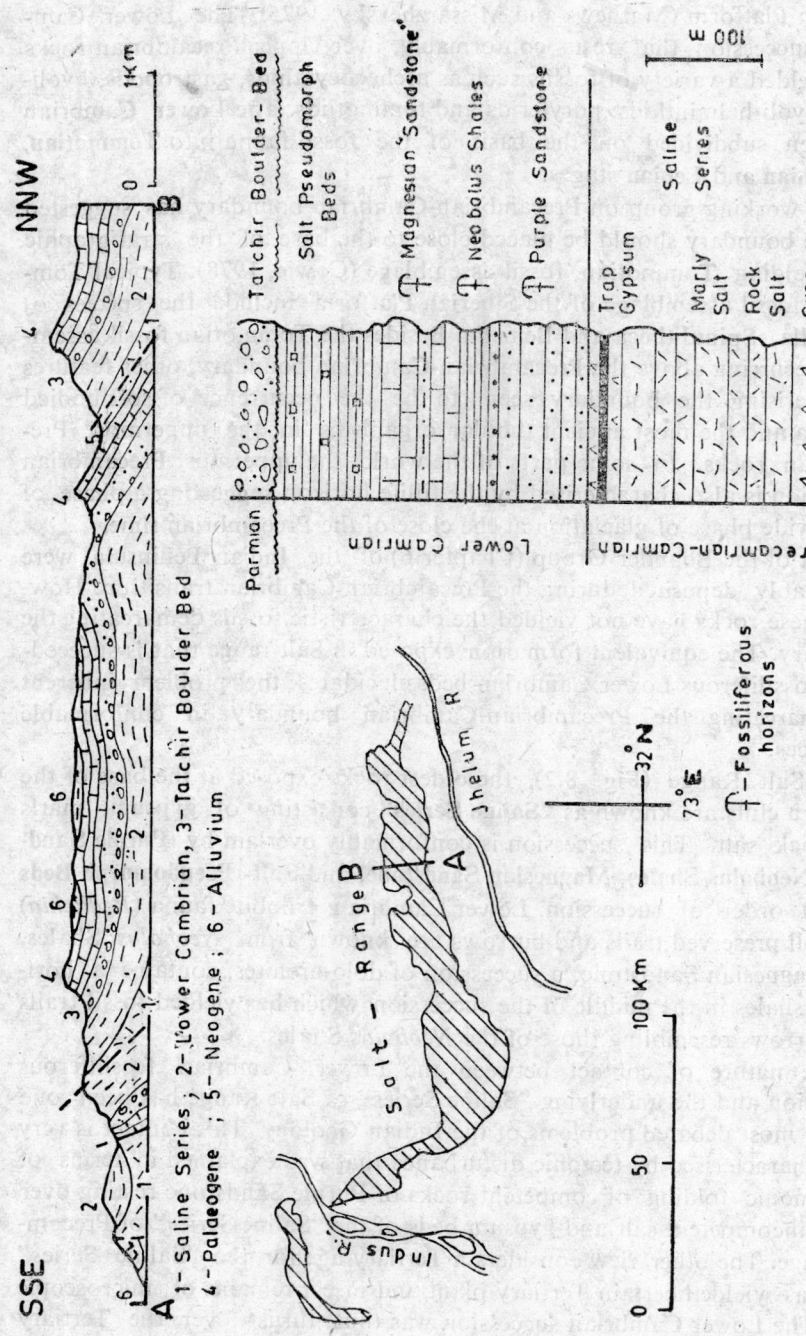

Fig. 8.2: Geological section and stratigraphical column of the Cambrian in Eastern Salt Range, Pakistan (based on Gansser, 1964)

"Saline Series" explained the presence of the Tertiary fossils on account of contamination by the ground water. The Tertiary fossils were presumed to have been leaked from the Tertiary rock formations exposed in the upper heights of the Salt Range.

Normal stratigraphic contacts between the "Saline Series" and the overlying Lower Cambrian succession have been carefully investigated by Schindewolf and Scailacher (1955). Recurrence of evaporitic conditions towards the top of the Lower Cambrian succession represented by the Salt Pseudomorph Beds further confirms the continuity in sedimentation from the "Saline Series" to Lower Cambrian succession. The evaporitic conditions of deposition of the "Saline Series" was followed by a brief duration of marine transgression through the shallow water (Purple Sandstone) to deeper water (*Neobolus* Shales and Magnesian Sandstone) facies. Withdrawal of the marine waters paved way for yet another phase of evaporitic conditions when the Salt Pseudomorph Beds were deposited. A prolonged phase of non-deposition and denudation ensued in this part of the Indian sub-continent until the Late Carboniferous Epoch. The base of the "Saline Series" is not exposed. It is presumed that the Precambrian-Cambrian time boundary is located somewhere in the exposed section of the "Saline Series".

Continuous successions of Precambrian-Cambrian rocks are exposed in various parts of the Tethyan Himalayan zone. In many sections, however, the Lower Cambrian rocks are unfossiliferous. Even in the fossiliferous sections, the fauna is poorly preserved for specific and generic identification.

In Kashmir, a rich assemblage of microbiota cryptarchs and algae of Late Precambrian (Late Riphean-Vendian) age have been reported from the Lower parts of the Lolab Formation (Kumar *et al*, in press). The upper parts of this formation has yielded trilobites and brachiopods of upper Lower Cambrian affinity. The Precambrian-Cambrian boundary is, thus, situated somewhere in the middle of the Lolab Formation which has so far not yielded any fossil record. In Spiti, the boundary is located in the middle part of the Kunzam la Formation. The fossil record of the Kunzam la Formation is similar to that of the Lolab Formation in that the lower parts have yielded acritarchs of Vendian affinity and the upper parts contain trilobites and trace fossils of upper Lower Cambrian affinity.

In the neighbourhood of Nandadevi, the fossiliferous Palaeozoic succession rests over about a 4 kilometre thick unfossiliferous sequence of argillaceous and calcareous fine grained rocks known as **Martoli Formation**. The formation is overlain by about 100 metre thick **Ralam Conglomerate** that constitutes the base of the fossiliferous Cambrian succession. The Martoli Formation has been assigned a Late Precambrian age and the Ralam Conglomerate presumably represents an unconformity demarcating the **Precambrian-Cambrian boundary**.

An unconformable contact between Precambrian and Lower Palaeozoic rock successions has been observed in Central Nepal (Fig. 7.2). The Precambrian schists, quartzites and marbles of the Bhimphedi Group are unconformably overlain by over 3 kilometre thick succession of unfossiliferous meta-sandstones and phyllites (**Tistung Formation**). The rocks of the Tistung Formation are conformably overlain by a succession of limestone and shale sequence which have yielded fossil fauna of Cambro-Ordivician age. The Tistung Formation has been, thus, assigned a Late Precambrian to Early Cambrian age. The time boundary between the Precambrian and Phanerozoic Eons is situated somewhere in the lower parts of this succession.

Tommotian fossil assemblage has been reported from the base of the Lower Tal Formation exposed in the Mussoorie Hills (Azmi and Pancholi, 1983; Bhatt et al., 1983). The assemblage comprising of conodonts and other shelly microfossils (hyoliths and bryozoan stems), has many similarities to the Tommotian fauna of the Siberian Platform. These fossil findings have upset the generally accepted age of Jurassic to Cretaceous for the Tal succession (Chapter IX). The upper part of the Tal succession contain well preserved Cretaceous fauna. In view of the reported occurrence of the Lower Cambrian fauna from the basal parts of Tal succession, a pronounced break in deposition is expected between the lower and the upper parts of the succession. However, the evidence for such an unconformity has not yet been established. Moreover, the formations underlying the Tal succession have yielded faunal and floral records of Late Palaeozoic-Triassic age. In view of these fossil findings suggesting contradictory ages for the Tal succession, the geology of the region needs a more careful study before any definite stratigraphy of the region can be worked out.

MARINE PALAEOZOIC FORMATIONS OF INDIA

The Himalayan region remained a marine basin for the most part of the Palaeozoic Era. The Tethyan Himalaya exposes some of classic Palaeozoic successions rich in fossil records (Fig. 8.3). The Lesser Himalayan Palaeozoic formations, however, are predominantly unfossiliferous except for sporadic and doubtful fossil records. On the evidence of this contrast in the fossil records, two separate basins, namely the 'Tethyan palaeo-sea' and the 'Himalayan palaeo-sea', have been postulated in the palaeogeographic reconstructions. These seas were supposed to have been separated by a land barrier or 'geanticline' which came into existence either during the Late Palaeozoic (Wadia, 1957) or Early Palaeozoic (Pande, 1967; Fuchs, 1968) or during the Late Precambrian (Saxena, 1971; Mehdi et al., 1972). The other view holds that the Himalayan palaeo-sea was separated from the Tethyan palaeo-sea by a vast open palaeo-sea which prohibited the intermigration of the Palaeozoic fauna. None of the explanations, however can satisfactorily explain the contrast in the facies of the Himalayan and the Tethyan rock formations of the Palaeozoic Era. The Palaeozoic rocks of the Peninsula comprising the continental facies consist of the lower part of the Gondwana sequence (Chapter 10).

Fig. 8.3.: The Phanerozoic basins of the Western Himalaya

Tethyan Regions

Kashmir: The Palaeozoic formations of Kashmir exposed along the Pir Panjal and Great Himalayan ranges (Fig. 8.4) rest either over unfossiliferous Dogra Slates or over the Precambrian crystalline rocks of the Salkhala Group. The **Dogra Slates**, first described by Wadia (1928) from the south western Kashmir and Poonch, consist of slate and phyllite succession with a few bands of quartzites and altered lava flows. When metamorphosed to higher grades, the rocks are indistinguishable from the rocks of the Salkhala Group.

Dogra Slates are conformably overlain by fossiliferous beds yielding poorly preserved organic remains (Tubicolous, Vermes, etc.) and a few specimens of *Microdiscus* and *Agnostus* of Early Cambrian Epoch. The Dogra Slates have been thus assigned as Late Precambrian to Early Cambrian age. Equivalent slate sequences are known as Hazara Slates and **Attock Slates in the western continuation of the Himalaya in Pakistan,**

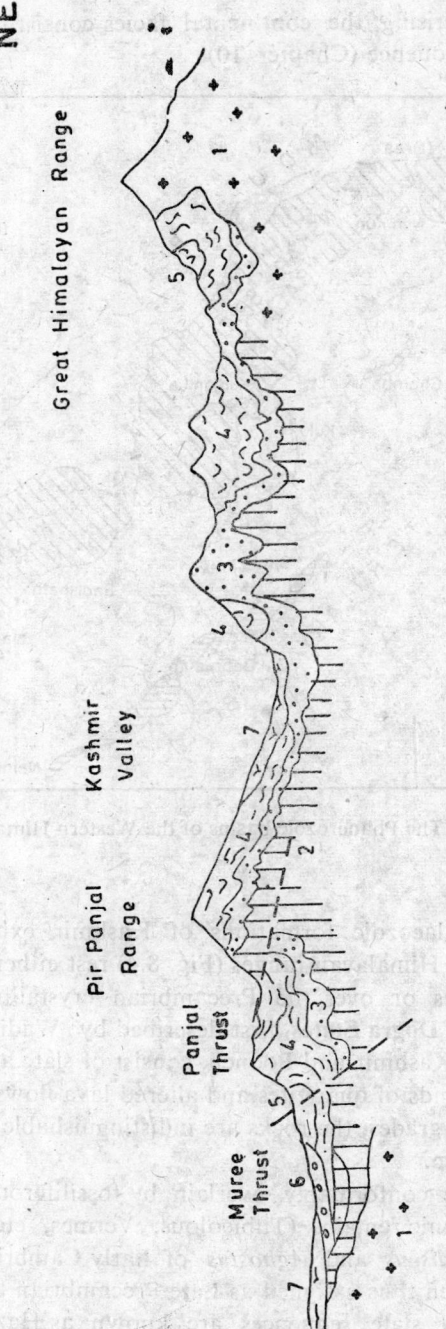

Fig. 8.4.: Geological section across the Pir-Panjal Range and Kashmir Valley (based on Wadia, 1934)

1– Crystalline rocks; 2– Purana (including Dogra slates; 3– Lower Palaeozoics; 4– Carboniferous–Trias
5– Cretaceous – Eocene; 6– Neogene; 7– Quaternary

Simla Group in Simla Hills (H.P.), Haimanta Group in Spiti (H.P.) and Garbyang Formation in northeastern Kumaun.

The Cambrian fossiliferous sequence of Kashmir comprises imperfectly cleaved clays, impure sandstones and greywackes with a few lenticular beds of limestone. The Lower Cambrian fauna is relatively scanty. *Redlichia* occurring in limited sections characterises the presence of Lower Cambrian. Trace fossils of arthropod origin have been reported from the beds lying below the *Redlichia* Zone (Shah, 1982). The middle and the upper parts of the Cambrian sequence have yielded well preserved fauna comprising trilobites, brachiopods, pteropods, crinoids and sponges. Typical trilobite assemblage includes *Ptychoparia, Solenopleura, Anomocare* and *Conocoryphe*. Most of these organisms are known to have inhabited deep turbid and muddy ocean bottoms. The fauna is typically endemic in character having no affinity to the Cambrian life of the adjacent regions (Cowper Reed, 1934). The Kashmir fauna shows an affinity with the Cambrian fauna of Indo-China, north Iran and North America.

Diversity in faunal assemblages in neighbouring regions is attributed to differing ecological conditions. The Cambrian succession of Kashmir represents an extra-cratonic, euxinic facies whereas equivalent formations of Salt Range experienced cratonic facies conditions and that of Spiti conditions of shallow water, platform facies.

The Upper Cambrian shales of Shamsh Abari Syncline in Baramula district are conformably overlain by a succession of ferruginous slates, greywackes and limestone yielding Ordovician fauna such as *Orthis, Leptelloidea, Leptaena, Strophomena, Refinesquina, Resserella, Sowerbyella, Raymondella, Antiella* and crinoid stems and joints (Cowper Reed, 1934; Suneja, 1971). In Liddar Valley of Anantnag district, an unfossiliferous shale, silt and limestone succession of probably Cambro-Ordovician age conformably underlies a fossiliferous shale horizon of Late Ordovician age known as **Gauran Beds** (Table 8.2). The shales have yielded well preserved bryozoans, brachiopods and cystoids.

The Ordovician rocks of Baramula district are conformably overlain by a succession of sandstones, slates and greywackes yielding typical Silurian (Wenlockian) fossil, *Pycnactis mitrata*. The Silurian rocks are disconformably overlain by Carboniferous succession. **Muth Quartzite** of Middle to Late Devonian age stands out conspicuously in the topographic feature of the northern parts of Kashmir (Table 8.2). The formation first described from the Spiti region of Himachal Pradesh was regarded as unfossiliferous until Middle Devonian fossils were reported from the Kashmir Valley (Gupta, 1966). The fossil fauna include representatives of brachiopods, corals, gastropods, pelecypods, trilobites and cephalopods.

In southwestern part of Kashmir, the Dogra Slates are conformably overlain by an unfossiliferous succession of phyllites, sandstones, massive quartzites, grits and conglomerates known as **Tanawals** (Wynne, 1878).

Wadia (1934) suggested that the unfossiliferous Tanawal succession bridges the gap between the Dogra Slates and the Upper Palaeozoic rocks in south and southwestern Kashmir.

Table 8.2: Silurian and Devonian succession of Anantnag District (after Gupta, 1969)

Stratigraphic Unit	Lithology	Characteristic Fossils	Age
Muth Quartzite	White & green ortho-quartzite	*Schizophoria, Arthyris, Eurispirifer, Calceola*	Middle to Late Devonian
"Naubug Beds"	Shales, sandy shales & siliceous limestone	*Illaenus, Calymene, Orthis, Triplica*	Silurian to Middle Devonian
"Harpatnar Beds"	Mudstones & shales	*Monograptus*	Silurian (Ludlovian)
Unfossiliferous	Shales & siliceous shales		Early & Middle Silurian
"Gauran Beds"	Shales	*Diplotrypa Prasopora, Monotrypa, Orthis*	Late Ordovician

In the northern parts of Kashmir, the Muth Quartzite is conformably overlain by *Syringothyris* **Limestone**, a succession of grey and dark blue limestones with a few interbeds of shales, quartzites and traps. The limestone is rich in *Syringothyris* suggesting an Early Carboniferous age for the formation. Three biostratigraphic zones namely the *Syringthyris* Zone, the *Productus* Zone and the Coralline Zone in ascending order, have been recognised in Anantnag district (Gupta, 1971). The formation exposed along the southern slopes of Pir Panjal Range near Banihal rests directly over the "Chamalwas Slates", a stratigraphic equivalent of the Dogra Slates.

The *Syringothyris* Limestone is overlain by a succession of unfossiliferous quartzites and shales, which, in turn, is conformably overlain by a 600 metre thick succession of inter-banded shales and quartzites with occasional beds of conglomerates. The shales contain abundant *Fenestella* after which the whole succession is named as *Fenestella* **Shales**. The shales have also yielded bryozoans, brachiopods, pelecypods, corals, trilobites and crinoids suggesting an Early to Late Carboniferous age for the formation. The *Fenestella* Shales are well exposed in the Anantnag district and near Banihal and Budil in Pir Panjal Range.

The Fenestella Shales are conformably overlain by a succession of slates, sandstones, quartzites, tilloids and conglomerates with a few bands of limestones. The succession named as **Agglomeratic Slate Series** (Middlemiss, 1910) is well exposed in the Pir Panjal Range, Baramula district, Lidar Valley, Anantnag district and Kishtwar in Jammu and Kashmir and Chamba in Himachal Pradesh. The polymictites consist of rock fragments derived from glacial erosion as well as from volcanic outbursts. The tilloids of glacial origin have been correlated with the Blaini Boulder Bed of the Simla Hills and Talchir Boulder Bed of the Indian Peninsula. The slate succession has yielded *Syringothyris, Spirifer, Productus* and *Eurydesma*

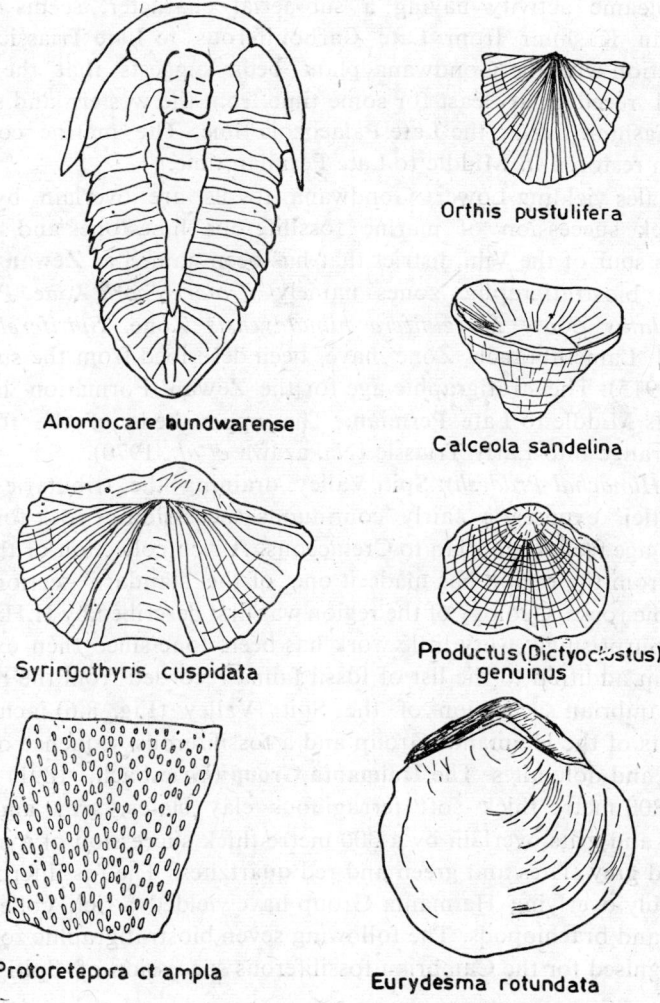

Fig. 8.5.: Palaeozoic fauna from Kashmir.

suggesting a Late Carboniferous to Early Permian age for the formation. The succession also contains a few interbeds of shales yielding Gondwana plants.

The "Agglomeratic Slate Series" is overlain and often intermixed with a thick succession of andesitic and basaltic traps known as **Panjal Volcanics**. The volcanics occupy the steep slopes and the high peaks of the Pir Panjal Range and higher reaches of the Lidar Valley. Andesites are without olivine and range in structure from massive to amygdaloidal varieties. The amygdules are often filled with secondary quartz and chlorite. Dolerite sills and dikes comprising the intrusive equivalents of the Panjal Volcanics have intruded the Fenestella Shales, Agglomeratic Slates and older successions.

The volcanic activity having a sub-aerial character, seems to have persisted in Kashmir from Late Carboniferous to Late Triassic epochs. Its association with the Gondwana plant beds suggests that the marine water had receded at least for some time from the western and southern parts of Kashmir during the Late Palaeozoic time. The marine conditions were again restored in Middle to Late Permian time.

The shales yielding Lower Gondwana fossils are overlain by a 240 metre thick succession of marine fossiliferous limestones and shales on the Zewan spur of the Vihi district that has been named as **Zewan Formation**. Five biostratigraphic zones namely *Protoretepora* Zone, *Productus semireticulatus* Zone, *Marginifera himalayensis* Zone, *Spiriferella rajah* Zone and Lamellibranch Zone have been described from the succession (Diener, 1915). The stratigraphic age for the Zewan Formation has been assigned as Middle to Late Permian. The upper beds of the formation probably range into Early Triassic (Nakazawa *et al*., 1970).

Spiti (*Himachal Pradesh*): Spiti Valley, draining the tributaries of the upper Sutlej, exposes a fairly continuous succession of rock formations ranging in age from Cambrian to Cretaceous. A rich collection of the fossils reported from the region has made it one of the standard sections of the Phanerozoic rocks. Geology of the region was first described by H.H. Hayden in 1904. Surprisingly, very little work has been done since then except for some recent addition to the list of fossil fauna recorded from the region.

The Cambrian succession of the Spiti Valley (Fig. 8.6) includes the upper parts of the Haimanta Group and a fossiliferous sequence of slates, quartzites and dolomites. The **Haimanta Group** (Griesbach, 1891) consists of about 800 metre thick soft ferruginous clay slates, grey and purple quartzites and grits overlain by a 300 metre thick succession of pyritiferous purple and grey slates and green and red quartzites. The fossiliferous beds conformably overlying Haimanta Group have yielded a rich collection of trilobites and brachiopods. The following seven biostratigraphic zones have been recognised for the Cambrian fossiliferous succession of Spiti (Gupta, 1973):

Palaeozoic History 137

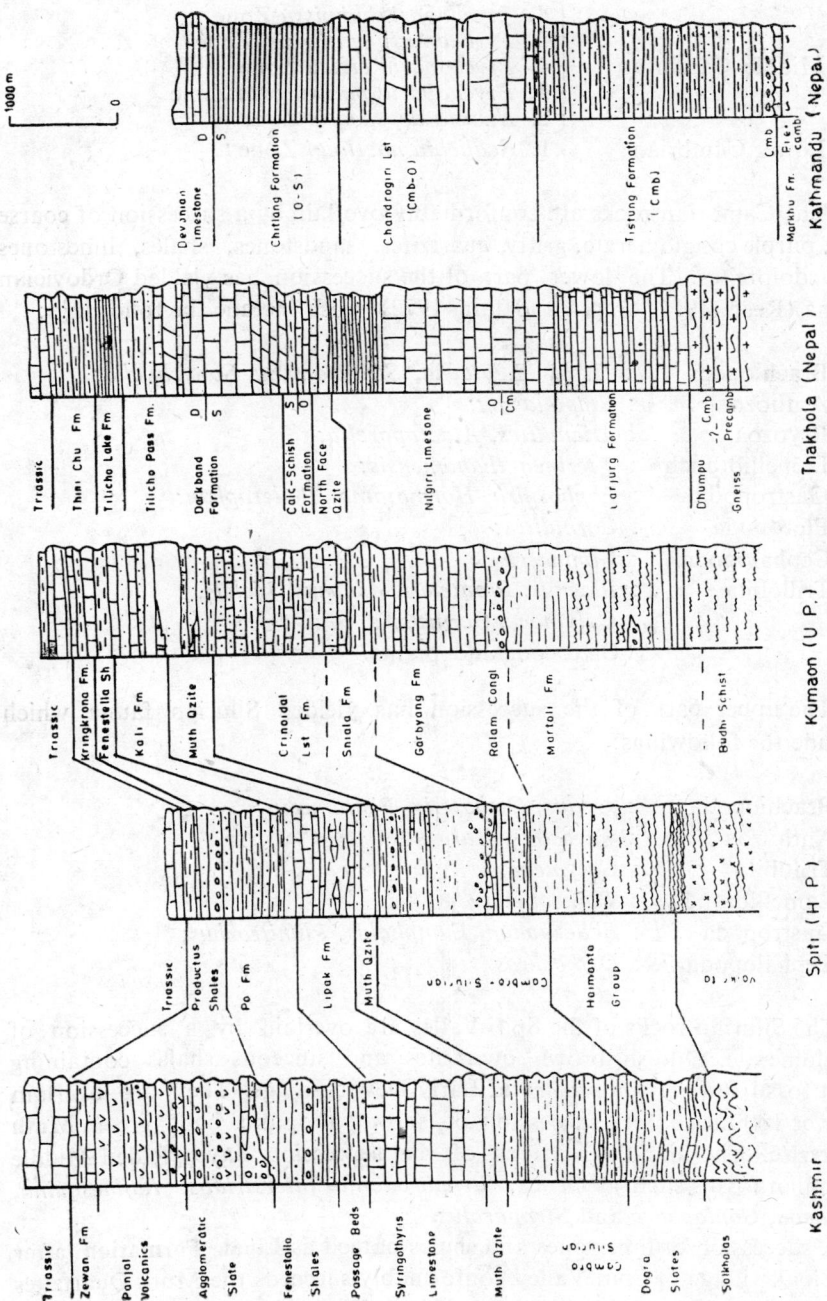

Fig. 8.6.: Stratigraphic columns showing the Palaeozoic formations of the Tethyan Himalaya.

138 Stratigraphy of India

Upper Cambrian	7. *Olenus haimantaensis* Zone
Middle Cambrian	6. *Ptychoparia admissa* Zone
	5. *Ptychoparia spitiensis* Zone
	4. *Ptychoparia civica* Zone
	3. *Ptychoparia peruvulgata* Zone
	2. *Nisusia depsaensis* Zone
Lower Cambrian	1. *Redlichia noetlingi* Zone

The Cambrian rocks are conformably overlain by a succession of coarse red, purple conglomerate, gritty quartzites, sandstones, shales, limestones and dolomites. The lower part of the succession has yielded Ordovician fauna (Reed, 1912; Gupta and Jain, 1972) which include the followings:

Brachiopods	: *Orthis, Leptaena, Strophomena, Sowerbyella;*
Anthozoa	: *Steptelasma;*
Bryozoa	: *Dianulites, Atactoporella;*
Lamellibranch	: *Pterinea thanamensis;*
Gastropoda	: *Lophospira, Hormotoma, Bellerophon;*
Pteropoda	: *Cornulites;*
Cephalopoda	: *Goniaceras;*
Trillobita	: *Illaenus, Brontenus, Asaphus, Lichas, Calymene, Cerurus;*
	Ostrocods and plants.

The upper part of the succession has yielded Silurian fauna which include the followings:

Brachiopoda	: *Orthis, Orthotetis, Pentamerus;*
Anthozoa	: *Favosites, Halysites;*
Trilobita	: *Calymene;*
Lamellibranch	: *Palaeoneilo;*
Gastropoda	: *Bellerophon, Eumphalus, Planitrochus;*
Cephalopoda	: *Orthoceras.*

The Silurian rocks of the Spiti Valley are overlain by a succession of conglomerate, reddish brown quartzites and siliceous shales containing plant fossil *Psilophyton princep* of Early Devonian age. They are overlain by over 100 metre thick succession of snow-white and light green **Muth Quartzite**. The middle portions of the succession have yielded Middle Devonian fossils such as *Orthis, Stropheodonta interstrialis, Rafinesquina, Leptaena, Goniophora* and *Straparolus*.

A succession of limestones and shales named as **Lipak Formation** after the Lipak village of Spiti Valley conformably succeeds the Muth Quartzites of Spiti Valley. The basal part of the Lipak Formation has yielded coral, *Cyathophyllum* ranging in age from Late Devonian to Early Carboniferous.

The middle and the upper parts of the succession contain characteristic Lower Carboniferous brachiopods, *Syringothyris cuspidata*. A rich assemblage of ostrocoda has also been reported from the limestones exposed in the upper Spiti Valley (Jain *et al.*, 1972).

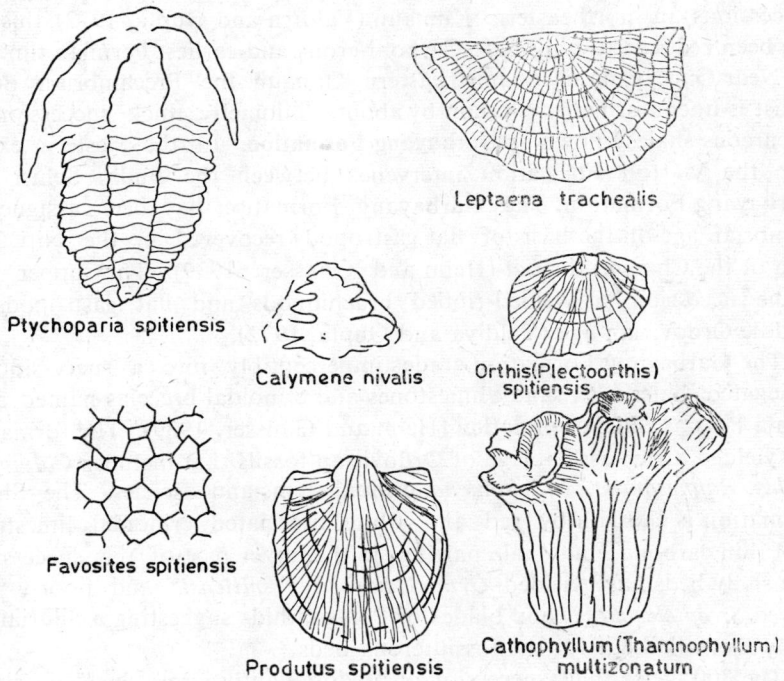

Fig. 8.7.: Palaeozoic fauna from Spiti.

The Lipak Formation is conformably overlain by a thick succession of shales and quartzites that has been named as **Po Formation** after the type area in Spiti Valley. The lower part of the succession known as Thabo Member has yielded Rhacopteris flora of the uppermost Early Carboniferous age. The upper part contains *Fenestella* and *Protoretepora* of Late Carboniferous age.

The Po Formation is conformably overlain by a succession of grits and conglomerates followed by calcareous sandstones and black shales. The lower coarse arenites are unfossiliferous whereas the calcareous sandstones have yielded *Spirifer*. The topmost black shales known as *Productus* **Shales** have yielded a rich fossil fauna including *Productus, Xenaspis, Cyclolobus* and *Marginifera himalayensis*. The fauna indicate a Permian age for the fossiliferous calcareous sandstones and black shales. The lower arenite sequence is probably equivalent of the Agglomeratic Slate Series of Kashmir which has been assigned a Late Carboniferous age.

Kumaun-Garhwal (*Tethyan Sequence*): The sections of the Kumaun and

Garhwal, in many respects, are similar to those of the Spiti Valley except for some variations in lithofacies characters. A large stratigraphic break in deposition was envisaged in the Kumaun Himalaya representing the whole of Carboniferous and lower parts of the Permian periods (Heim and Gansser, 1939). With the discovery of Lower and Middle Carboniferous successions in northeastern Kumaun (Valdiya and Gupta, 1972), this gap has been reduced to only Late Carboniferous and earliest Permian times.

Near Garbayang in the northeastern Kumaun, the Precambrian Budhi Schist is unconformably overlain by about 2 kilometre thick succession of calcareous shales known as **Garbayang Formation**. In the western extension, the Martoli Formation intervenes between the Budhi Schist and Garbayang Formation. The Garbayang Formation has been assigned a Cambrian age on the basis of flat gastropods recovered in the Niti Pass area of the Chamoli district (Heim and Gansser, 1939). The upper part of the succession has yielded ribbed brachiopods and flat gastropods of Middle Ordovician age (Valdiya and Gupta, 1972).

The Garbayang Formation grades imperceptibly into a succession of variegated shales with sandy limestones and crinoidal breccias named after Shiala Pass as **Shiala Formation** (Heim and Gansser, 1939). The formation has yielded a rich assemblage of Ordovician fossils that includes *Calymene, Orthis, Rafinesquina, Strophomena*, crinoid stems and oscicles. The Shiala Formation is overlain by red and white brecciated crinoidal limestones with thin carbonaceous shale partings in the upper part of the succession. The shaly beds have yielded *Orthis, Favosites spitiensis* and poorly preserved *Streptelesma* (?) and blades of Polygnathids suggesting a Silurian to Early Devonian age for the fossiliferous beds.

The 800 metre thick succession of predominantly cross bedded, ripple marked, white quartzites that overlies the crinoidal limestones has been correlated with the Muth Quartzite of Spiti Valley. The lower portion of the succession having gradational contact with the underlying formation is intercalated with dolomitic layers. The fossil assemblage of *Schellwienaella, Arthrophyens* (?) *Hostimella, Machaeraia, Salopina, Camaroboechia*, crinoid stems and poorly preserved bryozoans indicating Middle Devonian age has been reported from the northeastern Kumaun (Valdiya and Gupta, 1972).

The Muth Quartzite is conformably overlain by about 800 metre thick succession of grey siliceous limestones, rusty brown weathering dolomites and fine grained quartzites named as **Kali Formation** (Valdiya and Gupta, 1972). Lower Carboniferous fossils that have been reported from the succession include *Linoproductus, Chonetes, Reticularia, Caninophyllum, Caninia* and *Lophocarinophyllum*. The Kali Formation has been correlated with the Lipak Formation of the Spiti Valley and Syringothyris Limestone of Kashmir.

A succession of black carbonaceous shales interbedded with subordinate black limestones conformably resting over the Kali Formation has yielded Middle Carboniferous fossil assemblage which includes *Fenestella, Protoretepora, Spirifer* and *Productus*. This succession has been correlated with the *Fenestella* Shales of Kashmir and Po Formation of Spiti Valley. It has been disconformably overlain by the **Krinkrong Formation** that consists of a succession of dark ferruginous limestones interbedded with black micaceous shales. The shales have yielded *Spirifer, Marginifera, Costiferina, Strophalosia, Productus* and *Lyttonia* suggesting a Permian age for the formation. The fossil assemblage is well correlated with those of the Zewan Formation of Kashmir and Productus Shales of the Spiti Valley.

Nepal Himalaya: The Tethyan Palaeozoic rocks have been exposed in the Thakhola region of the upper Kali Gandaki valley (Bordet *et al.*, 1971 and 1975), the Dolpo regions in western Nepal (Fuchs, 1977) and the Kathmandu Nappe of the Central Nepal (Kumar, 1980).

The Lower Palaeozoic succession of the Kathmandu Nappe (Figs. 7.2 & 7.3) known as **Phulchauki Group** unconformably rests over the Precambrian rocks of Bhimphedi Group. The Phulchauki succession begins with about 3 kilometre thick unfossiliferous sequence of interbanded slates, calc-phyllites and metasandstones known as **Tistung Formation**. The formation grades upwards into the **Chandragiri Limestone** through a 200 m thick succession of a transitional formation known as **Sopyang Formation**. The Chandragiri Limestone consisting of 2000 to 2500 metre thick, thickly and thinly interbedded limestones with occasional quartzite beds has yielded *Caryocrinites, Codacystes, Lagynocystis, Dendrocystites* and *Himalayicalix* suggesting a Middle Ordovician age for the formation (Stocklin *et al.*, 1977; Gupta and Termier, 1978 and Gupta and Stocklin, 1978). Gupta and Chetri (1978) have recorded some additional fauna from the Chandragiri Limestone of Godavari Phulchauki section in the south of Kathmandu which include three species of *Orthis, Rafinesquina* and broken specimens of Strophomenids.

The Chandragiri Limestone is conformably overlain by a 100 metre thick slate succession known as **Chitlang Formation** that contains intercalated bands of quartzites in the lower parts and limestones in the upper parts. The upper parts of the formation is also characterised by the presence of hematitic rich beds in the Phulchauki hills. The formation has yielded fragmental brachiopods, trilobites and other fossils indicating a Silurian age for the formation. The youngest unit of the Phulchauki Group comprising a 300 metre thick succession of impure limestones and massive dolomites have yielded Devonian fossils which include *Ancyrodella, Icriodus, Palmatolepsis, Polygnathus* (three species), *Siphonodella* and *Bryantodus* and other Late Devonian conodonts (Gupta and Stocklin, 1978).

The rock succession of Thakhola region resembling the Tistung Formation has been transformed along with the Precambrian basement into high grade schists and gneisses broadly grouped as **Dhimpu Gneisses** (Bordet et al., 1975). A staggering thickness of about 5 kilometre of predominantly calcareous succession known as **Nilgiri Carbonate Group** lies in an apparent conformity over the Dhimpu Gneisses (Fig. 8.6). The Nilgiri Carbonate Group has been divided into **Larjung Formation** and **Nilgiri Limestone,** the former is differentiated from the latter by its higher grade of metamorphism. The upper part of the sequence consists of calcareous siltstones with current bedding structures. An Early Ordovician age for the higher parts of the succession is well established on the basis of faunal evidence which includes small ribbed brachiopods, *Aporthophylla* (nov. sp.) and a nautiloid.

In the Dolpo region, the Nilgiri Carbonate Group has been referred to as **Dhaulagiri Limestone** (Fuchs, 1977) consisting of about 4 kilometre thick sequence of interbanded calcareous sandstone, siltstone and dolomitic limestone. The rocks show well preserved sedimentary structure such as ripple marks, load-casts, convolutions and burrow casts of presumably turbidite origin. Increased amount of detrital components indicate a coastal nearness in this part of the basin. The Dhaulagiri Limestone containing gastropods, brachiopods, lamellibranchs, crinoids and orthoceratids is also exposed in the Lesser Himalayan nappes of the Jaljala Dhuri range in Western Nepal (Fuchs and Franck, 1970).

The Nilgiri Limestone of the Thakhola region grades upward into **North Face Quartzite** comprising about 500 metre thick succession of light coloured quartzitic arkoses and calcareous siltstones. The formation is overlain by a thin horizon of **Pitted Calc-Schist Formation** that has yielded fairly well preserved Llanvirnian fauna such as *Orthambonites* (nov. sp.), *Aprothophylla* (nov. sp.) and *Opekina* and echinoids. About 450 metre thick succession of dark grey limestones and dolomites, blackish calc-siltstones and shales that conformably overlies the Pitted Calc-Schist Formation has been named as **Dark Band Formation** for its typical dark colour attributed to disseminated carbonaceous matter. Occurrence of Llandoverian, Early Silurian graptolites and Early Devonian *Tentaculites* have been reported from the lower and upper parts of the succession (Egeler *et al.,* 1964; Bordet *et al.,* 1967).

The **Tilicho Pass Formation** comprising about 900 metre thick succession of micaceous and calcareous quartzites and arenaceous limestones marks a sharp change in litho-facies from that of underlying Dark Band Formation. Presence of sedimentary structures of turbidite origin is characteristic for the Tilicho Pass Formation. Westward, in the Dolpo region, the facies changes into dolomite rich formation that has been correlated with the Muth Quartzite of the Spiti Valley (Fuchs, 1977). Presence of reefal features in the formation indicates the proximity of strand line during its deposition. The formation has yielded a rich assemblage of corals, algal remains, crinoids, brachiopods and gastropods of Middle Devonian age.

The Tilicho Pass Formation is overlain with sharp conformable contacts by a succession of rhythmically bedded massive limestones and clayey limestones known as **Tilicho Lake Formation**. The formation contains a thin haematitic layer at its contact with the underlying formation. The formation has yielded a rich assemblage of fossil fauna which include corals (*Zapherentites?* sp., *Siphinophyllia* sp., *Caninia* sp., *Amplexus* sp., *Michelina megstoma* and *Aulopora*) and brachiopods (*Linoproductus pollex*, *Fussella mucronata* and *Syringothyris curzoni glaber*) of Early Carboniferous age.

Thini Chu Formation differs from the underlying Tilicho Lake Formation in the presence of coarser detrital contents. The succession comprises a 700 metre shick sequence of rhythmically bedded conglomerates, sandstones, slates and dolomitic limestones. A general trend of shallowing of the Tethyan basin seems to have set in during the deposition of the Thini Chu Formation. The ill-sorted sandstones and grits of the lower part of the succession containing angular fragments of quartz have been correlated with the Agglomeratic Slate Series of the Kashmir. The dolomitic limestones occurring as lenses in slates towards the upper part of the succession have yielded abundant brachiopods, corals, fenestellids, crinoids and some trilobites. The top of the succession is represented by sandstone beds conformably overlain by Lower Triassic sequence.

Lesser Himalayan Regions

The Lesser Himalayan formations lying to the south of the Central Crystalline Zone (Higher Himalaya) depict a totally different geological history. Stratigraphic positions of these formations are speculative in view of complex structure and largely unfossiliferous nature of the formations. Scanty fossil findings have been recently reported from the Upper Palaeozoic successions. Absence of fossils from the Lower Palaeozoic successions is still puzzling in view of rich collection of fossil fauna from the equivalent formations of the Tethyan regions.

Lower Palaeozoics: During the Early Palaeozoic time, the Lesser Himalayan sea presumably formed the northern part of the Vindhyan sea (Fig. 6.6). While the Vindhyan sea receded from the Peninsula at the dawn of the Palaeozoic Era, sedimentation seems to have continued in parts of the Lesser Himalaya. The upper part of the Simla Group (Chapter 7) was presumably laid down during the Early Palaeozoic time.

A complex association of limestones, quartzites, slates, volcanic rocks and ashes exposed in the Garhwal Himalaya was named as "Jaunsar System" by Oldham (1883). The succession was later divided into Mandhali, Chandpur and Nagthat "Stages" (Auden, 1934). The succession redefined as **Jaunsar Group** (Srikantia and Bhargava, 1974) shows a considerable lateral facies variations. The Mandhali and Chandpur Formations have been assigned Precambrian age (Chapter 7). The Nagthat Formation occupying the upper part of the Jaunsar Group comprises a thousand metre thick sequence of white to red current bedded quartzite with slate partings, slaty quartzites and volcanics. The volcanics are well exposed in the Bhowali-Bhimtal area in Nainital district (U.P.) where the succession has been named as **Bhimtal Volcanics** (Valdiya, 1980).

Gymnospermic wood of Lower Palaeozoic age has been reported from the upper phyllites of the Chandpur-Nagthat succession of Kumaon (Pawar and Phansalkar, 1971). A well preserved brachiopod *Salopina* indicating an Upper Silurian to Lower Devonian age has been reported from the Sirdang Quartzites that have been correlated with the Nagthat Formation. Radiometric dating of the basics from the Bhimtal volcanics (Varadarajan, 1977) gives a Permian age for the volcanics which, however, appears to be an apparent younger age in view of Permo-Carboniferous flora and fauna reported from the overlying rock successions.

A thick succession of green phyllites, schistose slates, psammitic and conglomeratic schists, talcose quartzites and limestones known as "Chail Series" (Pilgrim and West, 1928) has been described from the vicinity of the Chail town in the southeast of Simla (H.P.). The succession earlier assigned an Upper Precambrian age has been regarded as a part of the Jaunsar Group (Bhargava, 1972). Fuchs (1967) regards the Chails to be a molassic deposit of the Caledonian Orogeny. The stratigraphic age for the formation, however, still remains uncertain in view of its unfossiliferous nature and tectonic contacts with the fossiliferous units.

Upper Palaeozoics: Rock formations of the Upper Palaeozoic age are exposed in the Krol Belt of the Lesser Himalaya (Fig. 7.4). The Upper Palaeozoic succession of the Krol Belt comprises Blaini and Infra-Krol Formations.

The **Blaini Formation** in its type area of Baliana river near Solon consists of a succession of boulder beds, dark shales, siltstones and pink dolomitic and siliceous limestones. The thickness of the formation is highly variable ranging from a few tens of metres to about 200 metres. The boulder beds are composed of clasts of quartzites, slates, sandstones, siltstones, volcanics and granites. The clasts seem to have been derived from the underlying formations of the Simla and Jaunsar Groups which must have been partly eroded during the deposition of the Blaini Formation.

The age of the Blaini Formation has been one of the most debated problems of the Himalayan Geology. The age problem revolves around the tillitic nature of the Blaini Boulder Beds. First proposed by Oldham (1888), the "Blaini tillites" were correlated with the Talchir Tillites (Chapter 10) of the Indian Peninsula that represents a Permo-Carboniferous Epoch of world-wide glaciation. Holland (1908), on the other hand, correlated the Blaini tillites with the Late Precambrian Glacial Epoch. The origin of the Blaini Boulder Beds has been attributed by some geologists (Rupke, 1968; Valdiya, 1980) to the submarine slumping under turbid environments of unstable marine basins. Among the several other genetic models proposed for the origin of the Boulder Beds, the glacio-marine model, however, appears to be the most widely accepted (Bhatia, 1975).

Palaeontological evidence for the stratigraphic age of the Blaini Formation has been meagre. A wide range of microfossils reported from the type area by Prasad and Bhatia (1975), includes tracheids, foraminifers, dinoflagellates, algae, ostrocods and a radiolaria suggesting a Permo-Carboniferous age for the formation. The sporomorphs, reported by Shrivastava and Venkataraman (1975), suggest a Late Carboniferous age for the Blaini Formation. An Early Permian age has been suggested by Tewari and Singh (1981) on the basis of the presence of a large fusulinid foraminifera *Robustoschwagerina*, a cryptostome bryozoa *Protoretepora* and impressions of Productids from the calcareous beds of the Blaini Formation.

The base of the Blaini Formation demarcates an important datum line in the geological history of the Lesser Himalaya. It marks the beginning of a prolonged phase of sedimentation extending over a period from Late Palaeozoic until the closing of the Himalayan Geosyncline during the Tertiary Period. There are mainly two views regarding the nature of the unconformity at the base of the Blaini Formation. The one view holds that the Blainis were deposited over conformable and time transgressive contacts (Bhatia, 1975). The other view regards the unconformity to represent a time gap of over 100 million years as the middle part of the Palaeozoic succession has not been definitely established in the stratigraphic successions of the Lesser Himalaya.

The Blaini Formation is conformably overlain by a sequence of about 200 metre thick black carbonaceous shales and slates with intercalations of brown quartzites named as **Infra-krol Formation** in view of its underlying position below the Krol Formation (Chapter 9). The shales are often pyritiferous indicating an euxinic facies of deposition. Micro-floral remains (trilete spores) comparable with selaginellaceous types ranging in age from Carboniferous to Permian have been reported from the Infra-krol Formation exposed in Mussoorie hills (Ghosh and Srivastava, 1962). Tewari and Singh (1981) have reported the occurrence of Glossopteris flora from black shales of the formation exposed in the vicinity of Nainital.

The carbonate rich sequence of Krol Formation conformably overlying the Infra-krol Formation has been assigned a Permo-Triassic age (Auden, 1934). Valdiya (1980) has grouped the Blaini-Krol-Tal succession into **Mussoorie Group** assigning the whole succession a Late Palaeozoic age. A Permian age for the Tal succession was suggested on the basis of reported occurrence of fusulinids from the Duggada area which has, however, been doubted by some geologists (Chapter 9).

The Garhwal syncline exposes in the Duggada area of Garhwal district (U.P.) a sequence of slates and boulder slates forming the **Bijni Tectonic Unit.** The unit tectonically over-rides the Mesozoic-Tertiary sequence and is itself tectonically overlain by schists and gneisses of presumably Precambrian age (**Amri Tectonic Unit**). The shaly horizons of the Bijni Tectonic Unit have yielded six species of bryozoan genera suggesting a Middle to Upper Carboniferous age (Ganesan, 1972). Waterhouse and Gupta (1978) have reported the occurrence of nine species of brachiopods and five species of bivalves suggesting a Sakmarian (Lower Permian) age. The brachiopods and bivalves are comparable with the fauna of the Umaria marine (Chapter 10) beds of the Indian Peninsula.

Chapter 9
Mesozoic History

The term "Mesozoic" was introduced by John Phillips in 1840 for the rock formations containing remains of "middle forms" of life. The Mesozoic Era that began at about 230 million years ago and closed at about 65 million years ago represents less than half the duration of the Palaeozoic Era. The Mesozoic Era has been further subdivided into Triassic, Jurassic and Cretaceous periods (Table 9.1).

The **Triassic System** of rocks deposited during the Triassic Period was established in 1834 by the German scientist F. von Alberti. The system was named after its characteristic three-fold divisions in Germany. The Lower and the Upper divisions known as Bunstanstein (Bunter) and Keuper Series respectively consist of predominantly continental formations whereas the Middle division known as Muschelkalk Series is composed of mainly marine rocks. The Lower Triassic is made up of only one stage known as Scythian Stage. The Middle and Upper successions have been subdivided into two and three stages respectively. Type sections of most of the stages of the Triassic System are situated in the Alps.

The **Jurassic System** was established in Jura Mountains of Western Europe by the French scientist A. Brogniart in 1829 where the system is divisible into Lias, Dogger and Malm Series. Four stages each for the Lias and Dogger and three stages for the Malm have been established on the basis of their characteristic faunal assemblages. The Jurassic successions elsewhere are subdivided into three divisions known as Lower, Middle and Upper Jurassic. These divisions correspond to the three divisions of the type area except that the Aalenian stage of the Dogger is included in the Lower Jurassic. The type sections of the stages of the system are situated in England (Bathonian, Callovian, Oxfordian and Kimmeridgian), France (Hettangian, Sinemurian, Toarcian and Bajocian), West Germany (Pliensbachian) and U.S.S.R. (Tithonian).

The **Cretaceous System** was established by the Belgian scientist O. d'Halloy in 1822 after its typical lithology of chalk (known as Creta). The system is subdivided into Lower and Upper Cretaceous Series which are further subdivided into six stages each. The lower three stages of the Lower

Cretaceous are often grouped into a single unit known as Neocomian. The upper four stages of the Upper Cretaceous are likewise grouped into Senonian. Type sections of most of the stages except for Valanginian and Hauterivian (Switzerland) and Maastrichtian (Holland) are situated in France.

Table 9.1: Subdivisions of the Mesozoic Era

Period/System	Epoch/Series		Age/Stage
Tertiary (Cenozoic Era)	Palaeocene		Danian*
65 m.y.			
Cretaceous	Late/Upper	Senonian	Maastrichtian Campanian Santonian Coniacian
			Turonian Cenomanian
	Early/Lower		Albian Aptian Barremian
		Neocomian	Hauterivian Valanginian Berriasian
141 m.y.			
Jurassic	Late/Upper	Malm	Tithonian (Portlandian) Kimmeridgian Oxfordian
	Middle	Dogger	Callovian Bathonian Bajocian Aalenian
	Early/Lower	Lias	Toarcian Pliensbachian Sinemurian Hittangian
195 m.y.			
Triassic	Late/Upper	Keuper	Rhaetian Norian Carnian
	Middle	Muschelkalk	Ladinian Anisian
	Early/Lower	Buntstandstein	Scythian
230 m.y.			

*The Danian Stage is regarded by some as the uppermost stage of the Upper Cretaceous.

TECTONIC HISTORY

Perhaps the largest number of palaeogeographic reconstructions have been made for the Mesozoic Era. In many parts of the world, the Era began with a new phase of sedimentation. Pangaea, the supercontinent of the Palaeozoic Era was gradually torn apart during the Mesozoic Era. Fragmentation of the Pangaea began with the opening of proto-Atlantic and proto-Indian oceans. On the basis of the palaeo-magnetic evidence, it has been suggested that this break-up began with the separation of North America and Gondwana Land in Late Triassic Epoch. Dismemberment of the Gondwana Land began in Late Jurassic Epoch which led to the separation of India and Africa from Australia, Antarctica and South America.

Remarkably poor marine record of the Early Triassic indicates the continuation of world-wide marine regression that commenced at the close of the Palaeozoic Era. The Early Triassic was also a time of abnormal salinity which had a disastrous effect on the marine fauna of the time. The first marine transgression of the Era took place during the Ladinian Age of the Middle Triassic Epoch. This was followed by a climatic optimum and extensive deposition of carbonate rocks during the Carnian and Norian times. Yet another phase of marine regression at the close of the Late Triassic is recorded in many parts of the world. The regression is marked by a distinctive facies of sulphurous black shales, red beds and evaporites, undersized fauna and mass extinction of many faunal groups.

Deepening of the sea that began towards the close of the Early Jurassic is demonstrated by a widespread change in facies and appearance of new ammonitic fauna. A general eustatic rise in the sea level of Toarcian Age was presumably a direct consequence of emergence of a mid-Atlantic Ridge. Significant reef development and presence of oolites in the Bajocian rocks may indicate a climatic optimum. The most impressive of all the marine transgressions of the Era took place during Callovian and it was presumably caused by a sudden increase in the rate of ocean-floor spreading and further rise of the mid-Atlantic Ridge.

The Cretaceous seas were as oscillatory as those of the Jurassic period. The temperature reached its maximum during the Campanian time when the transgression was also maximum. The regression at the close of the Era was accompanied by a general cooling of the climate and widespread extinction of fauna that is almost as marked as the one at the beginning. Exciting achievements in space explorations during the last over two decades have led many scientists to believe that the earth's history has also been affected by the external factors. The discovery of high levels of iridium in thin beds of clay at the boundary of Mesozoic and Cenozoic formations has been attributed to asteroid impact over the earth's surface at the time of the deposition of clay beds. Such an impact might have cut out the sunlight for photosynthesis leading to starvation and extinction of a large number of faunal groups.

HISTORY OF MESOZOIC LIFE

Marine Forms

Only some of the Late Palaeozoic forms of life such as ceratites, productids and chonetids could survive the "salinity crisis" and inhospitable climate of Late Permian-Early Triassic time. New habitats were made available during the Middle Triassic marine transgression leading to a complete reorganisation of organisms. The marine forms of the Mesozoic Era include a great variety of ammonites and belemnites, greatly diversified bivalves, and gastropods, appearance of hexacorals in place of tetracorals, and appearance of bony fishes and many new orders of marine plants.

The Triassic ammonoids were rapidly diversified and dispersed through the Jurassic and Cretaceous periods making them the most important group of index fossils. Hexacorals became widely distributed in narrow equatorial belt during the Late Triassic and persisted through the Jurassic and Cretaceous periods. New orders of Bryozoans replaced the Palaeozoic forms. Terribratulids were also greatly diversified during the Mesozoic Era providing important index fossils. Pelecypods that had also adapted the brackish and fresh water basins replaced the dominance of brachiopods in shallow marine basins.

Crustaceans were the important arthropods of the Mesozoic Era. Echinoderms are represented by crinoids and echinoids constituting important index fossils of certain Mesozoic sections. Foraminifera, dominantly of microscopic sizes, proliferated during the Jurassic and Cretaceous periods. Unicellular planktonic plants, diatoms and cocoliths made their first appearance during the Mesozoic Era.

Mesozoic fishes are dominated by ray finned fish, the actinopterigians. Amphibians continued to decline during the Era. Frogs appeared in Jurassic and salamanders in the Cretaceous. These small amphibians are of little stratigraphical value although they are useful ecological guides.

Land Forms

An increase in the land area at the beginning of the Era facilitated the expansion of terrestrial fauna and flora. Nearly all the available land habitats were dominantly occupied by the reptiles and, thus, the Mesozoic Era is also known as the Era of reptiles. In spite of their success, a great majority of reptiles became extinct by the close of the Era.

Terrestrial reptiles included cotylosaurs, therapsids, thecodonts, dinosaurs, snakes, lizards and crocodiles. Therapsids that evolved from the cotylosaurs during the Permian became dominant reptiles during the Early

Triassic. They, however, rapidly declined and became extinct by the end of Triassic. Dinosaurs that evolved from the conodonts in Late Triassic greatly diversified during the Jurassic and Cretaceous periods. Competition for food between the land reptiles led to the evolution of some forms which adapted themselves to marine environments. Yet another group of reptiles soared the unchallengeable skies of the time.

Mammals, although appeared in Triassic, remained insignificant throughout the Era. The earliest known bird, Archaeopteryx, is known from the Jurassic deposits of Germany. The birds were well established by the Cretaceous period.

The flowering plants rose to dominance during the Mesozoic Era. The angiosperms that appeared in the Jurassic period had replaced the gymnosperms in abundance in Late Cretaceous. Of the two classes of angiosperms, the monocotyledons appeared in Jurassic whereas the dicotyledons made their first appearance in Early Cretaceous. Seed ferns, which were prominent in Permian, gradually declined through the Mesozoic Era. The gymnosperms of the Era are represented by conifers.

PERMIAN-TRIASSIC BOUNDARY

Demarcating the Permian-Triassic boundary is beset with problems due to incomplete stratigraphic records that is attributed to the widespread regression at the close of the Palaeozoic Era. In many parts of the world, the top of the Permian succession is characterised by the presence of red beds or a phase of non-deposition. Even when the highest Permian stage is immediately overlain by the lowest Triassic stage, there is often an evidence of uplift and erosion between them. Stratigraphic records from the Indian subcontinent comprise some of the few sections of the world that have preserved the imprints of this transitional phase of the Earth's history.

A fairly continuous succession of rock formations containing Permian and Triassic faunal assemblages have been observed in many parts of India. Fossil fauna of the lowermost Triassic rocks is markedly different from those of the uppermost Permian. Extinction of the brachiopod fauna of the *Productus* Shales is as abrupt as the appearance and predominance of Triassic cephalopods. Of the many ammonoids of the Lower Triassic rocks, only the genera *Medlicottia, Episageceras, Xenaspis* and *Xenodiscus* are found in the underlying Permian rocks.

The Guruyul ravine of Vihi District in Kashmir (Fig. 9.1) provides one of the best known sections where the Permian-Triassic boundary is defined both on faunal and lithological characteristics. The succession consists of an arenaceous sequence grading upwards through argillaceous rocks into a carbonate sequence. The lower part of the transitional facies of argillaceous rocks (Black Shales) has yielded a mixed faunal assemblage comprising productids of Permian age in association with the Lower Triassic pelecypod

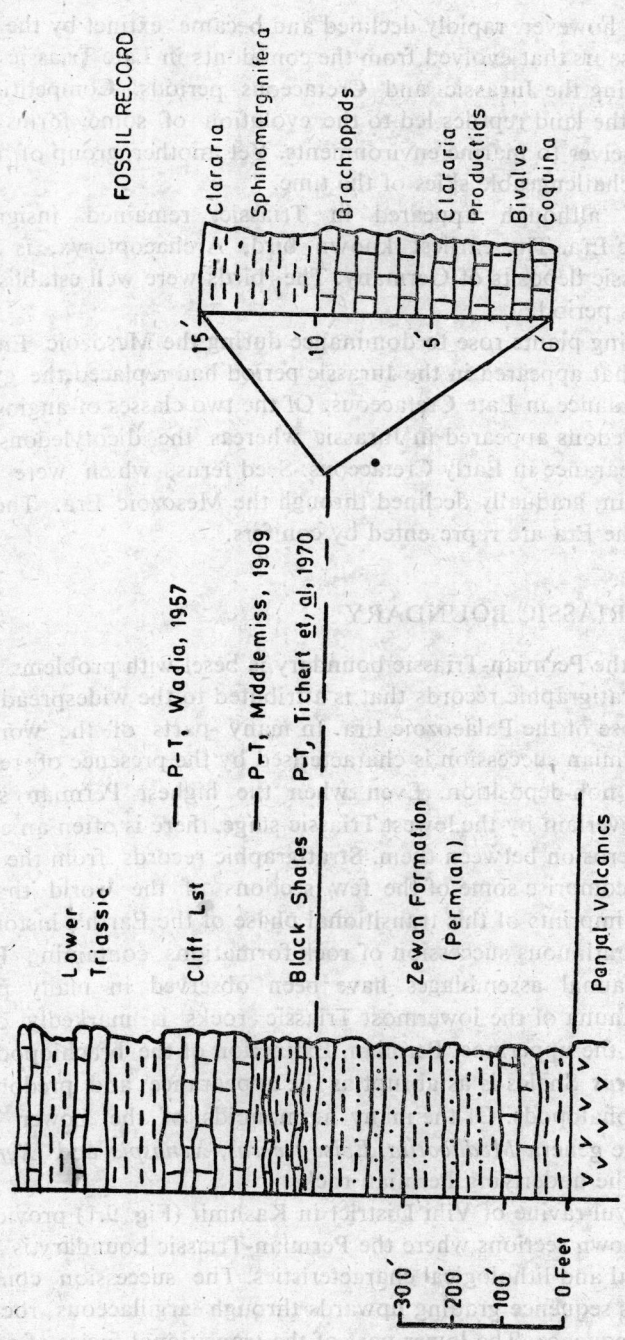

Fig. 9.1: Permo-Triassic boundary in Kashmir

Claraia. Permian productids are regarded to have survived until the Early Triassic Epoch (Teichert *et al.*, 1970). Earlier, this boundary was placed either at the top of the Black Shales (Middlemiss, 1909) or at the top of the overlying Cliff Limestone (Wadia, 1957).

Palaeontological data from Spiti were regarded to suggest that the first appearance of *Otoceras woodwardi* marks the beginning of the Triassic strata. This species is also reported from the strata of mixed Permian-Triassic fauna of Kashmir. The Triassic succession of Spiti begins with massive limestones (*Otoceras* Bed) overlain by flaggy limestone (*Ophiceras* Bed). The carbonate rocks rest over *Productus* Shale with an intervening thin layer of limonitised pebbly rocks (Fig. 9.2). The *Otoceras* Bed has yielded Permian conodonts whereas the first appearance of *Neospathodus*, a dominantly Triassic conodont, was recorded from the flaggy limestone of the *Ophiceras* Bed (Bhatt *et al.*, 1981). Similar observations have

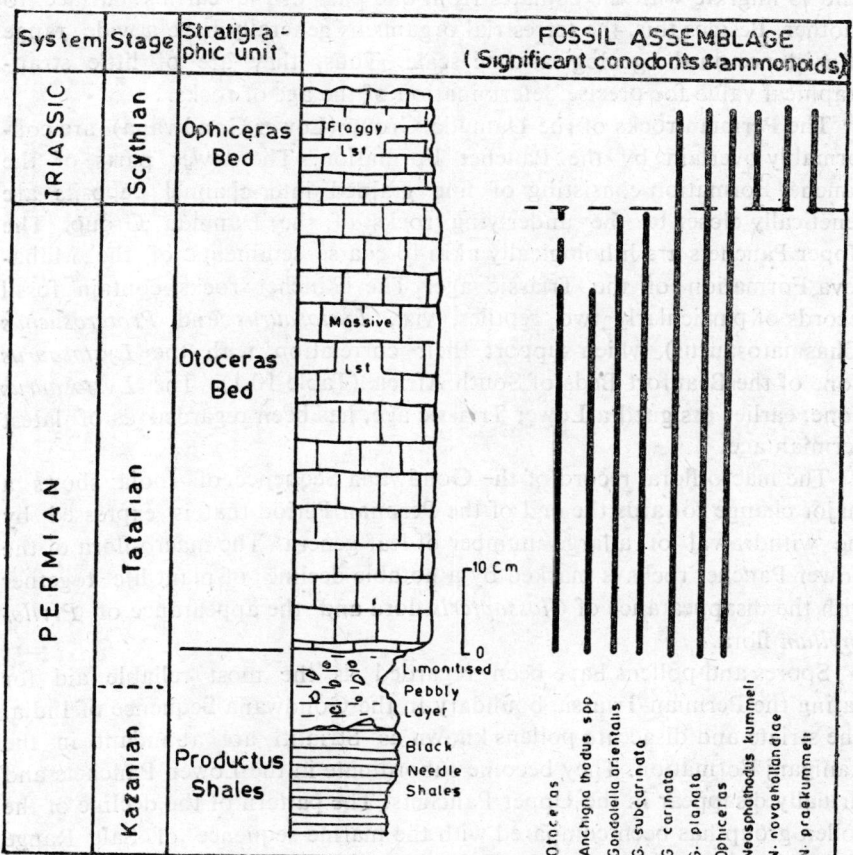

Fig. 9.2: Permo-Triassic boundary in the Lalung section, Spiti Valley (based on Bhatt *et al.*, 1981)

been recorded from Zanskar Range in the northwest and Kumaun region in the southeast.

The conodonts reported from Spiti, Zanskar and Kumaun establishes the affinity of the *Otoceras* Bed with the youngest stage of the Permian period. The mixed fauna of Kashmir reported from the Black Shales, thus, may appear to be truly the fauna of Late Permian alone. More so, in view of the fact that assigning an exclusively Early Triassic age to *Claraia* is doubtful.

An almost uninterrupted succession of terrestrial sediments of Permian-Triassic age is recorded in the Gondwana Sequence (Chapter 10) of the Indian Peninsula. However, delineation of Triassic Gondwana from the underlying Permian Gondwana is beset with great uncertainty in view of the limitations of terrestrial faunal and floral records. At any given time, the terrestrial organisms are susceptible to climatic conditions and they tend to migrate with the climates from one part of the earth's surface to another. Besides this, the terrestrial organisms generally have a wide range of existence on the geological time scale. Thus, they are of little stratigraphical value for precise determination of the age of rocks.

The Permian rocks of the Damuda Group (Lower Gondwana) are conformably overlain by the Panchet Formation. The lower parts of the Panchet Formation consisting of fine grained inter-channel deposits are genetically closer to the underlying rocks of the Damuda Group. The Upper Panchets are lithologically akin to coarse sediments of the Mahadeva Formation of the Triassic age. The Panchet rocks contain fossil records of particularly two reptiles, viz., *Lystosaurus* and *Proterosuchus* (Chasmatosaurus), which support their correlation with the *Lystosaurus* Zone of the Beaufort Beds of South Africa (Table 10.1). The *Lystosaurus* Zone, earlier assigned a Lower Triassic age, has been regarded as of latest Permian age.

The macro-floral record of the Gondwana Sequence of India shows a major change towards the end of the Permian Period that is expressed by the withdrawal of a large number of leaf genera. The macro-flora of the Lower Panchet rocks is marked by a notable decline of plant life together with the disappearance of *Glossopteris* flora and the appearance of *Ptyllophyllum* flora.

Spores and pollens have been regarded as the most reliable aid for dating the Permian-Triassic boundary in the Gondwana Sequence of India. The striate and disaccate pollens known as Striatiti are abundant in the Raniganj Formation. They become subordinate in the Lower Panchets and virtually disappear in the Upper Panchets. The pattern of the decline of the pollen group has been compared with the marine sequence of Salt Range (Pakistan) where the marine fossils are associated with micro-floral record. In the marine rocks, the striate disiccates suffer bulk termination at the

Permian-Triassic boundary. Thus, the Permian-Triassic boundary in the Gondwana Sequence of India lies somewhere in the middle of the Panchet Formation (Sarbadhikari, 1979).

MARINE MESOZOIC FORMATIONS OF INDIA

The Mesozoic rocks of marine facies described in the sequel are known from the Tethyan Himalaya, the Krol Belt of the Lesser Himalaya and northwestern and southern parts of the Indian Peninsula. The Mesozoic rocks of continental facies constitute the Middle and Upper Gondwana Sequences which are described in Chapter 10.

Tethyan Himalaya

A fairly continuous succession of rocks yielding Triassic, Jurassic and Cretaceous fossils is exposed in different parts of the Tethyan Himalaya (Fig. 9.3). Panjal Volcanism which had begun in Late Permian in Kashmir and other parts of the Himalaya continued until the Middle Triassic. The Triassic and Jurassic rocks of the Tethyan Himalaya are predominantly composed of carbonate facies. The Cretaceous rocks, at many places, show flyschoidal characters. The Tethys palaeo-sea became shallower during the Cretaceous period. Waters of the Tethys palaeo-sea was gradually withdrawn towards the Indian Ocean which widened during the Mesozoic Era as a consequence of the northward drift of the "Indian Plate". The emplacement of basic volcanics in association with the deep marine sediments during the Late Cretaceous resulted in the appearance of one of the most striking lineaments of the Himalaya known as the Indus Ophiolitic Belt of Ladakh.

Kashmir-Chamba: Permian rocks of Kashmir (Zewan Formation) are conformably overlain by a thick succession of limestones and shales yielding rich Triassic fauna. The **Lower Triassic** consists of about 100 m thick sequence of dark grey, compact and thick bedded limestones with a few beds of quartzites and shales. Four biostratigraphic zones have been established on the basis of typical ammonoid fauna of the Lower Triassic; they are, in ascending order, *Otoceras* Zone, *Ophiceras* Zone, *Meekoceras* Zone and *Hedenstroemia* Zone.

The Lower Triassic is conformably overlain by a 300 m thick succession of buff coloured, thin bedded, sandy limestones yielding a **Middle Triassic** faunal assemblage. The ammonoids dominated by the genus Ceratites include the genera: *Hungarites, Sibirites, Isculites, Pinacoceras, Ptychites, Gymnites* and *Budhaites*. The important genera of the nautiloids include: *Syringonautiloids, Gryphoceras, Paranautiloids* and *Orthoceras*. The lamellibranchs are represented by *Myophoria, Modiola, Anomia* and *Anodontophoria;* the brachiopods by *Spiriferina, Dielasma* and *Rhynchonella* and the gastropods by *Euomphalus* and *Conularia*.

156 Stratigraphy of India

The **Upper Triassic** consists of enormously thick succession of largely unfossiliferous limestones, dolomites and shales. A few fossiliferous horizons have yielded species of brachiopods, lamellibranchs, crinoids and corals. The *Spiriferina stratcheyi* bearing beds of Carnian age occur at the base of the succession. The upper beds often contain *Megalodon* of Early Jurassic age.

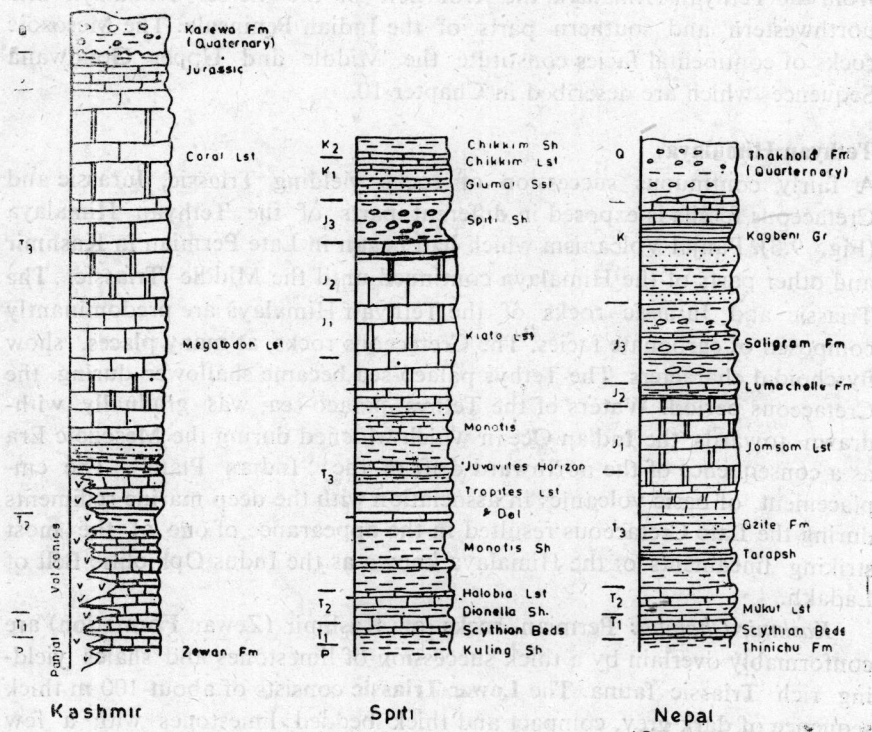

Fig. 9.3: Stratigraphic columns showing the Mesozoic formations of the Tethyan Himalaya.

The Triassic succession of Kishtwar and Chamba regions consists of a carbonate sequence known as **Kalhel Limestone**. The basal part of this succession has yielded *Claraia* in association with Permian fossils. The overlying rocks, although unfossiliferous, have been assigned a Triassic age on the basis of their supra-Permian stratigraphic relationship.

The outcrops of Jurassic rocks have a restricted distribution in Kashmir. A major part of the rocks is buried beneath the Quaternary sediments. Rocks yielding Jurassic cephalopods and lamellibranchs have been reported from the northern slopes of Pir Panjal Range, Baltal and Zozi-la areas. These rocks have conformable contacts with the underlying Triassic succession. The Cretaceous rocks have not been reported from the Kashmir Himalaya.

Zanskar-Spiti: The Triassic rocks of Zozi-la in the northern Kashmir represents a transitional facies between the Kashmir Basin in the southwest and the Zanskar-Spiti basin in the northeast (Kumar and Gupta, 1982). In the southern parts of the Ladakh Range of Ladakh, the Panjal Volcanics are overlain by an unfossiliferous slate phyllite sequence conformably overlain by a 60 m thick fossiliferous calcareous phyllite yielding *Ptychites* and fragments of other ammonoids and gastropods of Anisian Age. The rocks of Anisian Age are conformably overlain by a thick sequence of black shales and massive limestones which may range in age from Ladinian to Rhaetian.

The Triassic succession of the Spiti Valley (Fig. 9.3) known as **Lilang Group** begins with a 13 m thick massive dark limestone, shaly limestones and shales yielding a rich Lower Triassic faunal assemblage. Four biostratigraphic zones similar to the zones of Kashmir have been established in the Spiti section as well. The Lilang Group comprising a total of over 1000 m thick succession of limestone and shale intercalations topped by massive limestone has yielded few fossiliferous horizons of Middle and Late Triassic age.

The Middle Triassic succession conformably overlying the Lower Triassic rocks begins with a metre thick shaly limestone yielding *Rhychonella griesbachi, Novella kingi* and *Retzia himaica* suggesting a basal Muschelkalk age for the limestone bed. The limestone is overlain by a 20 m thick sequence of poorly fossiliferous nodular limestones. The overlying shaly limestones of Anisian Age have abundant cephalopods in the lower parts and prolific brachiopods in the upper parts. The Upper Muschelkalk consists of 50 m thick *Daonella* **Shales** yielding a rich fossil assemblage dominated by the lamellibranchs, *Daonella* and *Halobia*. The other important fossils of the horizon are Brachiopods (*Spiriferina* and *Spirigera*) and Cephalopods (*Ceratites, Ptychites, Trachyceras, Xenaspis, Monophyllites, Gymnites, Sturia, Proarcestes, Isculites, Hollandites, Dalmanites, Haydenites, Pinacoceras, Buddhaites, Nautilus* (sp. *spitiensis*), *Pleronautilus, Syringonautilus* and *Orthoceras*).

The *Daonella* Shales are conformably overlain by a 90 m thick sequence of dark coloured splintery limestones with a few interbeds of shales. The lower half of the sequence known as *Daonella* **Limestone** has yielded *Daonella indica* and *D. lamelli* and other fossils having affinities with both the Ladinian Stage of Middle Triassic and the Carnian Stage of Upper Triassic. The upper half of the sequence known as *Halobia* Limestone has yielded *Joannites thenamensis* and *Halobia* cf. *Comata* of a definite Carnian aspect.

The Upper Triassic consists of over 600 m thick sequence of mixed facies of calcareous, argillaceous and arenaceous rocks overlain by massive and pure limestones and dolomites. The rocks have yielded a rich assemblage of fossil fauna of Carnian and Norian ages.

The limestone and dolomite succession of Norian Age is conformably overlain by **Kioto Limestone** which consists of an impressive 750 m thick massive and bedded limestones and dolomites. The formation named after the Kioto village in Spiti Valley is also known as *Megalodon* **Limestone** after the most characteristic lamellibranch in the fossil assemblage of the succession. The lower part of the succession known as **Para "Stage"** has yielded Rhaetian fossils which include *Megalodon, Dicerocardium, Spirigera, Spiriferina, Lima, Pecten* and *Entoleum*. A greater middle part of the carbonate succession is devoid of fossils. The upper part known as **Tagling "Stage"** has yielded a few fossil forms of Lower Jurassic to Middle Jurassic age.

The overlying euxinic facies of **Spiti Shale** consists of pyritiferous and splintery black shales with a few interbeds of impure limestones. The thickness of the formation varies from about 100 to 300 m. The shales contain numerous calcareous concretions (*Saligram*, the deity of devout Hindus) enclosing well preserved ammonites and other fossils. The Spiti Shales forming the most characteristic stratigraphic unit of the Tethyan Himalaya are exposed in many parts, from Hazara mountains (Pakistan) in the west to Sikkim in the east. The Spiti Shale is also well known for their great faunal wealth. The fossil assemblage of the formation dominated by abundant ammonites, a few pelecypods and some gastropods indicates a Late Jurassic to Early cretaceous age for the formation.

The Spiti Shale has been subdivided into a Belemnite Shale overlain by Chidamu and Lochambal "stages" (Diener, 1895). The Belemnite Shale has yielded *Belemnopsis gerardi* and *Mayaites* of Late Oxfordian age. The Chidamu "Stage" has yielded a rich ammonite fauna dominated Perisphinctids ranging in from Kimmeridgian to Tithonian. The Lochambal "Stage" has yielded a fossil assemblage of Tithonian to Valanginian age. The assemblage includes *Spiticeras, Blanfordiceras, Neocomites* and *Holeostephanus*.

The Lower Cretaceous sequence named as **Giumal Sandstone** after the type area, Diumal (Giumal) village in Spiti Valley, consists of about 100 m thick yellow coloured silliceous sandstones and quartzites. The fossil assemblage indicating a Hauterivian to Albian age that has been reported from the formation includes the followings:

Lamellibranchs:	*Cardium, Ostrea, Gryphea, Pecten, Tellina, Pseudomonotis, Arca, Opis, Corbis, Cucullaea* and *Tapes*.
Ammonoids:	*Holocostiphanus, Acanthodiscus, Perisphinctes* and *Hoplites*.

The Giumal Sandstone is conformably overlain by about 33 m thick light grey, massive and fine-grained foraminiferal limestone succession named as **Chikkim Limestone** after a peak near the Chichim (Chikkim) village in Spiti Valley. A Late Cretaceous age was assigned to the limestone formations on the evidence of fragments of *Rudistes*. On the basis of species of *Globotruncana* obtained from the limestone succession, Kohili and Sastry (1956) subdivided the succession into three units of Cenomanian, Turonian and Senonian ages. The Chikkim Limestone is overlain by **Chikkim Shale** which has yielded Maastrichtian microfauna (Jain and Gupta, 1973.)

Ladakh: The Mesozoic rocks of the Zanskar Basin are thrust northward over the rocks of 'Indus Belt' in the Ladakh region. The oldest rock succession known as **Namikla Flysch** is exposed in the southern parts of the Belt. The flyschoidal succession consists of soft dark argillites alternating with thin beds of sandstones, siltstones and limestones. The rocks have yielded bryozoa, crinoid stems and corals of Triassic age. The upper parts of the succession presumably range up to Jurassic in age. The flyschoidal rocks were deposited in a geosynclinal basin bordering the shelf sea of the Zanskar-Spiti region.

The Namikla Flysch are also thrust northwards over a sequence comprising predominantly volcanics with minor sedimentary associations known as **Dras Volcanics.** The igneous complex of Dras Volcanics comprising over 3000 m thick succession of well differentiated rocks varying in composition from ultra-basic to acidic varieties constitutes a part of the 'Indus Ophilitic Belt' which extends across Ladakh in west-north-west direction up to Burzil and Nanga Parbat and further beyond. A major portion of the ophiolitic suite is represented by basic lavas and agglomerates. The serpentinites and dunites usually occur at the basal part of the succession. The volcanics laterally grade eastward into a sequence of phyllites and shales with interbeds of quartzites, limestones and greywackes. The lower part of this volcano-sedimentary sequence has been assigned Aptian to Cenomanian age on the basis of the presence of a number of species of Orbitolines. Fossil algae reported from the beds suggest a Campanian to Maastrichtian age. The uppermost horizons of the sequence have yielded Eocene foraminifers.

Nepal: The Triassic succession of western Nepal (Fig. 9.3) rests conformably over the argillaceous rocks of the Thini-Chu Formation (Chapter 8) yielding Permian fauna. The Lower Triassic consists of 15 to 30 m thick thinly bedded limestones with subordinate shales showing ochre weathering. The beds have yielded fossil assemblage of Scythian age which includes *Anchignathodus minitus, A. latidentatus, Gondolella carinata, G. subcarinata, G. planata* and *G. orientalis*.

Scythian rocks are overlain by a 300 m thick calcareous succession known in western Dolpo as **Mukut Limestone**. The carbonate beds yielding Middle Triassic fauna has been conformably overlain by **Tarap Shales** containing Ladinian and Carnian fauna. The calcareous beds in the lower parts of about 100 m thick **Quartzite Formation** that conformably overlies the Tarap Shales have yielded *Monotis salinaria* of Norian age.

The Quartzite Formation is conformably overlain by a thick succession of massive dolomites and limestones known as **Jomsom Limestone** after its type area in the upper Kali Gandaki Valley. The succession has yielded a number of species of gastropods, pelecypods, brachiopods, crinoids, algae, corals and bryozoans. Rhaetic fauna is reported from the lower part of the succession whereas the upper parts have yielded Lower Jurassic forms. The limestone succession has preserved several sedimentary features indicating a shallow and unstable condition of deposition.

The Middle Jurassic is represented by **Lumachelle Formation** comprising 1100 m thick alternation of calcareous sandstone, lumachelle limestone and calcilutites. The formation yielding bivalves of Lower Dogger age is topped by a less than 7 m thick horizon of fossiliferous beds that contain fauna of Callovian to Oxfordian age. The reduced thickness of the formation has been attributed to a deeper water condition of deposition.

Lesser Himalaya (Krol Belt)

The Mesozoic rocks of the Krol Belt (Fig. 7.4) are classified into a carbonate predominant **Krol Formation** overlain by a flyschoidal **Tal Formation**. The contact between the two formations (Fig 9.4) is characterised by the presence of phosphorite-bearing beds which are well exposed in the east of Mussoorie. The phosphatic horizon represents the evaporite facies indicating a partial withdrawal of marine conditions. The phosphate-bearing beds have been generally regarded as the basal part of the Tal succession. Views, however, differ whether there was any break in deposition between Krol and Tal formations. Evidence of erosional unconformity as well as perfect Krol-Tal transition has been reported from different sections of the Krol Belt. These observations presumably suggest variable conditions of deposition in different segments of the Krol Basin.

Krol Formation: The marine transgression which engulfed parts of the Lesser Himalaya during the Late Palaeozoic persisted for a major part of the Mesozoic Era. A thick sequence of dolomites, limestones and shales were laid down in quiet, shallow water shelf sea during the Triassic and part of Jurassic periods. This succession was named as "Krol Series" by Medlicott (1864) after a prominent hill near Solon (H.P.). The Krol Formation is often separated from the underlying Infra-Krol Formation of Permian age (Chapter 8) by about 7 m thick yellow coloured soft sandstones known as **Krol Sandstone Member**.

The Krol Formation has been divided into a Lower and an Upper

Limestone Members intervened by a Red Shale Member (Medlicott, 1864). Auden (1934) classified the succession into Krol-A, Krol-B, Krol-C, Krol-D and Krol-E Members. The upper three members of Auden correspond to the Upper Limestone Member whereas Krol-A and Krol-B represent the Lower Limestone and Red Shale Members respectively.

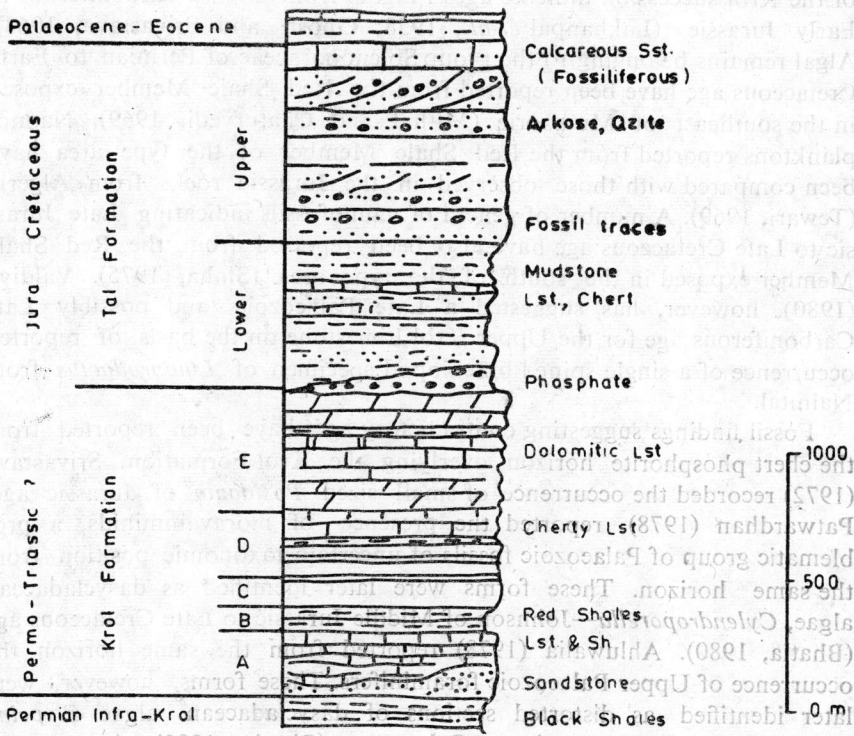

Fig. 9.4: Stratigraphic succession of the Krol Belt (based on Auden, 1934).

The Lower Krol Limestone Member consists of interbedded limestones and calcareous shales showing current and graded bedding structures. The Red Shale Member comprises red and green shales with a few interbeds of grey limestones and layers of gypsum. The Upper Limestone Member comprising massive dolomites, limestones and cherty beds contain pockets of barytes and gypsum in its upper parts. Oolitic and algal structures are common in the Upper Limestone Member. The entire Krol succession represents a tidal flat facies with the development of evaporite facies conditions towards the top of the succession. According to Bhattacharya and Niyogi (1971), the lower part of the Krol sequence was deposited in a near shore, high energy environment while the upper part accumulated in a shallow and stable marine basin.

The age of the Krol Formation has been highly disputed in view of lack of definite fossil evidence. The underlying Infra-Krol Formation of Permian age fixes the upper age limit of the Krol Formation. The upper parts of the overlying Tal succession have yielded Upper Cretaceous fossils. Thus, a Triassic-Jurassic age has been generally accepted for the Krol Formation. Certain groups of palynomorphs occurring in different parts of the Krol succession indicate ages ranging from Permo-Carboniferous to Early Jurassic (Lakhanpal et al., 1958; Ghosh and Srivastava, 1963). Algal remains belonging to the group Solenoporaceae of Permian to Early Cretaceous age have been reported from the Red Shale Member exposed in the southeast of Mussoorie (Mithal and Chaturvedi, 1969). Nannoplanktons reported from the Red Shale Member of the type area have been compared with those observed in the Jurassic rocks from Algeria (Tewari, 1969). A number of genera of nannofossils indicating Late Jurassic to Late Cretaceous age have also been reported from the Red Shale Member exposed in the south of the type area (Sinha, 1975). Valdiya (1980), however, has suggested a Late Palaeozoic and possibly Late Carboniferous age for the Upper Krol Limestone on the basis of reported occurrence of a single spined brachiopod specimen of *Linoproductus* from Nainital.

Fossil findings suggesting contradictory ages have been reported from the chert phosphorite horizon overlying the Krol Formation. Srivastava (1972) recorded the occurrence of small sized *Posidonia* of Jurassic age. Patwardhan (1978) reported the presence of moravamminids, a problematic group of Palaeozoic fossils of uncertain taxonomic position from the same horizon. These forms were later identified as dasycladacean algae, *Cylendroporella*—Johnson of Middle Jurassic to Late Cretaceous age (Bhatia, 1980). Ahluwalia (1978) reported from the same horizon the occurrence of Upper Palaeozoic foraminifers. These forms, however, were later identified as distorted sections of dasycladacean algae *Clypeina* ranging in age from Jurassic to Palaeocene (Bhatia, 1980). Azmi et al. (1981) have reported the presence of Cambro-Ordovician conodonts from the chert phosphorite horizon. Reported occurrence of Tommotian fauna may even suggest the existence of Precambrian-Cambrian boundary at the base of the horizon (Chapter 8). However, similar forms recovered from the same locality have been referred to as annelids of either Early to Late Cretaceous age (Singh and Shukla, 1981) or of Permian-Triassic age.

Tal Formation: The Tal Formation first described from the southwestern Garhwal (Medlicott, 1864) is exposed in the central and northern parts of the Krol Belt in Sirmur district of Himachal Pradesh and Mussoorie district of Uttar Pradesh. The succession consists of black pyritiferous cherty, calcareous, arenaceous and argillaceous flyschoidal rocks in the lower part and sandy oolitic and shelly limestones in the upper part of the succession. Poorly preserved remains of corals, belemnites, lamellibranchs and gastropods suggesting a probable Jurassic age were reported from the basal part of the Upper Tal succession (Middlemiss, 1885). The uppermost shelly limestones have yielded Lower Cretaceous bryozoans and foraminifers (Tewari and Kumar, 1967). Bhatia (1980) suggested that the presence of certain echinoid spines in the upper Tal grainstones may indicate a Late Cretaceous-Palaeocene age for these beds.

The age of the lower part of the Tal succession has been variously regarded as either Jurassic or Permian. Those favouring a Permian age envisage a major hiatus in deposition between the lower and upper parts of the Tal succession for which the evidence has been lacking. The Permian age for the lower part of the succession was first suggested on the basis of the discovery of Permian invertebrate fossils from the Boulder Slate Sequence (Chapter 8) exposed near Jogira in Garhwal (Ganesan, 1972). This locality is very close to Gajwar from where Middlemiss (1885) first recorded the probable Jurassic fauna from the Tal succession. The fossiliferous beds of Jogira were regarded by some as part of the Lower Tal succession. This view was strengthened by the reported discovery of 'fusulinids' from the oolitic limestones in the neighbouring locality (Kalia, 1972). Identification of these so called 'fusulinids' was later contested. According to Bhatia (1975 and 1980), the specimens represent deformed oolites and certain algae.

Fossil assemblage of the bioclastic grainstone forming the uppermost part of the Tal succession indicates a Maastrichian to Danian age (Bhatia, 1980). The fossil assemblage consists of calcareous algae, hydrozoans, bryozoans, foraminifers ostracodes, spines and tubercles of echinoids and some unidentifiable gastropods and bivalves.

Indian Peninsula

Except for marine beds of Manendragarh and Umaria (Chapter 10), the Indian Peninsula is devoid of marine Palaeozoic rocks. However, the Mesozoic rocks of marine facies have been extensively recorded from the northwestern Peninsula, the central India and southeastern coast (Table 9.2). The coastal areas of the northwestern India came under the marine influence in Jurassic Period. The marine conditions persisted for the remaining part of the Mesozoic Era. The southeastern coast of India was submerged by the transgressing sea during the Late Cretaceous Epoch. This sea soon spread northeastward over the coastal regions of the Andhra Pradesh, Orissa and southern parts of Assam.

Kutch-Saurashtra: In Kutch region (Fig. 9.5), the Jurassic rocks comprising the Patcham, Chari and Katrol formations are exposed in three anticlinal chains of ridges trending in east-west direction. The northern chain comprises the islands of Patcham, Khadir, Bela and Chorar. The middle chain consists of the most prominent ridge extending for about 193 km from Habo in the east to Lakhpat in the west. The southernmost chain forms a 64 km long Katrol-Charwar range in the south of Bhuj. The

Stratigraphy of India

Table 9.2: Correlation of Mesozoic formations of Peninsular India

Kutch	Saurashtra	Western Rajasthan	Narmada Valley	Cauvery Basin (South India)
			Lameta Beds	Ariyalur Formation
			Bagh Beds	Trichinopoly Formation
		Abur Formation		Uttatur Formation
Bhuj Formation	Wadhawan Group			Dalmipuram Formation
Umia Formation		Parihar Formation		
Katrol Formation	Dharangadhara Group	Badesar Formation		
Chari Formation		Baisakhi Formation		
Patcham Formation		Jaisalmer Formation		
		Lathi Formation		

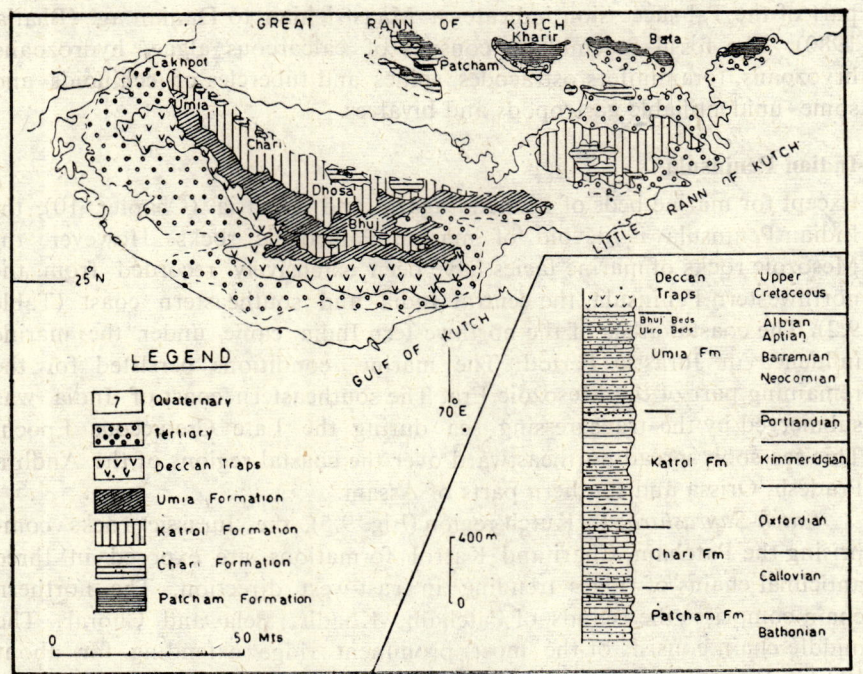

Fig. 9.5: Geological map and stratigraphic succession of Kutch Region (based on Biswas and Deshpande, 1970).

Jurassic rocks are overlain by the Umia and Bhuj formations exposed along the flanks of the anticlinal ridges.

The Mesozoic succession of the Kutch region has been intruded by various sills and dikes which are genetically related with the overlying Deccan Traps (Chapter 11). The basal conglomerate of the Patcham Formation contains pebbles of crystalline rocks suggesting a Precambrian Basement for the Mesozoic succession of the region.

Table 9.3: Mesozoic succession of Kutch Region (based on Sastry and Mamgain, 1971)

Formation	Subdivisions	Characteristic fossils
	Deccan Traps	
	----- unconformity -----	
Bhuj Formation	Umia plant beds	*Ptillophyllum* flora
Umia Formation	Ukra beds (calcareous shales)	*Australiceras* sp.
	Sandstones and shales	unfossiliferous
	Trigonia beds	*Trigonia*
	Umia Ammonite beds	*Virgosphinctes* sp.
Katrol Formation	Upper Katrol Shale	*Hildoglochiceras*
	Upper Katrol Sandstone	mainly unfossiliferous
	Middle Katrol Sandstone	*Torquasphinctes* sp. and *Katroliceras*
	Lower Katrol Shale	ammonites
	Belemnites marls of Jurum	*Belemnites*
	Kantkote Sandstone	*Euapidoceras, Taramelliceras* sp.
Chari Formation	Dhosa Oolite	*Mayaites* and *Epimayaites*
	Athleta beds	*Metapeltoceras, Peltoceras, Reineckei*
	Anceps beds	*Kinkeliniceras, Hubertoceras, Indosphinctes*
	Rehmani beds	*Reineckeia tyranniformis, R. rehmanni*
	Macrocephalus beds	*Macrocephalites, Dolikephalites*
Patcham Formation	Coral beds	*Macrocephalites, Sivajiceras, Procerates*
	Shelly Limestone	*Macrocephalites*
	Kuar Bet beds	*Corbula lyrata, Protocardia, Pseudotrapezium*
	----- unconformity -----	
	Precambrian Basement (not exposed)	

The **Patcham Formation** representing a neritic facies of transgressive sea comprises a 300 m thick succession of dark pisolitic limestones and olive green shales overlain by nodular fossiliferous cherty limestones and marls. The rocks of the formation are best exposed in the Patcham, Kharir and Bela islands of the Great Rann of Kutch. Lower part of the succession known as Kuar Bet beds has yielded Bathonian pelecypods such as *Corbula*, *Eomiodon*, *Protocardia* and *Pseudotrapezium*, corals, ammonoids, foraminifers and plant fossils. The upper beds have yielded a rich assemblage of corals, brachiopods, pelecypods and ammonites including *Macrocephalites triangularis* and *Sivajiceras congener* indicating a Callovian age for the beds.

The **Chari Formation** conformably succeeding the Patcham Formation consists of about 400 m thick succession of sandy limestones, marls, calcareous and sandy shales and oolitic limestones. The formation is best exposed near village Habo that is very close to the Chari village in the main land of Kutch. The lithology and fauna of the formation of relatively deeper water facies represent a change in the environment of deposition from that of the transgressive facies of the Chari Formation. The basal beds have yielded a rich collection of Callovian ammonites (*Macrocephalites macrocephalus*) and gastropods (*Nucula* and *Astrate*). The overlying 'golden oolites' coated with ferric oxide have yielded ammonoid *Indocephalites diadematus*. Beds overlying the 'golden oolites' have also yielded a rich assemblage of ammonoids, a few terribratulids and *Trigonia*. The Chari succession is topped by 'Dhosa oolites' yielding Oxfordian faunal assemblage.

The **Katrol Formation** succeeds the Chari Formation presumably with a break in deposition demonstrated by the presence of pebbly beds in the basal part of the succession (Rajnath, 1932). The microfaunal record, however, suggests an uninterrupted phase of sedimentation across the boundary of the two formations. The Katrol Formation named after the east-west trending Katrol-Charwar range in the south of Bhuj comprises about 750 m thick shallow marine succession of shales, limestones, sandstones and grits with lenticular beds of gypseous sandy shales. Lower beds of the succession have yielded *Trigonia* and concretions of ammonoids (*Oppelia* and *Haploceras*). The middle part of the succession contains Oppelids and Perisphinctids. The upper part, generally devoid of fossil record, is topped by grity sandstone yielding Portlandian fossils (*Virgatosphinctes, Hildoglochiceras, Auloosphinctes* and *Trigonia*). The Katrol succession has also yielded mioflora and other plant remains.

The **Umia Formation** succeeding the Katrol Formation with slight unconformity is composed of mixed rock facies of marine and continental origin. The lower part of the succession is composed of a 15 m thick ferruginous sandstone bed overlain by about 510 m thick sequence of oolitic sandstones, sandy shales and marls yielding a rich fossil assemblage. The assemblage indicating a Tithonian age includes ammonites (*Virgosphinctes, Aulacaosphinctes, Ptychophylloceras, Umiatites* and *Microcanthoceras*) Belemnites and *Trigonia*.

The upper part of the Umia Formation consists of **Ukra Beds**. The Ukra Beds comprise about 23 m thick succession of greenish sandstone, calcareous shales, grits and marls containing ammonoids (*Australiceras, Cheloniceras* and *Tripocium*) of Aptian age. The overlying **Bhuj Formation** (also known as Umia Plant Beds) has yielded abundant plant fossils which include Filicales, Cycadophyta, Conifers and Incertae. The saurian remains of *Plesiosaurus indica* have also been reported from these beds. The plant fossils are closely related to those reported from the Jabalpur Formation (Chapter 10) of post-Aptian age.

The northern parts of Saurashtra have exposed rock formations ranging in age from Tithonian to Cenomanian ages. The formations have been grouped into a lower Dharamgadhara Group and an upper Wadhawan Group (Table 9.4). Rocks of the Dharangadhara Group were laid down under deltaic environment whereas the rocks of Wadhwan Group were deposited under shallow marine conditions.

Table 9.4: **Classification of Mesozoic rocks of Saurashtra**

Group	Formation	Lithology
DECCAN TRAPS		
Unconformity		
WADHWAN GROUP (Albian to Cenomanian)	Khamisara Formation	Sandstone
	Malachimata Formation	Fossiliferous limestones
	Kukuda Formation	Maroon coloured sandstones
DHARANGADHARA GROUP	Ranipat Formation	Ferruginous and white sandstone & quartzite
	Surajdewal Formation	Red shales & sandstones
	Than Formation	Grey shales & sandstone
Unconformity		
PRECAMBRIAN BASEMENT		

Western Rajasthan: The Mesozoic rocks of the western Rajasthan exposed in the Jaisalmer region (Fig. 9.6) have been grouped into Lathi, Jaisalmer, Baisakhi, Badesar and Parihar formations.

The **Lathi Formation** exposed from Barmer through Jaisalmer to Lathi comprises a 360 m thick succession of conglomerates, coarse sandstones, arkoses, lithic arenites and siltstones. In the outcrops around Badhaura, the Lathi Formation rests unconformably over the Lower Gondwana rocks (Badhaura Formation, Chapter 10). However, in the Jaisalmer region, the rocks of the Lathi Formation rest directly over the Precambrian Basement. The Lathi succession represents a typical transgressive facies with the lower beds of terrestrial and deltaic origin and the upper beds of shallow marine origin. The terrestrial beds contain abundant fossil woods (gymnosperms), indeterminable fish teeth, gastropods and rich assemblage of pollens and

168 *Stratigraphy of India*

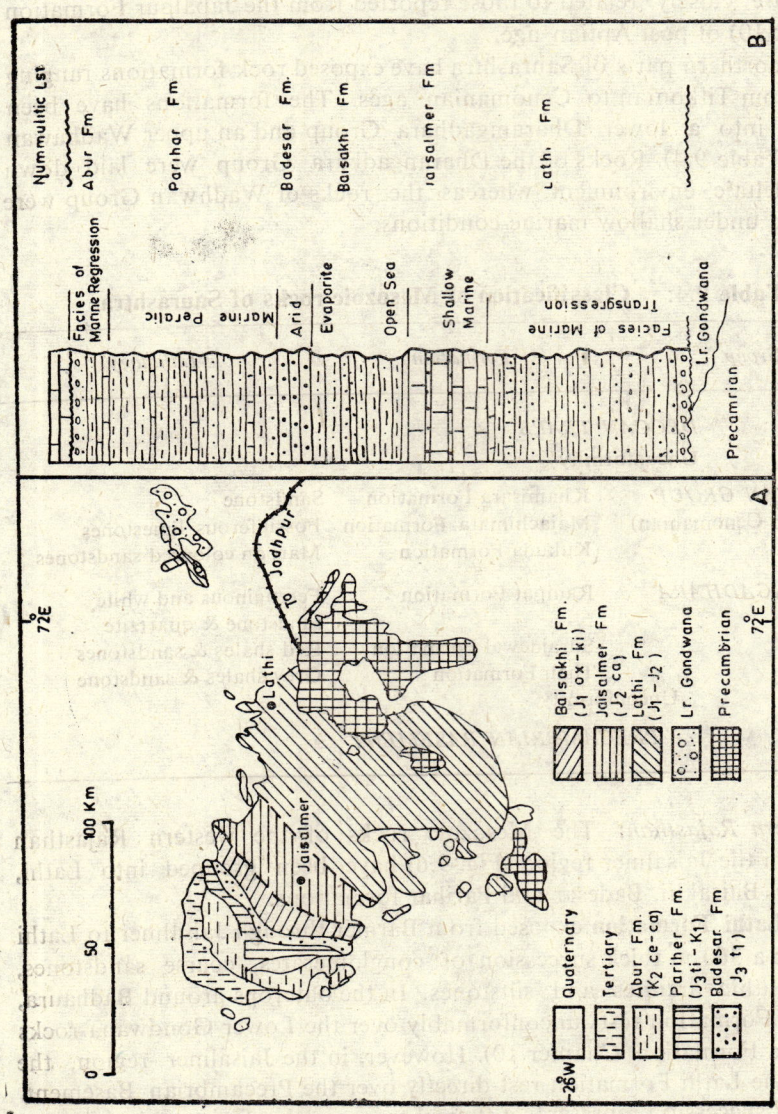

Fig. 9.6: Geological map and stratigraphic succession of Jaisalmer Region (based on Poddar, 1964).

spores indicating an Early to Middle Jurassic age for the formation (Poddar, 1964).

The **Jaisalmer Formation** conformably overlying the Lathi Formation consists of about 130 m thick succession of marlstones, fossiliferous oolitic limestones, calcareous sandstones and conglomerates. Limestones form conspicuous scarps close to the town of Jaisalmer. The succession depicts current bedding, ripple marks and other sedimentary features typical of littoral neritic facies deposited on a slightly unstable shelf. Fossiliferous limestones and calcareous beds have yielded a fossil assemblage of Callovian age. The fossil assemblage includes *Terebratula biplacata, Corbula lyrata, Idiocycloceras singulare* and *Sindeites sindensis.* The upper parts of the formation have yielded some ammonites and foraminifers of Late Callovian to Oxfordian age (Bhalla, 1983).

The **Baisakhi Formation** consists of about 150 m thick succession of brownish silty shales, calcareous sandstones and concretionary sandy limestones with gypseous and bentonitic clay intercalations. The basal oolitic limestones conformably overlying the Jaisalmer Formation have yielded cephalopods and micro-forms of Oxfordian to Kimmeridgian age. The upper parts of the succession is devoid of fossil record. Litho-facies characters and abundant fossil record in the lower parts indicate shallow sea conditions with life supporting moderate climate. The arid climate that gave rise to abnormal salinity conditions explains the paucity of the records of life in the upper part of the succession.

The **Badesar Formation** having gradational contact with the underlying Baisakhi Formation consists of 50 to 60 m thick extremely hard ferruginous sandstones intercalated with thin layers of red calcareous sandstones. The formation typically represents a succession of rocks deposited by a regressing sea. The lower beds of marine origin contain Tithonian ammonites such as *Pachysphinctes* aff. *bathylocus* and *Virgatosphinctes.* Occurrence of fossil wood and current bedding structures in the upper part of the succession indicates a near shore marine conditions of deposition.

The **Parihar Formation** overlaps the Badesar Formation and, at places, it rests directly over the Baisakhi Formation. The Parihar Formation comprises about 300 m thick succession of unfossiliferous soft white felspathic sandstones, coarse sandstones and calcareous grits of peralic facies. The succession represents the commencement of a phase of marine transgression. The arenaceous succession exposed near Barmer town (**Barmer Sandstone**) is considered as an eastern extension of the Parihar Formation. The Barmer Sandstone contains plant fossils (dicotyledonus wood and unveined leaves), casts of *Unio* and fragments of Cardium. Poddar (1984) assigned a Valengian to Barremian age to the Parihar Formation while Sastry and Mamgain (1971) considered a Tithonian to Neocomian age more probable.

The overlying **Abur Formation** overlaps the Parihar and Badesar

Formation indicating an expansion of the transgressive sea. The Abur Formation comprises a 60 m thick succession of sandy bioclastic limestones with abundant fossils overlain by a band of white and reddish well sorted quartzites and silts. The lower part was deposited in open shelf sea while the upper part of the succession represents off lap facies associated with marine regression. The basal calcareous beds of the Abur Formation have yielded Aptian fauna (Spath, 1933) comparable with those found in Ukra Beds of Kutch. The Abur Formation has yielded nine species of foraminifers suggesting a Cenomanian to Coniacian age for the formation.

Narmada Valley: Marine rocks of Cretaceous age known as **Bagh Beds** occur as small detatched outcrops along the Narmada Valley over a distance of about 350 kilometres from Rajpipla in the west to the neighbourhood of Indore in the east (Fig. 9.7). The rocks are exposed as inliers of

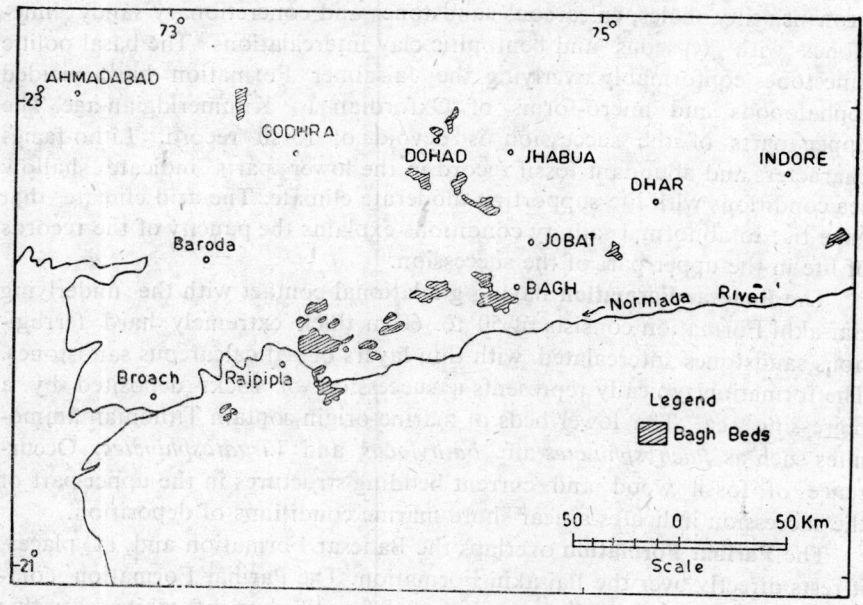

Fig. 9.7: Distribution of Bagh Beds in western India (based on Jain, 1970).

older succession surrounded by Deccan Traps of a younger age. The Bagh Beds have been named after the Bagh town in the Dhar district of Madhya Pradesh. The town itself is situated over the Archaean gneisses and schists forming the basement for the deposition of the Cretaceous rocks. The Bagh Beds exposed a few kilometres east in the Man river section consists of about 30 m thick succession of unfossiliferous sandstone and conglomerates known as **Nimar Sandstone** overlain by fossiliferous **Nodular Limestone** and **Coralline Limestone** (Fig. 9.8).

Mesozoic History 171

The Nimar Sandstone was deposited in fresh water basins. The Sandstone has yielded plant fossils of Late to Early Cretaceous (Hauterivian) age. The marine transgression began towards the close of the deposition of the Nimar Sandstone. Presence of certain Oyster beds in the upper parts of the sandstone succession indicates a periodic oscillation of the sea level at the commencement of the transgression.

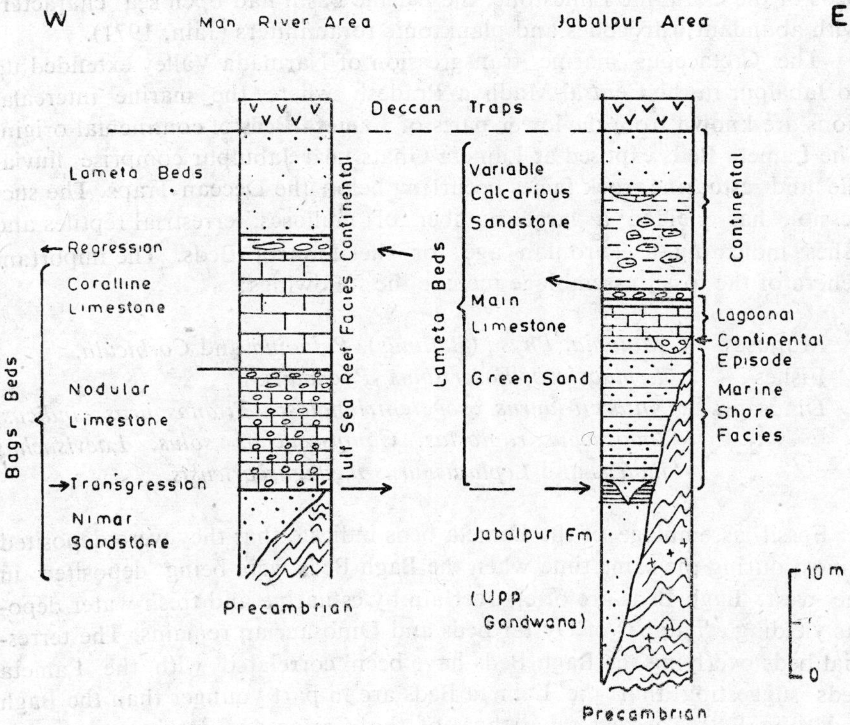

Fig. 9.8: Turonian marine transgression along the Narmada Valley (after Robinson, 1967).

The marine succession begins with a 12 m thick sequence of fossiliferous, white, argillaceous and compact Nodular Limestone. The fossil assemblage of the limestone succession includes *Hemiaster fourteavi, Parastantoceras mintoi, Proplacenticeras santoni* var. *bolli, Coilopoceras scindiae* and *C. bosei*. The overlying Coralline Limestone consists of about 10 m thick succession of red or yellow carbonate rocks that contain abundant small fragments of bryozoa such as *Cerinopora dispar* and shells of lamellibranchs, gastropods, brachiopods, corals and echinoids. The two limestone successions are often intervened by a 3 m thick succession of *Deola Marl* which also contain abundant fossils.

The marine Oyster beds in the upper parts of the Nimar Sandstone are more frequently observed in the western outcrops of the Bagh Beds suggesting that the marine transgression came from the west. Marine conditions were firmly established during the deposition of Nodular Limestone and Coralline Limestone. Faunal evidence suggests that shallow marine conditions predominated during the deposition of Nodular Limestone and the lower parts of the Coralline Limestone. During the deposition of the upper beds of the Coralline Limestone, the marine basin had open sea characters with abundant ostrocodes and planktonic foraminifers (Jain, 1971).

The Cretaceous marine transgression of Narmada Valley extended up to Jabalpur in the Central Madhya Pradesh where the marine intercalations are known from the lower parts of **Lameta Beds** of continental origin. The Lameta Beds exposed at Lameta Ghats near Jabulpur comprise fluviatile and estuarine rock facies occurring below the Deccan Traps. The succession has yielded a large number of mulluscs, terrestrial reptiles and fishes indicating a Turonian age for the Lameta Beds. The important genera of the fossil assemblage include the followings:

Mulluscs : *Malania, Physa* (*Bullinus*), *Paludina* and *Corbicula*.
Fishes   : *Lepidosteus, Eoserranus, Pycnodus*.
Dinosaurs : *Antarctosaurus sepetentrionales, Titanosaurus indicus, Indosuchus raptorius, Compsosuchus solus, Laevisuchus indicus* and *Leplatasaurus medagascariensis*.

Fossil assemblage of the Lameta beds indicate that they were deposited almost during the same time when the Bagh Beds were being deposited in the west. Bagh Beds are often overlain by estuarine and fresh water deposits yielding a number of Oyster beds and Dinosaurian remains. The terrestrial beds overlying the Bagh Beds have been correlated with the Lameta Beds suggesting that the Lameta Beds are in part younger than the Bagh Beds (Fig. 9.8). The regressive phase of the Cretaceous marine transgression was apparently quicker as evidenced by a rapid change from open sea facies of the Coralline Limestone to estuarine-fluviatile facies of the overlying beds.

Cauvery Basin: Marine Cretaceous rocks of Cauvery basin attaining a maximum thickness of about 2500 m are well exposed in the vicinity of Tiruchchirapalli (Formerly spelled Trichinopoly), Vridhachalam and Pondichery (Fig. 9.9). The fossil fauna and lithofacies of the Mesozoic succession ranging in age from Aptian to Maastrichtian indicate at least four phases of marine transgression and regression paving way for the conditions of depositions of the four successive formations known as Dalmiapuram, Uttatur, Trichinopoly and Ariyalur formations.

The **Dalmiapurum Formation** (Bhatia and Jain, 1969) comprising a 27 to 55 m thick sequence of pyritiferous grey shales and limestones has yielded ostracodes, ammonites, small foraminifers and plant fossils. The

Mesozoic History 173

fossil assemblage indicates an Aptian-Albian age for the formation. Occurrence of pyrite indicates a reducing environment of deposition commonly associated with the euxinic facies of partly enclosed marine basins. Certain normal sized ammonites and several ostrocode genera reported in the fossil assemblage were transported to the area of deposition from open sea of the time. The Dalmiapur Formation rests unconformably over the Precambrian Basements or Upper Gondwana plant beds.

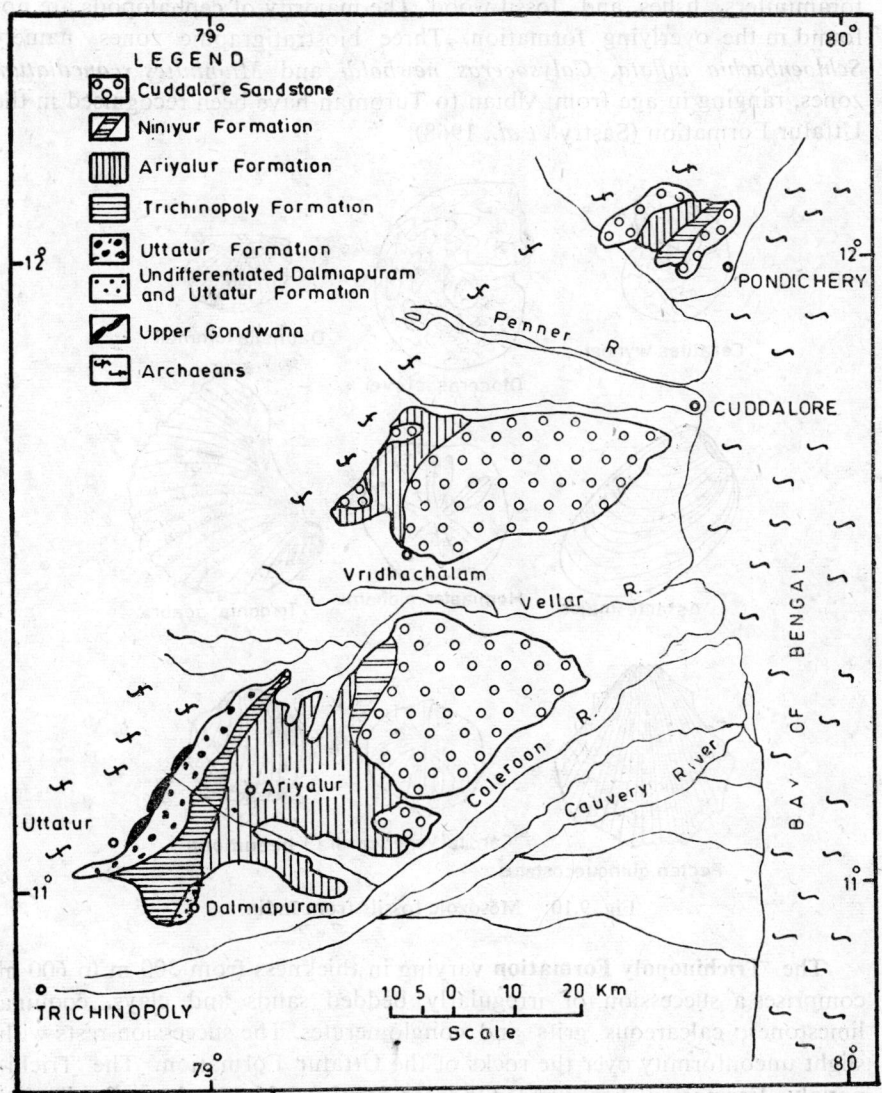

Fig. 9.9: Geological map of the Cauvery Basin (based on Sastry and Rao, 1964).

174 Stratigraphy of India

The **Uttatur Formation** comprises a 700 m thick succession of fine silts, calcareous shales and sand clays. The formation rests either over the Dalmiapur Formation, Upper Gondwana sediments or over the Precambrian Basement. The basal beds often comprise coral limestones which are deposited in shallow and warm tropical sea. The uppermost beds contain gypseous and phosphatic nodules indicating evaporitic conditions of a regressing sea. The Uttatur Formation has yielded more than 300 taxa of cephalopods, pelecypods, gastropods, brachiopods, echinoids, corals, smaller foraminifers, fishes and fossil wood. The majority of cephalopods are not found in the overlying formation. Three biostratigraphic zones, namely *Schloenbachia inflata*, *Calysoceras newboldi* and *Mammites conciliatum* zones, ranging in age from Albian to Turonian have been recognised in the Uttatur Formation (Sastry *et al.*, 1968).

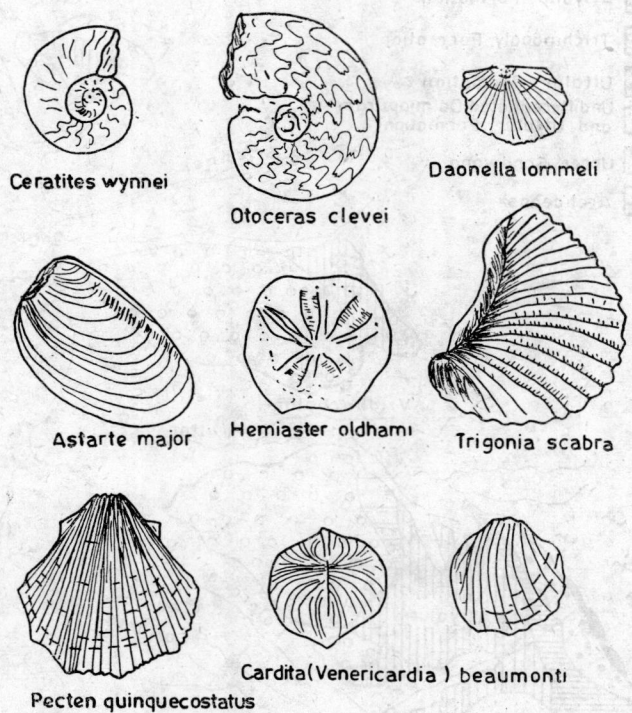

Fig. 9.10: Mesozoic fossils from India.

The **Trichinopoly Formation** varying in thickness from 300 m to 600 m comprises a succession of irregularly bedded sands and clays, coquina limestones, calcareous grits and conglomerates. The succession rests with slight unconformity over the rocks of the Uttatur Formation. The Trichinopoly Formation has yielded a large number of invertebrate fossils and tree trunks indicating a Turonian to Lower Senonian age for the formation.

Presence of coquina limestones indicates an environment of deposition of mud-flat bottoms with depths of 7 to 8 fathoms. The coquina limestones are interbedded with clastic bands of shallow littoral environments. The succession of intercalated rocks of neritic and littoral facies indicates rapid fluctuation of sea level of the basin. Sastry *et al.* (1968) have divided the Trichinopoly Formation into three biostratigraphic zones, they are in ascending order *Lewesiceras vaju, Kossamaticeras theobaldianum* and *Placenticeras tamulicum* zones.

The **Ariyalur Formation** unconformably overlying the Trichinopoly Formation comprises a 1200 m. thick succession of sandstones with some marly clays, calcareous shales and limestones (Ramanathan, 1968). The lower parts of the succession have yielded some well preserved fossils of organisms inhabiting the shallow and quiet sea. The upper parts of the succession are generally unfossiliferous and show lacustrine conditions of deposition. The fossil assemblage of the lower parts includes several genera of ostracods, foraminifers, mulluscs, echinoids, brachiopods, corals and bryzoans. Sastry *et al.* (1968) have divided the Ariyalur succession in to two biostratigraphic zones on the basis of the species of *Globotruncana*. The lower zone has been assigned a Campanian age whereas the fossils of the upper zone have a Maastrichtian affinity. During the closing phases of the Ariyalur sedimentation, the northern parts of the Cauvery Basin seem to have undergone a gradual uplift as indicated by a gradual coarsening of the sediments in the northern outcrops and the presence of land vertebrates in the uppermost parts of the succession.

Meghalaya: The Cretaceous rocks exposed along the sourhern slopes of the Shillong Plateau have been grouped in to **Mahadek Formation** that is comprised of conglomerates and glauconitic sandstones. The formation unconformably overlies either the Sylhet Traps of Jurassic age or a Precambrian Basement. The Mahadek Formation has yielded a fairly rich faunal assemblage which has a close affinity with the Ariyalur assemblage of the Cauvery Basin. The assemblage includes *Stigmatopygus elatus, Alectryonia ungulate, Nerita (Ostostoma) divaricata, Lyria crassicostata* and *Eubaculus vagina* which indicate a Senomanian age for the formation. The upper parts of the Mahadek Formation has yielded *Globotruncana stuarti, Guembelina plummerae, Orbitoides, Siderolites calcitropoides, Globogerina pseudobulloides* and *G. triloculinoides* suggesting a Maastrichtian to Danian age (Bhalla, 1983). The Mahadek Formation is conformably overlain by the **Langpar Formation** that comprises a succession of calcareous shales and sandy limestones. The Langpar Formation is known to contain abundant foraminifers of Danian affinity (Nagappa, 1959).

Chapter 10
Gondwana Sequence of India

The term "Gondwana" was coined by Medlicott (1872) for a 6 to 7 kilometre thick succession of fluviatile and lacustrine deposits with a glacial deposit at the base. The sequence represents a long episode of continental sedimentation that began in Permian Period and closed during the Cretaceous Period representing a time span of over 180 million years. These rocks are now exposed in three river valley grabens, viz., the Narmada-Son-Damodar, the Mahanadi and the Godavari grabens (Fig. 4.3). Views differ regarding the original aerial extent of the Gondwana deposits outside the present basin limits. According to one view, the sedimentation was confined to the river valley grabens and it was contemporaneous with faulting. The other view holds that the Gondwana basins were originally wider and the rocks have been preserved only in the down-faulted grabens.

The Gondwana sediments of India has a typical association of lithology and floral remains. The floral assemblages do not help in precise determination of chronostratigraphic boundaries. Vertebrate remains are sporadic in occurrence. They are mainly restricted to the middle part of the succession. Intercalations of marine fossiliferous rocks are known from the basal and uppermost parts of the succession. The lower and upper boundaries of the succession determined by the marine faunal records do not coincide with the boundaries of the Standard Time Scale. Nevertheless, the Gondwana succession has a time connotation indicating a certain period of the earth's history. The succession has been informally designated as **Gondwana Sequence** in the Standard Stratigraphic Terminology.

The classification of the Gondwana Sequence has been one of the disputed problems of the Indian geology. Some would prefer a two-fold classification while others favour a three-fold classification. In a two-fold classification, adopted by Oldham (1893), Cotter (1917), Fox (1931) and others, the lower and upper units are separated by a break in stratigraphy and plant life.

Equisetaceous plants prevail in the lower, cycades and conifers in the upper while ferns are common to both. The *Glossopteris* flora flourished during the deposition of the lower division while the predominance of the

Ptilophyllum flora is characteristic for the upper division. The change in floral characters has been attributed to the climatic changes through the successive periods of the earth's history.

The three-fold classification advocated by Feistmantal (1880) and later followed by Vredenberg (1910) and Wadia (1926) is based on the lithological similarity of the upper parts of the Lower Gondwana Sequence with those of the lower parts of the Upper Gondwana Sequence. These transitional formations were separated and grouped as Middle Gondwana Sequence. The tripartite subdivisions were fitted into the International Standard Scale, so that the Lower, Middle and Upper Gondwana sequences correspond roughly to the Permian, Triassic and Jurassic systems respectively.

Robinson (1967) favoured a two-fold classification of the Gondwana Sequence based on a marked floral break between the Permian and Triassic Gondwana rocks. Similar break has also been observed in continental succession of western Australia that formed a part of the **Gondwanaland** (Fig. 10.1). Robinson's two-fold classification differs from the earlier two-fold classification adopted by Oldham (1893) and others. While in the earlier classification, the floral break demarcating the boundary between the two divisions was placed at the top of the Panchet Formation (Lower Triassic), Robinson (1967) brought down the boundary at the base of the Panchet Formation. Thus, in this scheme of classification, the Lower Gondwana consists of predominantly Upper Palaeozoic rocks whereas the Upper Gondwana comprises essentially the Mesozoic rocks.

SEDIMENTATION AND PALAEOCLIMATES

Beginning of the Gondwana sedimentation is characterised by the deposition of glacial sediments. Study of the till fabric, boulder trains and pavement striae indicate that the glacier advance was in the form complex lobes of ice-sheets (Robinson, 1967). The most prominent centre of glaciation was located in the south-west of the present Godavari Valley from where the glaciers moved in northerly and northwesterly directions. In the Damodar Valley region, the highlands were located in the north and northwest resulting in a southerly glacial advance. Records from the east coast of India suggest that the ice moved from an upland that existed in the east of the present eastern shore-line of India (Datta et al., 1983).

With the retreat of the glaciers at the close of glacial epoch, the irregular topography of the Indian Peninsula was filled in by swamps rich in vegetative matter. The conditions were favourable for the deposition of coaliferous sediments. The Gondwana basins of the Damodar Valley region show a centripetal dispersal pattern of sediments deposited by the streams flowing in from the surrounding uplands. The basins of the Godavari Valley region, however, show a longitudinal dispersal pattern of sediments.

After the deposition of the coal bearing sediments, the Gondwana basins were effected by a variable condition of deposition. A thick sequence of ironstone shales were deposited in the eastern and the southern Gondwana basins. The shales were deposited in closed basins under reducing environments favourable for the precipitation of iron carbonates. The sedimentation in the Satpura basins of Central India took place in an oxidising environment giving rise to the deposition of variegated clays and ferrugenous (iron oxides) sandstones.

The coal forming conditions were again ushered in during the deposition of the Upper Permian rocks. The younger coaliferous sediments were laid in almost all the Gondwana basins. The Raniganj basin of the Damodar Valley region, however, shows the maximum development coaliferous beds which contain several superior quality coal measures. The deposition of the superior quality coal-seams has been attributed to repeated formation of extensive back swamps which were deep and yet received less inflow of fresh water (Datta *et al.*, 1983).

The Triassic Gondwana is characterised by cyclic alternations of arkosic sandstones and red shales whereas the Jurassic and Lower Cretaceous Gondwana formations consist of quartz-arenite, pebble-sandstone and red siltstone association. The Triassic sedimentation took place in streams bearing a moderate to high channel sinuosity characteristic of gentle slopes. On the other hand, the younger sediments were deposited by mainly a network of braided streams characteristic of steeper palaeoslopes (Cashyap, 1979).

The lower part of the Gondwana Sequence is dominated by arkosic sediments representing a cold palaeo-climate. The overlying coal bearing sediments were deposited in humid sub-tropical palaeo-climate. A great change in the climatic conditions has been noticed at the beginning of deposition of the Triassic Gondwana. In contrast to the humid palaeo-climates of the Permian time, the Triassic sediments were laid down in dry climatic conditions. The dry climates were responsible for the disappearance of the *Glossopteris* flora. Milder temperatures and wet conditions were again set in during the later parts of the Triassic Period during which the *Ptilophyllum* flora established itself and flourished during the later part of the Gondwana sedimentation.

LOWER GONDWANA SEQUENCE

Talchir Formation

The Talchir Formation having a fairly wide distribution were first described from Talchir in Orissa. The formation rests unconformably over the basement of either Archaean gneisses and schists or rocks of Proterozoic age. The formation comprises a 20 to 30 metre thick **Talchir Boulder Beds** overlain by a succession of green silty shales, mudstones, fine soft sandstones with a few calcareous bands. The sandstones often contain abundant undecomposed felspar representing a cold climate at the time of their deposition. The Boulder Beds, also known as **Talchir Tillites,** have abundant evidence to suggest that they were laid by the continental glaciers. Similar tillite beds are known from several other continents of the southern hemisphere (Table 10.1). The southern continents were joined together during the Palaeozoic Era. In palaeogeographic reconstructions, it has been shown that such a supercontinent known as **Gondwanaland** was located nearer to the South Pole of the Permian Period (Fig. 10.1).

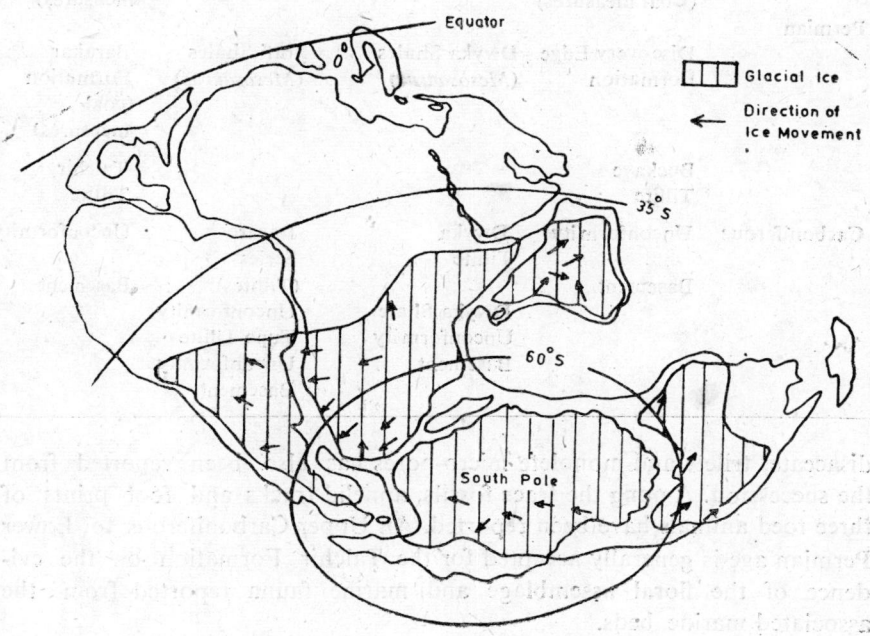

Fig. 10.1: Distribution of ice-sheet over the Gondwanaland during the Late Carboniferous and Early Permian.

The Talchir Boulder Beds comprise usually well rounded pebbles and boulders varying in size up to blocks of 4 to 5 metres and rarely up to 10 metres across. The boulders, embedded in fine grained matrix, consist of gneisses, granites, quartzites, slates and amphibolites which can be very well matched with the lithology of the Precambrian Basement. Characteristic features of the glacial transport can be seen in the form of smoothened, polished and striated boulders and polished and grooved "glacial pavements" beneath the Talchir deposits.

The Talchir Formation contains a few plant fossils of seed ferns *Gangamopteris cyclopteroides* and *Glossopteris indica* and its characteristic stem *Vertebraria indica*. A rich assemblage of monosaccates and a few

Table 10.1: Upper Palaeozoic-Lower Triassic succession of the southern continents comprising the "Gondwanaland"

Period	Antarctica	South Africa	South America	India
Lower Triassic	Beacon rocks	Beufort Series	Santa Maria Formation	Panchet Formation
Permian	Mount Glossopteris Formation (Coal measures)	Ecca Series (Coal measures)	Estrata nova	Raniganj Formation (Coal measures)
	Discovery Edge Formation	Dwyka Shales (*Mesosaurus*)	Irati Shales (*Mesosaurus*)	Barakar Formation (coal measures)
	Buckaye Tillite			Talchir Tillite
Carboniferous	Unconformity	Dwyka Tillite	Itarare Series (Tillites)	Unconformity
	Basement		Unconformity	Basement
		Dwyka Shale Unconformity Basement	Tupe Tillite Unconformity Basement	

disaccates trilete and monolete micro-pores has also been reported from the succession. Among the trace fossils, annelid tracks and foot prints of three toed animals have been reported. An Upper Carboniferous to Lower Permian age is generally accepted for the Talchir Formation on the evidence of the floral assemblage and marine fauna reported from the associated marine beds.

Marine Intercalations

Marine beds in association with typical Lower Gondwana rocks are known to occur at Umaria in Madhya Pradesh (Reed, 1928) and at Manendragarh (Ghosh, 1954) about 150 kilometres southeast of Umaria. The Umaria Marine Bed exposed at Narsahra Railway cutting comprises a single three metre thick bed of shelly limestone containing fossil shells of *Productus, Spiriferina, Reticularia* and others. The limestone bed occurs in the Karharbari Formation which was earlier regarded as the upper part of the Talchir Formation. The fossil assemblage is suggestive of Lower Permian age for the limestone bed. The faunal assemblage is also suggestive of a warmer climate which appears to be in conformity with the view that the Talchir succession witnessed glacial conditions at the beginning and warmer climate during the later parts of its deposition.

The marine beds of Manendragarh occur in several patches attaining a thickness of up to 5 metres. The marine rocks occur in the basal parts of the Talchir succession. The fauna include *Protoretepora, Spirifer, Aviculapecten, Eurydesma, Hyperammina, Glomospira.* Total absence of Productids and abundance of Eurydesmids make these marine beds distinctly different from those at Umaria. A slightly older and cooler climatic conditions are inferred on the basis of the faunal assemblage of the marine beds at Manendragarh (Bhatia and Singh, 1959).

The discovery of the marine beds within the Lower Gondwana sediments showed that the sea had transgressed over the Peninsula in two phases of marine transgressions. The first marine transgression recorded at Manendragarh was probably an extended arm of the Tethys sea through Sikkim to Central India (Fig. 10.2). The second phase of marine transgression recorded at Umaria took place along the Narmada Valley which appears to be a tectonically active region since the Late Precambrian time. The two marine transgressions were probably separated in time by a few million years. Ahmed (1962) suggests that an open sea existed in the

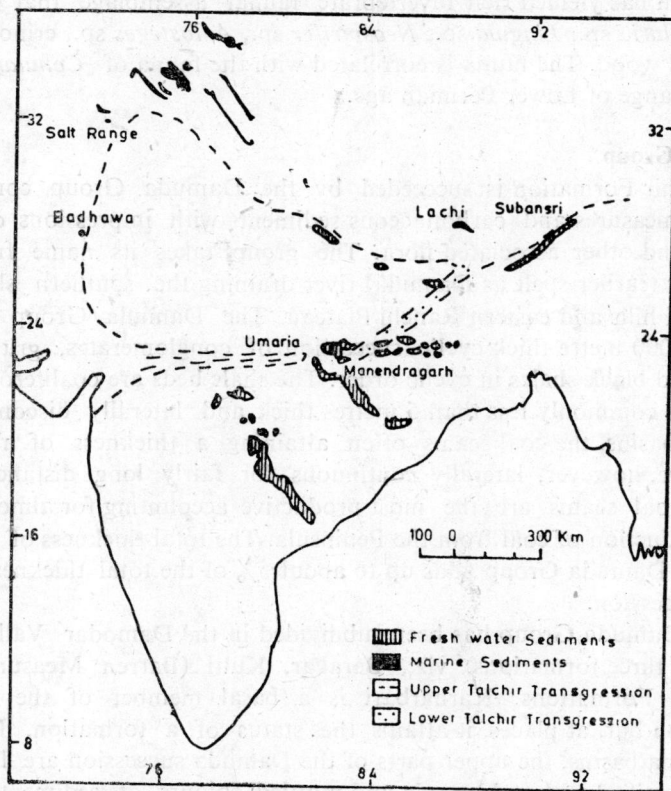

Fig. 10.2: Permian marine transgression in Peninsular India (after Sastry *et al.*, 1964).

southeast of the Central India during the Permo-Carboniferous time and that the marine transgressions of Umaria and Manendragarh were attributed to successive phases of coastal inundations of this palaeo-sea. In either case, one would expect to find evidence of marine beds along the route which has not so far been found.

Bap and Badhaura Formations

Nearly a century ago, Oldham (1886) while traversing around Bap some 150 km northeast of Jaisalmer, noticed glacial boulder beds resting over Vindhyans. These boulder beds were correlated with Talchir Boulder Beds as they are also associated with glacio-fluvial conglomerate, sandstones and varved clays. The succession, ranging in thickness from 50 m to 160 m has been named as **Bap Formation**. The formation has yielded a rich and well preserved invertebrate fauna that include the characteristic fossil *Eurydesma* (Ranga Rao et al., 1979). The **Badhaura Formation** (Misra et al., 1961) consisting of about 350 km thick succession of alternating sandstone and clay beds, rests conformably over the Bap Formation. The Badhaura Formation has yielded rich invertebrate faunal assemblage that includes *Paraconularia* sp., *Lingula* sp., *Neospirifer* sp., *Aulosteges* sp., crinoid stems and fossil wood. The fauna is correlated with the fauna of *Conularia* beds of Salt Range of Lower Permian age.

Damuda Group

The Talchir Formation is succeeded by the Damuda Group comprising the coal measures and carbonaceous sediments with impressions of *Glossopteris* and other associated flora. The group takes its name from the Damodar (earlier spelt as Damuda) river draining the southern slopes of Rajmahal hills and eastern Ranchi Plateau. The Damuda Group consists of over 2000 metre thick cyclic succession of conglomerates, grits, sandstones and black shales in cyclic order. The shale beds are coaliferous. Coal seams are commonly less than 5 metres thick and laterally discontinuous. In some basins, the coal seams often attaining a thickness of about 40 metres are, however, laterally continuous for fairly long distances. The thicker coal seams are the most productive accounting for almost entire total production of coal from the Peninsula. The total thickness of all coal seams of Damuda Group adds up to about 5% of the total thickness of the rock succession.

The Damuda Group has been subdivided in the Damodar Valley type area into three formations, viz., Barakar, Kulti (Barren Measures) and Raniganj Formations. Karharbari is a basal member of the Barakar Formation but, at places, it attains the status of a formation. In other Gondwana basins, the upper parts of the Damuda succession are lithologically diversified with red, purple and mottled colours of sediments representing oxidising environments. Carbonaceous matter in these areas thus

Table 10.2: Biostratigraphic classification of Lower Gondwana Sequence of India (after Sastry et al., 1979)

	Mega-floral Zones	Damodar Valley	Narmada Valley	Mahanadi Valley	Godavari Valley
Upper Permian	*Glossopteris conspicua—G. retifera* Zone	Raniganj Fm (2, 3, 4, 5)	Bijori Fm (2, 5)	Kamthi Fm (2)	Kamthi Fm (2, 3, 6)
	Glossopteris Assemblage Zone				
	Cyclodendron Zone	Kulti Fm (Barren Measures) (2, 3, 4)	Motur Fm (2)	?	Motur Fm (5)
Lower Permian	*Glossopteris walkomiella* Zone	Barakar Fm (2, 3)	Barakar Fm (2, 3)	Barakar Fm (2, 3, 4)	Barakar Fm (2, 3)
	Gandwanidium Buriadia Zone	Karharbari Fm (2, 3)	Karharbari Mb (2, 3)	Karharbari Fm (2, 3)	
	Gangamopteris Zone			Umaria Marine Beds (1, 7)	
	Noeggerathiopsis-Paranocladus Zone	Talchir Fm (2, 3)	Talchir Fm (2, 3)	Talchir Fm (2, 3)	Talchir Fm (2)
				Manendragarh Marine Beds (1)	

Fm—Formation; Mb—Member; 1—marine invertebrate, 2—mega-flora, 3—mioflora, 4—fresh water invertebrate, 5—vertebrate, 6—Estherid, 7—phytoplankton.

becomes sparse or absent. In these sections, the Damuda Group is divisible into Barakar, Motur and Kamathi or Bijori formations (Table 10.2).

The sedimentation of the Damuda succession from Karharbari through Barakar to Raniganj Formations went uninterruptedly and the basin of deposition enlarged progressively in areal expanse (Cashyap, 1979). The sedimentation began with the retreat of the Talchir ice-sheet which led to the uplift of parts of the Peninsula due to isostatic adjustments. The pebbly arkosic sands of Karharbari and basal Barakars were deposited in channels of braided streams. With the decline of gradient through the time, the braided streams were replaced by the meandering streams which laid the cyclic sandstone-shale-coal sequence of the upper parts of Damuda succession. Backswamps favoured the coal forming environments of Karharbari, Barakar and Raniganj Formations and a muddy environment favoured the deposition of fine red clay of Kulti and Motur Formations. The Damuda succession shows a progressive decrease upwards in grain size of sediments and thickness of cross bedded units and increase in mineral maturity. These characters are correlated with a gradual decline in the relief of the source area of sediments during the Damuda sedimentation.

Karharbari Formation: The lowermost subdivision of the Damuda Group consists of sandstones, grits and conglomerates with coal seams. Shales occur only in association with the coal seams. Much of the sediments seem to have been derived from the Talchir Formation part of which must have been uplifted and eroded during the deposition of the Karharbari Formation. The sandstones of the formation contain angular to subangular fragments of quartz and felspar which are in contrast to particularly well rounded sediments of the overlying Barakar Formation. The *Gondwanidium-Buriadia* megafloral assemblage of this formation has much similarity with the *Noeggerathiopsis-Paranocladus* megafloral assemblage of the Talchir Formation. However, the reported occurrence of a slight unconformity at the base of the Karharbari Formation has led it to separate it from the Talchir Formation and to include it with the Damuda succession.

Barakar Formation: The formation resting conformably over the Karharbari Formation comprises a 1700 m thick cyclic succession of conglomerates, grits, sandstones, shales and coal seams. The formation derives its name from a river in the Raniganj coal field in Bihar. The pebbles of conglomerate consist of largely quartzites and granites derived from the Precambrian Basement. The sandstones are coarse, light in colour, often current bedded and highly felspathic. Unlike the Talchir sandstones, the felspars of Barakar sandstones are usually kaolinised. The china-clay deposits of the formation are often of economic significance. The thickness of the formation is highly variable being maximum in Raniganj area. The thickness is reduced to less than 100 metres in the Godawari graben. The floral assemblage of the formation contains abundant *Glossopteris* along with *Walkomiella, Pseudoctenis* and *Barakaria*.

In the south Karanpura coal field, the Barakar Formation consists of two sub-facies (Banerjee, 1963). The coal-shale sub-facies is laterally persistent representing a quieter inter-channel condition of deposition. The sandstone-siltstone sub-facies occurs in the form of lenses representing the shifting bars of a meandering river system.

Kulti Formation: The formation also known as Barren Measures, succeeds conformably the Barakar Formation. The formation is devoid of workable coal seams although streaks of carbonaceous matter are reported from about 500 metre thick succession of sandstones with intercalations of clay beds. In the coal fields of eastern India, this formation is represented by 300 to 400 metre thick succession of predominantly red brown shales known as **Ironstone Shale Formation**. The formation contains a few deposits of sideritic iron ores.

In Narmada Valley, the Kulti Formation is referred to as **Motur Formation** comprising white sandstones with intercalations of red, yellow and carbonaceous shales. The Motur Formation of Godavary Valley Graben has a much reduced thickness measuring about less than 100 metres. The Kulti and Motur formations have yielded scanty mega-flora. The floral assemblage comprises *Cyclodendron, Glossopteris indica* and *G. conspicua*.

Raniganj Formation: The formation overlying the Kulti Formation consists of about 1000 m thick cyclic succession of sandstones, shales and coal seams. The sandstones are finer grained than those of the Barakar Formation. The coal seams of economic significance comprising a part of the Raniganj Formation occur only in the Raniganj coal field. The Raniganj coal contains a higher content of volatiles but some of them are superior quality coal. The floral assemblage of the Raniganj Formation is characterised by the acme of the *Glossopteris* flora. The Raniganj Formation has been correlated with **Kamthi Formation** of the Godavari Graben. The Kamthi Formation named after the type area near Nagpur comprises over 500 metre thick succession of red and grey argillaceous sandstones and conglomerates with interstratified shales. The micaceous fine sandstone and shale sequence of Chindwara district (Narmada Valley Graben) is known as **Bijori Formation** which has also been correlated with the Raniganj Formation on the basis of their floral records.

Lower Gondwana of Eastern Himalaya

Rock successions that have been correlated with the Lower Gondwanas of the Peninsula are exposed in the Darjeeling district (West Bengal), the Rangit Valley (Sikkim) and the Siang district (Arunachal Pradesh). Patches of the Lower Gondwana elements are known from the Lesser Himalaya of Nepal (Kumar and Gupta, 1981). The succession of Darjeeling and Sikkim Himalayas comprises a basal **Rangit Pebble Slate Formation** overlain by **Damuda Group**. The basal formation named after the Rangit Valley in Sikkim (Acharya, 1971) consists of pebbly to gritty slates and lithic wackes, quartzites, pyritous and carbonaceous argillites, rhythmites, volcanoclasts and marls. The diamictite horizon is associated with **Abor Volcanics** in Arunachal Pradesh which is comparable with the Agglomeratic Slate Sequence and Panjal Volcanics of Kashmir (Chapter 8). The Rangit Pebble Slate has yielded marine fauna characterised by the dominance of bivalve *Eurydesma* and absence of *Productus*. A fenestillid bryozoan dominated assemblage comprising *Fenestella, Protoretepora* cf. *ampla,? Polypora megastoma, Geinitzella* sp., brachiopods, bivalves and crinoid stems have been reported from Kameng district, Arunachal Pradesh (Acharya *et al.*, 1975).

The Damuda Group of the Eastern Himalaya rests conformably over the Rangit Pebble Slate in the Rangit window zone of Sikkim. Elsewhere, the contact is often tectonised leading to reversal in stratigraphy. The arenaceous and carbonaceous succession of the Damuda Group has been classified into three formations in Arunachal Pradesh. The older **Khelong Formation** in Kameng district comprises interbedded pyritous and carbonaceous shale, siltstones, marl, calcareous sandstone and felspathic wacke. The succession contains abundant floral record which includes *Glossopteris indica, G. communis* var. *stenoneura, G. damudica, Gangamopteris cyclopteroides, Vertebraria indica, Phyllotheca* sp., *Schizoneura* sp. The formation, at places, contains some beds with marine fauna. The lithofacies and fossil contents of the formation indicate a peralic environment of deposition.

In Subansiri and Siang districts, the Khelong Formation is represented by **Rilu Formation** comprising pyritous and calcareous siltstones and sandstones with rich faunal assemblage. The assemblage indicating a Late Asselian to Lower Sakmarian age is characterised by Linoproductid brachiopods such as *Steponoviella, Linoproductus* etc., together with *Cyretella*, tetropod *Paraconularia* and goniatitic ammonites.

The **Bhareli Formation**, overlying the Khelong Formation in Kameng district, is a typical fluvial sequence of coarse arenaceous and coal bearing rock strata. The floral assemblage is characterised by the predominance of *Glossopteris* indicating an Upper Permian age.

UPPER GONDWANA SEQUENCE

The close of the Lower Gondwana sedimentation in the Satpura Hills is marked by a phase of tectonic deformation producing gentle folds and faults (Crookshank, 1936). In this region, the Upper Gondwana Sequence rests unconformably over the Lower Gondwana rocks. In other Gondwana basins, the sedimentation seems to have been continuous with a gradual passage from the Lower to the Upper Gondwana sequences. The floral

Table 10.3: Classification of the Upper Gondwana Sequence of India (based on Sastry et al., 1979)

Stratigraphic Age	Mega-floral Zones	Damoder Valley Rajmahal Hills	Satpura (Narmada) Region	Son-Mahanadi Valley	Godavari-Prahnita Valleys
Lower Cretaceous	Weichselia-Onichiopsis Zone	R A J Traps			
Upper Jurassic	Pagiophyllum-Brachyphyllum Zone	M Fm. Nipania A Beds H (1, 2) A Traps	J A B A L Jabalpur Beds (1)		Gangpur Fm (1)
Lower-Middle Jurassic	Dictyzamites Pterophyllum Zone	L Lower plant beds (1, 2, 4)	P U R Gr. Chaogaon Beds (1)		Kota Fm (1, 3, 4)
		Dubrajpur Fm (1, 2)			
Upper Triassic	Dicrodium-Noegerathiopsis Zone	Supra Panchet (Mahadeva Fm)	M A H Unconformity A Bagra Congl.	Parsora Fm (2) Pali Fm (2, 5)	Dharmaram Fm (1, 4) Maleri Fm (4)
Middle Triassic	Flora III Flora II	Unconformity	A Gr. (3) D Denwa Clays E (4)		Bhimasaram Fm Yerapalli Fm (4)
Lower Triassic	Flora I	Panchet Fm (1, 2, 3, 4)	V Panchmarhi A Sst. (1)		Mangli Beds (4, 5)
			Unconformity	Kamthi Fm	Kamthi Fm
Permian	Lower Gondwana	Raniganj Fm	Bijori Fm		

Gr.—Group; Fm.—Formation; Congl.—Conglomerate; Sst.—Sandstone; 1—megaflora; 2—mio-flora; 3—fresh water invertebrates; 4—vertebrates; 5—Estherids.

record, however, shows a rapid change with the elimination of the *Glossopteris* flora during the Permian-Triassic transition.

The floral assemblage of the Upper Gondwana Sequence consists of a *Lepidopteris-Dicrodium* flora characterising the lower part of sequence and a *Ptilophyllum* flora characteristic of the upper part of the sequence. Sastry et al. (1979) have distinguished seven mega-floral zones (Table 10.3) in the Upper Gondwana Sequence.

The sedimentation of the Mesozoic Gondwana seems to have continued with several interruptions in different basins of the Gondwana Sequence of India. Correlation of various stratigraphic units of the Gondwana basins is beset with several limitations in view of wide geological ange of floral records.

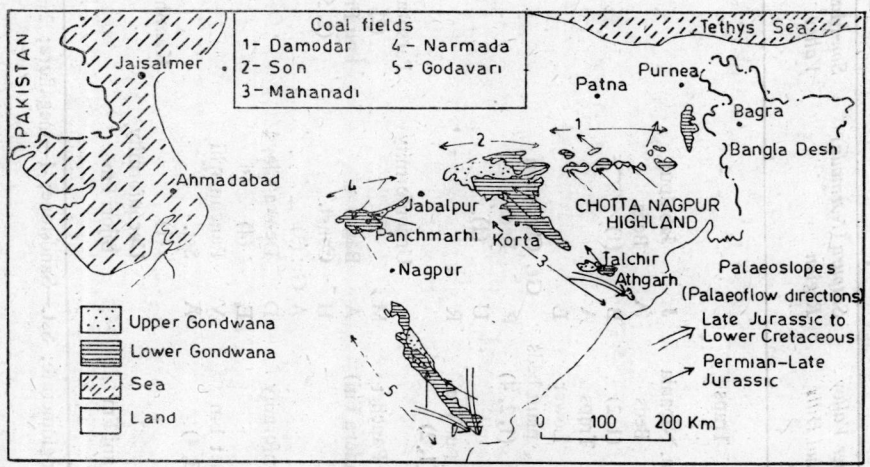

Fig. 10.3: Palaeogeography and palaeoslopes of the northern Indian Peninsula during the Gondwana sedimentation (based on Cashyap, 1979, and Pareek, 1979; Political boundaries are approximate).

Damodar Valley Basin

The Raniganj Formation of the Damodar Valley basin is overlain with a slight unconformity by the **Panchet Formation**. Often the Raniganj Formation is missing and the Panchet Formation rests directly over the older unit of the Damuda Group. The Panchet Formation named after the Panchet hill in the Raniganj coal field was distinguished from the underlying Gondwana rocks by its distinct lithology, fauna and flora (Blanford, 1861). The formation comprises a succession of alternating beds of coarse felspathic, micaceous and cross bedded sandstones, thin beds of green shales and variegated clay beds. The formation is thicker in the Raniganj area (about 700 m) and becomes thinner in the west (about 250 m). The Panchet Formation comprises red bed facies of fluviatile origin.

The Panchet Formation has yielded a rich assemblage of mega-flora, mio-flora, vertebrates and freshwater invertebrates. The megafloral assemblage includes *Glossopteris, Schizoneura, Pecopteris, Cyclopteris, Samorpsis*, sp. and *Taeniopteris*. The vertebrates are represented by fishes (*Amblypterus*), amphibians (*Pachygonia* and *Gonioglyptus*), reptiles (*Dicynodonts* and *Lystosaurus*) and theodont (*Proterosuchus*). The invertebrate include esthiids, *Cyzicus, Estheriella* and insect remains. The Lower Triassic age of the formation has been inferred mainly on the basis of its vertebrate fauna.

The Panchet Formation is often overlain by a succession resembling in lithology and fossil content with those of Mahadeva Group of the Satpura Basin.

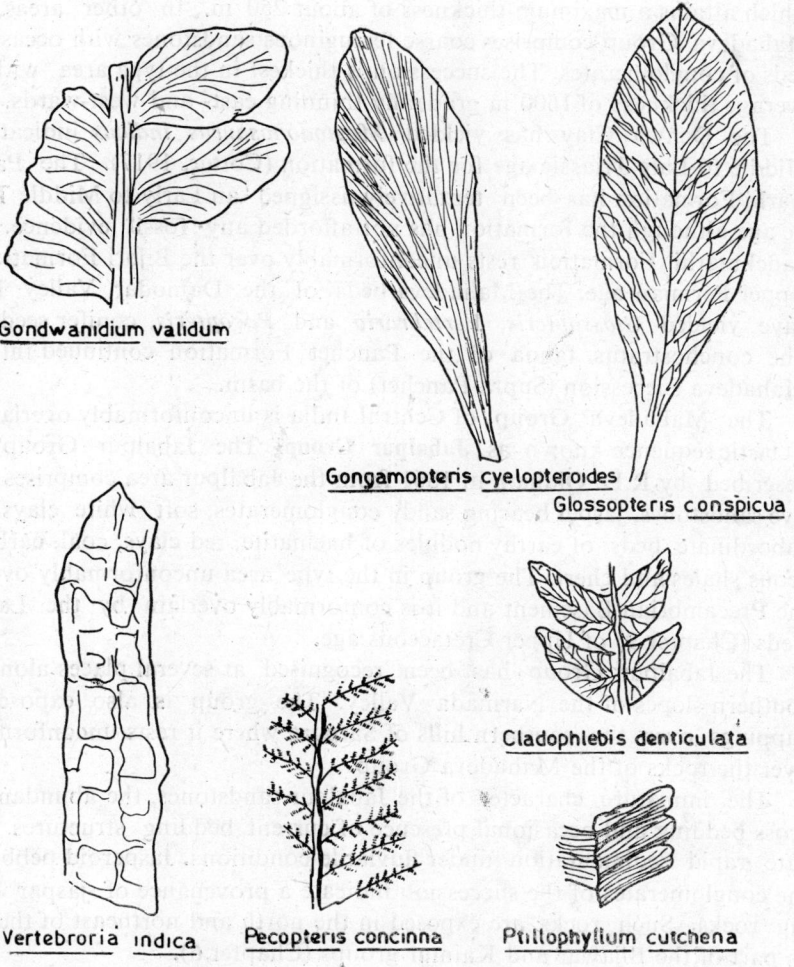

Fig. 10.4: Gondwana flora from India.

Satpura Basin

The Kamthi Formation of the Lower Gondwana is unconformably overlain by a thick succession of continental red beds which were grouped into **Mahadeva Group** named after the Mahadeva Hills near Panchmarhi (Oldham, 1856). The group is generally divided into Panchmarhi Sandstone, Denwa Clays and Bagra Conglomerate (Table 10.3). The three formations generally lying one over the other often grade laterally into one another. The **Panchmarhi Sandstone** consists of about 750 m thick coarse, white and cross-bedded sanstones with characteristic ferruginous partings. The overlying **Denwa Clay** is interstratified with discontinuous and subordinate bands of white sandstones. Locally the sandstones are conglomeratic. Eastward, the Denwa Clay grade upwards into the **Bagra Conglomerate** which attains a maximum thickness of about 250 m. In other areas, the Mahadeva Group comprises coarse ferruginous sandstones with occasional beds of conglomerates. The succession is thickest in the type area with an average thickness of 1600 m gradually thinning east- and west-wards.

The Denwa Clay has yielded *Mastodonasaurus indicus* indicating a Middle to Late Triassic age for the formation (Cotter, 1917). The Panchmarhi Formation has been tentatively assigned an Early to Middle Triassic age although the formation has not afforded any fossil evidence. The Panchmarhi Formation rests unconformably over the Bijori Formation of Upper Permian age. The Mahadeva beds of the Damodar Valley Basin have yielded *Glossopteris, Vertebraria* and *Pacopteris,* conifer seeds etc. The conchostratus fauna of the Panchet Formation continued into the Mahadeva succession (Supra-Panchet) of the basin.

The Mahadeva Group of Central India is unconformably overlain by a clastic sequence known as **Jabalpur Group**. The Jabalpur Group first described by R.D. Oldham in 1861 from the Jabalpur area comprises massive sandstones, jasper bearing sandy conglomerates, soft white clays and subordinate beds of earthy nodules of haematite, red clays, coal, carbonaceous shales and chert. The group in the type area unconformably overlies the Precambrian Basement and it is conformably overlain by the Lameta Beds (Chapter 9) of Upper Cretaceous age.

The Jabalpur Group has been recognised at several places along the southern slopes of the Narmada Valley. The group is also exposed as cappings over the southern hills of Satpura where it rests unconformably over the rocks of the Mahadeva Group.

The immature character of the Jabalpur sandstones, the abundance of cross bedding and occasional presence of current bedding structures indicate rapid sedimentation under fluviatile conditions. Jasperoid pebbles in the conglomerates of the succession indicate a provenance of jaspar bearing rocks. Such rocks are exposed in the north and northeast of the area as part of the Bijawar and Kaimur groups (Chapter 6).

The Jabalpur succession has been subdivided into Chaugaon Beds,

Jabalpur Beds and Bansa Beds. The subdivisions having similar lithology have been established on the basis of their characteristic floral assemblages. The floral record, however, does not give a precise age correlation with the standard stratigraphic scale. The Chaugaon and Jabalpur Beds have been assigned variable ages on the evidence of floral record from Middle Jurassic to Early Cretaceous. The Bansa Beds have yielded well preserved Wealdean flora indicating an Early Cretaceous age for the beds.

Rajmahal Hills

The Barakar Formation is unconformably overlain by a 150 m thick succession of sandstones and conglomerate known as **Dubrajpur Formation** after a village in Rajmahal hills of Bihar. The Dubrajpur Formation has been assigned a Late Triassic age indicating a break in deposition from Late Permian to Middle Triassic.

The Dubrajpur Formation is conformably overlain by trap flows with intercalated grits and carbonaceous shales named as **Rajmahal Formation** (Oldham, 1861). The **Rajmahal Traps** are plateau basalts characterised by the absence or rare occurrence of olivine. The traps are predominantly composed of fine grained to coarsely crystalline dolerite with irregularly distributed phenocrysts or aggregates of felspar and pigeonite. The ground mass comprises labrodorite, pigeonite, augite, magnetite and glass. Some flows are vescicular, the cavities filled with calcite, chalcedony and analcite.

The plant bearing intertrappean beds informally known as **Rajmahal plant beds** are composed of white and grey shales, carbonaceous shales, sandstones and quartzites. More than 15 intercalations of 1.5 to 8 metre thick such intertrappean beds are known from the 150 to 600 m thick volcanic succession of the type area. On the floral evidence, the plant beds have been grouped into **Lower Plant Beds** and **Nipania Plant Beds**.

The Rajmahal plant beds have yielded one of the richest floral records of the world consisting of more than 45 genera and 116 species. The Lower Plant Beds and the upper parts of Dubrajpur Formation have yielded *Dictyozamites-Pterophyllum* assemblage characterised by broad-based cycads, some fish scales and unionids. The Nipania Plant Beds contain *Pagiophyllum-Brachyphyllum* assemblage dominated by pteridophytes and conifers. The Rajmahal Formation has been assigned Early Jurassic to Early Cretaceous age on the basis of the floral assemblage. Radiometric (K-Ar) dating of the Rajmahal Traps has given an Albian age for a part of the trap sequence.

Mahanadi-Son Valley Basin

The Kamthi Formation of the Mandla Plateau in Madhya Pradesh ranging in age from Late Permian to Early Triassic is conformably overlain by about 1500 m thick succession of cross bedded, immature, felspathic sandstones and arkoses with abundant amount of brick red shales and siltstones. The succession earlier referred to as Pali Beds (Hughes, 1881) and Tiki Beds (Fox, 1931) has been named as **Pali Formation**. In the Son Valley, the formation rests over the Damuda coal measures. Robinson (1967) included the Pali Formation as a part of the Lower Gondwana as the Formation has yielded *Glossopteris communis, G. indica* and *Vertebraria indica* constituting the typical Glossopteris flora of the Lower Gondwana. The fossil fauna of the formation consists of fishes, amphibians, reptiles and crocodiles. Typical reptilian remains like Hyperdopedon and Parasuchus and Uniods have been reported from the red clays exposed near Tiki. The floral record of the Pali Formation which comes mostly from the lower part of the succession shows a Kamthi affinity. The faunal record reported largely from the upper part of the succession is closely comparable with that of the Maleri Formation of Late Triassic age (Table 10.3).

The **Parsora Formation** comprising about 450 m thick succession of medium to coarse sandstones of violet and red shades rest conformably over the Pali Formation. The contact between the two formations has been defined on the basis of a gradual disappearance of felspars which are characteristic of the underlying Pali Formation. The sandstones of the Parsora Formation are mineralogically mature composed of quartz with little or no felspar. The formation has yielded a typical *Dicrodium* floral assemblage showing an Upper Gondwana affinity. On the basis of lithology, the rocks of the Parsora Formation have been correlated with the Mahadeva Group of the Satpura Basin.

Pranhita-Godawari Basin

A fairly continuous succession of Middle and Upper Gondwana rocks is exposed in the Pranhita Godawari Basin. The succession conformably resting over the Lower Gondwana Sequence (Kamthi Formation) has been subdivided into six formations (Table 10.3).

The "Maleri Stage" of King (1881) conformably overlying the Kamthi Formation has been subdivided into three litho-units. The lowermost **Yerrapalli Formation** comprising soft red and green clays contains a distinct vertebrate faunal assemblage that includes fishes, amphibians, reptiles, cynodonts, proterosuchians and pseudosuchians. The fossil assemblage indicates a Middle Triassic age for the formation. The Middle unit of the "Maleri Stage" named as **Bhimasaram Sandstone** (Sengupta, 1970) is composed of poorly sorted cross bedded coarse sandstones with intercalations of red clay beds. The formation devoid of any fossil record has been conformably overlain by the clay formations yielding rich vertebrate fauna.

The **Maleri Formation** comprising the uppermost unit of the "Maleri Stage" consists of bright red clay beds intercalated with minor sandstone beds (Hislop, 1864). The formation has yielded the characteristic "Maleri fauna" which include fragments of conifer-wood, dipnoian fish, amphibians

(Metaposaurs), reptiles (Rhynchosaurs, Phytosaurs, Pseudosuchians and Coelurosaurs). The fossil assemblage indicating a Late Triassic age for the formation has been correlated with the fossil assemblage of the Denwa Clay of Mahadeva Group.

The **Dharmaram Formation** conformably overlying the Maleri Formation comprises alternation of pale coloured sandstones and red clay beds (Kutty, 1969). The sandstones are cross bedded and in places contain abundant lime pellets and clay galls. The formation has yielded platosaurid, codontosaurid, prosauropods and archosaurs of Upper Norian to Rhaetic age. The Dharmaram Formation has been correlated with the Parsora Formation of the Mahanadi-Son Valley Basin.

The **Kota Formation** named after the village Kota near the confluence of the Pranhita and the Godavari rivers comprises grey white and occasionally ferruginous gritty and pebbly sandstones with the basal beds and lenses of clay. A few beds of fossiliferous limestones have been reported from the upper part of the succession. The Kota Formation rests over the Dharmaram Formation with a slight unconformity inferred on the basis of the faunal records of the two formations. The faunal assemblage of the Kota Formation indicating an Early to Middle Jurassic age consists of Coicus (Estheria), insects (Blattoidea, Coleoptera and Hemiptera), fishes, reptiles, crocodilea and dinosaurs. The sauropods, typical of Kota Formation, have not been reported from the underlying formation. Jurassic fresh water ostrocodes have also been reported from the formation. The floral assemblage which includes *Equisetites, Sphenopteris, Hansmanica, Coniopteris, Otozamites, Pagiophyllum, Arancorites* and *Elatocladus* also indicate a Lower Jurassic age for the formation.

The **Gangapur Formation** named by Kutty (1969) after Gangapur in Adilabad district of Andhra Pradesh consists of a sequence of mudstones, siltstones, sandstones and conglomerates. The Chikiala Sandstone exposed along the eastern margin of the Basin has been correlated with the Gangapur Formation. The Gangapur Formation resting unconformably over the Kota Formation has yielded a rich assemblage of mega-flora having mixed characters of the eastern Rajmahal flora and the Jabalpur flora of the west. A rich assemblage of spores and pollens resembling that of the Nipania Plant Beds of the Rajmahal Formation has been reported from the carbonaceous shales of the Gangpur succession. A Late Jurassic to Lower Cretaceous age has been assigned to the formation on the basis of floral evidence.

Chapter 11

Cenozoic History

The term "Cenozoic" ("Cainozoic" meaning new life) was introduced by John Phillips (1840-41) for the era that followed the Mesozoic. The Cenozoic Era comprises the shortest and the most recent phase of the earth's history embracing a total duration of about 70 million years. The era is generally divided into **Tertiary Period** and **Quaternary Period**. The Tertiary System of rock formations that were laid down during that period has been subdivided into a lower **Palaeogene** succession and an upper **Neogene** succession. In view of contrasting fossil assemblages of the two subdivisions, some geologists prefer to assign them independent status of "system". Thus, in a three-fold division, the Cenozoic Era comprises the Palaeogene, the Neogene and the Quaternary periods.

The beginning of the Tertiary Period is placed at about 70 million years and the period came to an end at about 1.5 million years ago. Charles Lyell (1833) subdivided the Tertiary System into three series known as **Eocene** (dawn of the recent), **Miocene** (middle of the recent) and **Pliocene** (major of the recent). Later, **Palaeocene** (older than Eocene) and **Oligocene** (younger than Eocene) were carved out from the Eocene succession. The Palaeocene, the Eocene and the Oligocene epochs constitute the Palaeogene Period whereas the Miocene and the Pliocene epochs comprise the Neogene Period. The various epochs of the Tertiary Period have been further subdivided into successive ages (Table 11.1) on the basis of characteristic assemblages of flora and fauna.

The Quaternary Period beginning at about 1.5 million years ago comprises the **Pleistocene** (most recent) and the **Holocene** (Recent) epochs. The Pleistocene Epoch covering the major part of the Quaternary Period forms one of the most interesting epochs of the earth's history. The *Homo sapiens* evolved to a position of dominance over all other kinds of life during the Pleistocene Epoch. The Pleistocene Epoch is also characterised by a marked fluctuation of climate. Glacial ice covered over one-third of the continents during the glacial ages of the Pleistocene Epoch. Five such glacial ages have been recorded in the Pleistocene succession. They are in order of their succession, **Biberian, Danubian, Gunjian, Mindelian, Rissian** and **Wurmian** ages. The glacial ages were interrupted by interglacial phases during which many glaciers receded and others disappeared.

Table 11.1: Subdivisions of the Tertiary Period

Period/System	Series/Epoch		Stage/Age
Neogene	Pliocene	Late	Astian
			Plaisanchian
		Early	Pannonian
	Miocene	Late	Sarmatian
		Middle	Vindobonian
		Early	Burdigalian
			Aquitanian
Palaeogene	Oligocene	Late	Chattian
		Middle	Stampian
		Early	Sannoisian
	Eocene	Late	Priabonian
		Middle	Lutetian
		Early	Ypresian
	Palaeocene	Late	Landenian
		Early	Montian
			Danian

TECTONIC HISTORY

The physiographic features of India were shaped during the Cenozoic Era. The era began with the outbursts of lava flows over the Indian Peninsula, a marine transgression over the northwestern Peninsula and tectonic instability of the Himalayan basin. The Palaeocene sea that formed a part of the continuous sea connecting the Arabian sea and the Bay of Bengal across the northern parts of the subcontinent had receded towards the west and east during the Early Eocene (Fig. 11.1). This trend of marine regression continued till the final emergence of the Himalayan mountains by the end of the Tertiary Period. The Eocene Epoch is marked by marine invasion over the parts of the western, eastern and southern India. After a brief period of marine regression during Oligocene the western and eastern Peninsula was again inundated by the shelf sea at the beginning of Neogene Period (Fig 11.2). The Pliocene Epoch witnessed a general withdrawal of the sea from the Indian sub-continent.

Rise of the Tertiary Mountains

The rise of the Himalayas was accomplished in a series of five or more impulses punctuated by intervals of comparative quiescence (Krishnan, 1968). The first movement took place during the Late Cretaceous-Early Eocene time. It was accompanied by the emplacement of the Dras Volcanics (Indus Ophiolites) along the northern borders of the mountains. This phase of extensive volcanism represents a palaeo-island arc system that was formed at the commencement of the Himalayan orogeny (Kumar and Gupta, 1982). This tectonic phase was also accompanied by intense deformation, regional metamorphism and emplacement of granitic gneisses in the deeper parts of the orogen. Some of the granitic gneisses represent remobilised Precambrian Basement of the basin.

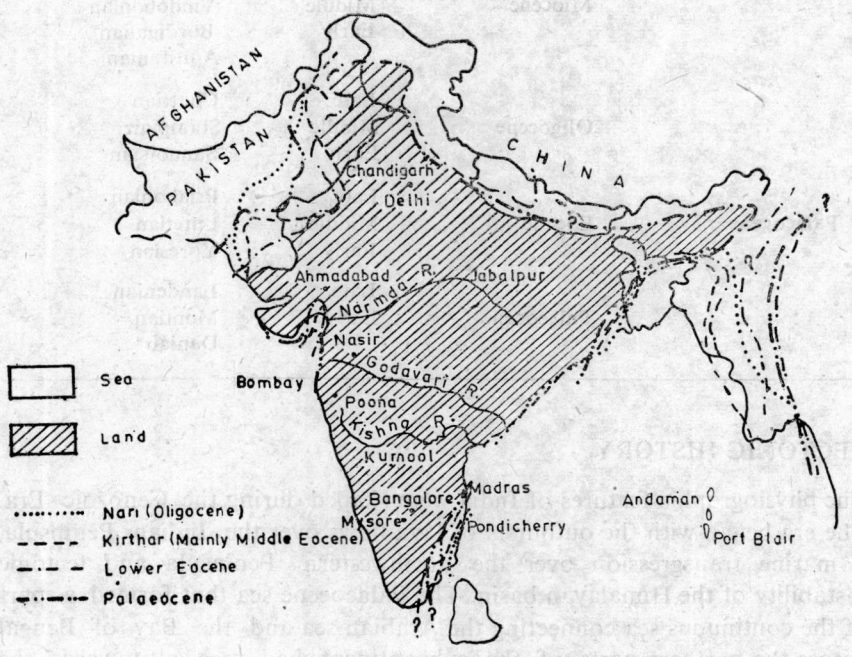

Fig. 11.1: Palaeogeography of the Indian sub-continent during the Paleogene Period (based on Sahni and Kumar, 1974; political boundaries are approximate).

The Tethys sea was furrowed into longitudinal ridges and basins as a consequence of the first tectonic movement of the Himalayan Orogeny. The basins were the sites of thick accumulation of flyschoidal sediments during the Palaeogene time. The Palaeogene flyschoidal successions are exposed both in the north (Ladakh) and in the south (southern Lesser Himalaya) of the Higher Himalayan mountains. The trans-Himalayan region (Ladakh and Karakoram) witnessed a phase of batholithic intrusion of tonalitic granitoids during the Palaeocene-Early Eocene resulting in the rise of the Trans-Himalayan ranges.

The second major upheaval took place during the Late Eocene time when the Tethyan Himalayan Zone was uplifted as a land mass. This movement was accompanied by the emplacement of tourmaline-granites in the metamorphites and granitic gneisses that comprise the Higher Himalayan Zone. The Lesser Himalayan basins became shallower with the partial withdrawal of marine water. Brackish water sediments were laid in these basins during the Late Eocene-Oligocene time.

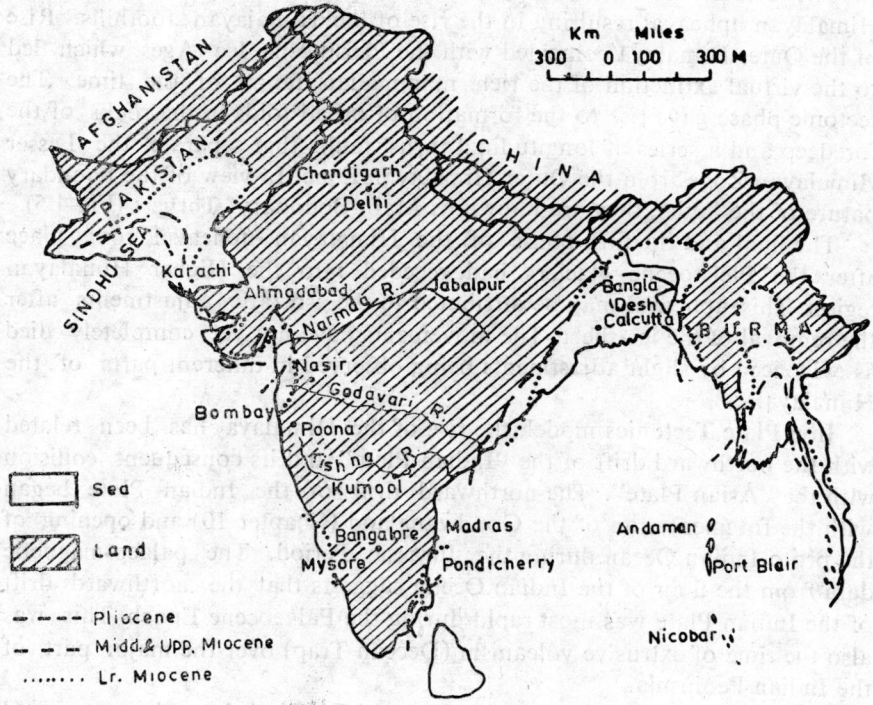

Fig. 11.2: Palaeogeography of the Indian sub-continent during the Neogene Period (Political boundaries are approximate).

The third upheaval that took place during the Middle Miocene time was the most pronounced of all the phases of the Himalayan Orogeny. Rocks of the Lesser Himalayan Zone were deformed into broad folds trending parallel to the Himalayan chain of mountains. Thrust sheets originating from the northern parts were piled one over the other in a southward translational movement. The nappes were further folded and thrust faulted due to continued south directed couple movement of the rising Himalaya.

The Middle Miocene upheaval of the Himalaya resulted in the formation of a Foredeep between the rising Himalaya and the northern edge of the Peninsula. The foredeep was filled with the fluviatile and lacustrine sediments derived from the freshly exposed Himalayan rocks. The molassic succession of the foredeep is characterised by finer low energy sediments in the basal part overlain by coarse high energy sediments in the upper part. The upper part of the succession progressively contain increasing proportion of clasts of metamorphic and granitic rocks which were exposed to weathering at successive phases of the Himalayan uplift.

The Pliocene-Pleistocene Epochs witnessed the fourth phase of the Himalayan upheaval resulting in the rise of the Himalayan foothills. Rise of the Outer Himalaya coincided with the Pleistocene Ice Ages which led to the virtual extinction of the rich mammalian fauna of the time. The tectonic phase gave rise to the formation of broad folds in the rocks of the foredeep and a series of longitudinal thrust faults that separate the Lesser Himalayan Zone from the Outer Himalayan Zone. In view of the boundary nature of the faults, they are known as Main Boundary Thrusts (Fig 4.5).

The final and the fifth phase of the Himalayan upheaval took place after the Pleistocene glaciers had receded into the Higher Himalayan region. This upheaval may be attributed to the isostatic adjustments after the removal of the ice-sheet. The movements have not yet completely died as evidenced by slight adjustments being observed in different parts of the Himalaya.

In a Plate Tectonics model, the rise of the Himalaya has been related with the northward drift of the "Indian Plate" and its consequent collision with the "Asian Plate". The northward drift of the Indian Plate began with the fragmentation of the Gondwanaland (Chapter 10) and opening of the proto-Indian Ocean during the Jurassic Period. The palaeo-magnetic data from the floor of the Indian Ocean suggests that the northward drift of the Indian Plate was most rapid during the Palaeocene Epoch. This was also the time of extrusive volcanism (Deccan Trap) over the major part of the Indian Peninsula.

The "collision" of the Indian Plate with the Asian Plate occurred during the Early Eocene. The "collision" retarded the pace of the northerly drift of the Indian Plate. The drift was resumed at the beginning of the Oligocene with a slightly changed direction of translation. The Indian Plate began its rotational movement giving rise to the formation of syntexial bend at the north-western extremity of the Himalayan mountains.

The Arakan mountains of the eastern India and their continuation into the Andaman-Nicobar islands were also formed during the Tertiary diastrophism. The rocks of the Arakan ranges were laid down during Late Cretaceous-Tertiary time in a sedimentary basin that was formed by the closing of the Indian and Burmese "Plates". Six phases of diastrophic movements having a widespread impact on the sedimentation pattern have been recognised in the Assam-Arakan region. The first two tectonic disturbances that occurred during the post-Cretaceous and post-Eocene times were confined to the eastern parts of the region. The third movement of post-Oligocene age was widespread affecting the Assam shelf and the Arakan mountains alike. The post-Miocene movements gave rise to the deposition of conglomerate predominant sediments of Pliocene age. The Late Pliocene movements were responsible for the rise of the Arakan mountains that supplied the coarse sediments filling molassic troughs. The last folding movements occurred towards the close of the Pleistocene Epoch although small scale uplifts and warping still continue.

HISTORY OF CENOZOIC LIFE

The organic forms underwent yet another phase of pronounced changes at the close of the Mesozoic Era. On land, the reptiles and other groups of animals that were so conspicuous in Mesozoic Era rapidly declined during the Mesozoic-Cenozoic transition. They were replaced by the abundance of mammals, birds and insects. The land mammals were provided with plentiful supply of food with a widespread growth of angiosperms. Nine new orders of mammalian fauna were added during the Palaeocene Epoch to the three orders, Multiberculata, Marsupalia and Insectivora that existed during the Cretaceous Period. Some of the mammalian orders reached the zenith of their evolution and became extinct during the Neogene Period. The Pleistocene Epoch witnessed the appearance of *Homo sapiens* and their steady rise to a position of dominance over other groups of animals.

In marine environments too, many typical Mesozoic invertebrate rapidly died out before Palaeocene, paving way to the appearance of many new forms. Some Palaeocene invertebrates have been recognised on al continents whereas others have shown a restricted distribution on account of their limited environmental tolerance. The warm Palaeocene and Eocene marine basins were swarmed with predominantly agglutinated (arenaceous) and calcareous foraminifers. Abundance of these micro-organisms in the fossil records has facilitated a detailed classification of rock successions into stages and zones. Larger foraminifers of the Palaeocene and Eocene Epochs include *Nummulites* and *Discocyclina*. *Lepidocyclina* of the family Orbitoididae which replaced the *Discocyclina* in Late Eocene became abundant in Early Miocene and survived through the Late Miocene. The smaller foraminifers represented mainly by the planktonic Globigerinidae (genus *Globigerina*) are widely distributed in all parts of the Cenozoic marine successions. The other tiny benthonic foraminifers of importance include the family Buliminidae (genus *Siphogenerina*).

Among the other marine invertebrates, the gastropods and the pelecypods achieved conspicuous radiations through the Cenozoic Era. The gastropod *Turritella* thrived the fine silty neritic floors of the oceans sheltered from wave action. The pelecypods included free moving giant venericards, sedentry oysters, fan-shaped pectens and burrowing types. The nautiloids became abundant in some environments during the Palaeocene and the Eocene epochs and diminished during the Miocene and the Pliocene epochs. Echinoids of the era included the regular and irregular types. Bryozoans and corals are also frequently encountered in the Cenozoic rock formations. Corals that first declined during the Palaeocene had a steady slow growth throughout the era (Fig. 11.3).

Appearance of modern groups of birds, tarriers and lemurs characterise the Palaeocene vertebrate record. Reptiles though rare in Palaeocene are represented by crockodilians and turtles. All the modern orders of mammals are known since the Eocene Epoch. Primitives apes and monkeys, saber toothed cats, cats and dogs had appeared during the Oligocene Epoch. The Miocene Epoch witnessed the rise and rapid evolution of the grazing mammals. The mammals achieved the peak of their evolution during the Pliocene Epoch. The man-like creature that appeared during the Pliocene evolved to be replaced by the *Homo sapiens* during the Pleistocene Epoch. The marine vertebrates of the Cenozoic Era included the bony fishes, sharks and highly specialised pelagic mammal whale.

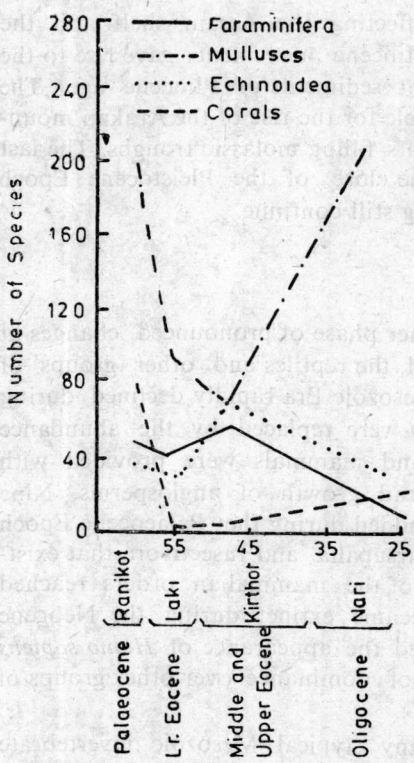

Fig. 11.3 : Graph showing the fluctuation in number of species during the Palaeogene (after Sahni and Kumar, 1974).

BOUNDARY PROBLEMS

Stratigraphic boundaries are determined by one or more of geological events such as volcanic activity, sedimentation, tectonism, palaeo-environments and evolution of life. Faunal records have played major role in determining the boundaries of the Phanerozoic units. The other geological events are dated on the evidence of fossil records. With the advent of radiometric dating methods and palaeomagnetism, phases of geological events have been identified independent of fossil record.

The problem of demarcating the boundary line between the Cretaceous and the Tertiary successions has two aspects. The one pertains to locating this boundary in a particular region on the evidence of some fossil record and the other concerns with the "fixing" of this boundary in the internationally accepted "Standard Stratigraphic Scale". Boundaries between the Phanerozoic units are determined on the faunal evidence with the presumption that the organic forms have periodically undergone profound changes at certain periods of the earth's history. In general, this assumption has been substantiated by the fossil records. In detail, however, various groups of animals did not have synchronous phases of crisis in their evolutionary history. Demarcating the stratigraphic boundaries with the help of the fossil records of diverse groups of animals often gives contradictory results.

The Maastrichtian Stage of the Upper Cretaceous is overlain by Danian, Montian and Thanetian Stages. The Montian and Thanetian Stages are grouped into the Palaeocene Series (Table 11.1). However, the opinion is sharply divided as regard to whether the Danian Stage should be included with the Palaeocene or with the Upper Cretaceous Series. Yet others have suggested an independent status of "series" to Danian and some have strangely doubted the very existence of such a stratigraphic unit.

The planktonic foraminiferal biostratigraphy of the Gulf of Atlantic coastal plains (Loeblich and Tappan, 1957) favours the inclusion of the Danian Stage with the Tertiary System. A significant palaeontological "break" has been noticed in the foraminiferal succession at the close of the Maastrichtian Age. Some other co-existing faunal groups, however, do not show such a profound break in the faunal succession. The foraminiferal biostratigraphy of the Danian Stage in the type area in Denmark also suggests such a marked "break" at the base of the Danian Stage warranting its inclusion in the Tertiary System. The other view held by Hofker (1966) and others suggests that the change from the Cretaceous oraminiferal fauna into the Tertiary fauna was gradual during the Maastrichtian and Danian time. Wherever this change seems to be an abrupt one, the same may be attributed to a gap in sedimentation. Although the problem remains yet to be resolved, the Danian Stage has been provisionally included as the lowest unit of the Tertiary System.

In many parts of the world, stratigraphic breaks have been observed between the Maastrichtian and Montian beds. The post-Maastrichtian—pre-Montian time interval is represented in other parts of the world by "passage beds" (Rama Rao, 1972) which is known as Danian Stage. Rocks above the "passage beds" contain fauna that have a definite Tertiary aspect while the rocks underlying them have a definite Cretaceous affinity. There are three regions in the Indian subcontinent which have exposed such "passage beds" (Rama Rao, 1964). They are the Sind area in Pakistan, the Tiruchchirappalli area in south India and the Assam region in the north-eastern India.

The geology of the Sind area was first described by Blanford (1880) which was later improved by Vredenberg (1909). The basic divisions of the stratigraphic units representing the Cretaceous-Tertiary transition are given below (Table 11.2).

Table 11.2: Upper Cretaceous-Palaeocene succession of the Sind area, Pakistan (based on Rama Rao, 1964)

Stratigraphic Units		Characteristic fossils	Age
Ranikot Formation	Upper	Miscellenia miscella, Nummulites (Ranikothalia) nuttali, N (R) sindensis, Discocyclina ranikotensis, Lokhartia haemei, Globigerina and Guembilina	Palaeocene
	Lower	Unfossiliferous	
	Basal	Unfossiliferous	
Cardita beaumonti beds		Cardita (Venericardia) beaumonti mainly unfossiliferous	Danian
Pab Sandstone Limestone Formation		Orbitoides media, O. macropora, Siderolites calcitrapoides	Maastrichtian

The Pab Sandstone which is mainly unfossiliferous contain Maastrichtian fossils in some thin beds at the base and the top of the sandstone succession. Since the *Cardita beaumonti* beds overlie the uppermost Maastrichtian succession, they have been assigned a Danian age and the *Cardita (Venericardia) beaumonti* has been recognised as an index fossil of the Danian on the Indian subcontinent. Of the three sub-divisions of the Ranikot Formation, the basal and lower sub-divisions are composed of practically unfossiliferous sandstones of fluviatile origin. These sub-divisions are often intervened by subaerial vescicular traps which constitute the local representative of the Deccan Trap (see later). It is evident that the Sind area witnessed a marine regression concurrent with a phase of volcanism immediately after the deposition of Danian rocks. The upper sub-division of the Ranikot Formation contain abundant marine Palaeocene fauna that indicates that the marine regression was short lived.

The Upper Cretaceous rocks of the Tiruchchirappalli area have been classified into Uttatur, Trichinapoly and Ariyalur formations (Chapter 9). Blainford (1865) divided the Ariyalur succession into lower, middle and upper units. The upper unit was later recognised as forming a distinct stratigraphic unit named as Niniyur Formation (Rama Rao and Ria, 1936). The Niniyur Formation has yielded a rich assemblage of lamellibranchs, corals and gastropods. The important forms of the fossil assemblage are *Cardita* (*Venericardia*) *jaquinoti*, *Lucian percrassa*, *Cerethium* and *Nautilus indicus*. The *Cardita* (*Venericardia*) *jaquinoti* is identical with the well known *Cardita* (*Venericardia*) *beaumonti* of the Sind area. The underlying Ariyalur Formation has yielded a foraminiferal assemblage of Maastrichtian age. The assemblage consists mainly the species of *Lepidoobitoides*, *Siderolites*, *Robulus*, *Lenticulina*, *Nonian*, *Rotalia* and *Operculina*.

The Upper Cretaceous rocks of Assam region comprising the Mahadek Formation (Chapter 9) is conformably overlain by a marine succession known as Langpar Formation. The Langpar Formation has yielded a Danian foraminiferal assemblage (Nagappa, 1959). The marine rocks are overlain by a thick continental sequence of coal bearing sandstones known as Therria Formation. Biswas (1962) has suggested that both the Langpar and the Therria formations were deposited at the same time, the former being a marine facies equivalent of the continental Therria Formation. Rama Rao (1964) grouped both the formations into the "passage beds" representing the Cretaceous-Tertiary transition. Sah and Singh (1977) have recorded the presence of a mio-floral break between the Mahadek and Langpar formations representing the Cretaceous-Tertiary boundary in Assam.

The Neogene-Quaternary boundary in India (Fig. 11.4) has been investigated in the continental deposits of the Outer Himalayas and the Kashmir Valley and the marine deposits of Andaman-Nicobar islands. The basic criteria for the recognition of this boundary in the continental deposits are the first appearance of certain group of mammals such as *Elephus*, *Leptobos* and *Equus* and disappearance of *Hipparion*. The continental succession contains evidence for the first climatic deterioration leading to the glacial phases at the commencement of the Quaternary deposits. The boundary

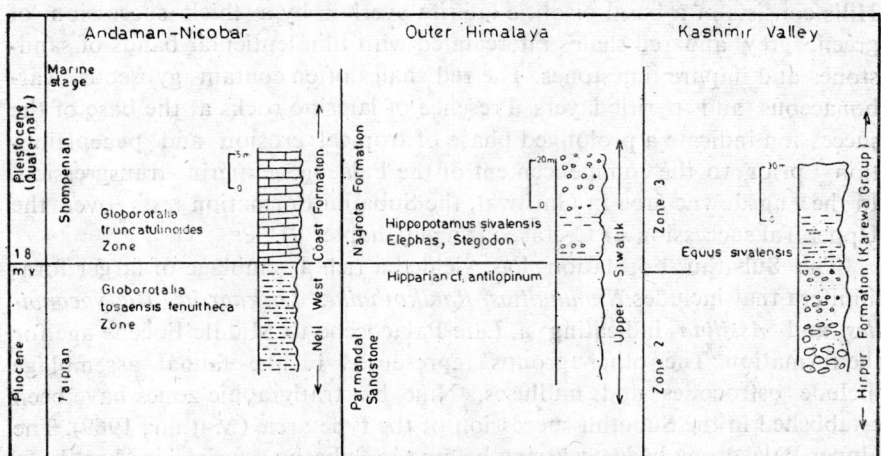

Fig. 11.4: Neogene-Quaternary boundary in India.

in the marine succession has been demarcated on the basis of the appearance and extinction of certain planktonic foraminifera.

INDIAN CENOZOIC FORMATIONS

The Tertiary rocks are exposed in the Himalayan and Arakan mountains and the shelf basins of Kutch-Saurashtra, Western Rajasthan, Tiruchchirappalli-Pondicherry and western coast of Kerala. The Himalayan successions are exposed in two parallel belts, a northern belt of the Palaeogene rocks and a southern belt of Neogene rocks. The two belts are separated from each other by the Main Boundary Thrust (Fig. 4.5). Tertiary rocks of miogeosynclinal facies overlain by rocks of the molassic facies are also known from the northern fringe of the Indus Ophiolitic Belt in Ladakh. The Tertiary rocks of the Arakan Orogenic belt laterally grade into the shelf facies of the Assam region. The Quaternary sediments fill the immense thickness of the Indogangetic Plain and some intermontane basins of the Himalaya.

Himalayan Palaeogene Succession

Simla Hills: The Palaeogene rocks of the Lesser Himalaya of Himachal Pradesh are exposed in a NW-SE trending belt bounded in the north by the Krol thrust and in the south by the Main Boundary Fault (Fig. 7.4). The succession comprising Subathu, Dagshai and Kasauli Formations rests unconformably over the Simla Group of Late Precambrian age (Chapter 7). The Subathu Formation has yielded definite marine fossils whereas the overlying Dagshai and Kasauli Formations contain an admixture of estuarine and freshwater sediments.

The **Subathu Formation** named after the Subathu town of the Simla Hills consists of a basal pisolitic laterite overlain by a thick succession of green, grey and red shales intercalated with thin lenticular bands of sandstones and impure limestones. The red shales often contain gypseous, carbonaceous and pyritic layers. Presence of lateritic rocks at the base of the succession indicate a prolonged phase of tropical erosion and peneplaination prior to the commencement of the Palaeogene marine transgression. In the Lansdowne area in Garhwal, the Subathu Formation rests over the Upper Tal succession of Cretaceous age (Chapter 9).

The Subathu Formation has yielded a rich assemblage of larger foraminifera that includes *Nummulites, Ranikothalia, Lockhartia, Dictyoconoides* and *Assilina* indicating a Late Palaeocene to Middle Eocene age for the formation. The other groups represented in the faunal assemblage include ostrocodes and mulluscs. Nine biostratigraphic zones have been established in the Subathu succession of the type area (Mathur, 1969). The Upper Palaeocene beds occurring below the Subathu succession has been grouped into a formation known as "Kakara Series" (Srikantia and

Bhargava, 1967). The formation comprising shales and limestones has yielded *Globorotalia, Globogerina, Operculina, Rotalia* and other fossils of Upper Paloeocene age.

The **Dagshai Formation** named after the Dagshai town of Simla Hills comprises a 600 metre thick alternating sequence of grey and purple clay and hard, fine grained greenish impure sandstones. Except for some trace fossils (fucoid markings and worm burrows), the formation has not yielded any fossil record. A Late Eocene to Oligocene age is generally assigned to the formation. The Subathu-Dagshai contact has been regarded by some to represent an unconformity. Some of the pebbles occurring in the basal conglomerate of the Dagshai succession were derived from the erosion of locally uplifted parts of the basin of deposition (Chaudhri, 1968). The conglomerates having intraformational characters are in perfect conformity with both the underlying and the overlying beds.

The **Kasauli Formation** named after a hill station of that name in the southern Himachal Pradesh consists of about 2100 metres thick alternation of massive pale grey to buff micaceous sandstones and purple and grey clays. Lithology of the Kasauli Formation differs from that of the underlying Dagshai Formation in having a relatively smaller proportion of "red beds". The upper part of the Kasauli Formation contain plant fossils that include *Sabal major,* fossil wood and monocot and dicot leaf impressions. The succession has also yielded a large number of *Unio* shells. The formation has been generally assigned a stratigraphic age ranging from Late Oligocene to Middle Miocene.

The Subathu, Dagshai and Kasauli formations are repeatedly exposed in the southern ranges of Himachal Pradesh. The repetition of beds is generally attributed to the presence of structures such as folds and faults. However, Raiverman and Raman (1971) have shown that the green and red facies represented by the Subathu and Kasauli formations are in fact repeated in a single continuous sequence of rocks that were deposited during a prolonged phase of sedimentation in alternating marine-brackish-fresh water conditions. Raiverman *et al.* (1979) divided the Tertiary sequence of the region into a number of energy sequences representing successive impulses of uplifts and subsidence related to the Himalayan Orogeny. The eight energy sequences named as En-seq 1 to En-seq 8 are characterised by their typical association of rocks, colour, stratification, texture, structure and mineralogy.

The Subathu Formation has been exposed in a number of outliers in the Jammu region in the west and Central Nepal in the east. The rock formations of the northwestern Himalaya that have been correlated with the Dagshai and Kasauli Formations have been grouped into the Dharmshala and Murree Groups.

The **Dharmshala Group** comprises an alternation of fine to medium grained grey and maroon sandstones and variegated clays. The succession is often divisible into a lower and an upper formations. The lower unit contains certain definite marine features whereas the upper unit is predominantly of continental origin. The succession has yielded pollen, fossil wood and leaf impressions indicating an Oligocene to Early Miocene age.

The type area of the **Murree Group** is situated in Pakistan. The succession extensively exposed on the Jammu-Srinagar highway around Batot consists of a basal conglomerate bed overlain by intercalations of bright reddish purple clay and green sandstones. The succession rests with a possible unconformity over the rocks of the Subathu Formation. The Murree rocks contain plant fossils (*Sabal major*) and bivalves. Vertebrate fossils indicating an age range of Middle Eocene to Oligocene have been reported from Kalakot area in Jammu (Sahni and Khare, 1973; Rangarao and Obergfell, 1973).

Himalayan Neogene Succession

The Neogene rocks of the Himalaya representing the molasse facies are exposed in the Siwalik hills and other foothills comprising the Outer Himalaya. The Neogene rocks are also known from several intermontane basins occurring in different parts of the Himalayan Orogen. The Outer Himalayan rocks have been grouped into the Siwalik Group whereas the intermontane molasse best exposed in the Kashmir Valley comprise the Karewa Formation.

Siwalik Group: The succession named after the Siwalik Hills near Hardwar is best exposed in the Tawi Valley of Jammu and Hartalyangar area in Himachal Pradesh. The basement of the succession is generally not exposed. The **Nahan Formation** of the southeastern Himachal Pradesh which has a gradational contact with the underlying Kasauli Formation has been correlated with the lowermost unit of the Siwalik Group. The sediments of the Siwalik Group were derived from the rising mountains in the north and they were laid down in the alluvial plains of a series of rivers or a single river system named as "Indo-Brahm" river (Pascoe, 1919).

The lower parts of the Siwalik succession contain fine to medium grained, relatively matured sediments indicating a low relief for the provinance. Greywackes indicating a marine influence have also been reported from the lower part of the succession (Sikka *et al.*, 1961). With the rapid uplift of the Himalayan mountains in succeeding tectonic impulses, the sediments deposited in the upper parts of the Siwalik succession acquired increasingly coarse and immature characters.

The Siwalik Formation has yielded a rich assemblage of vertebrate fauna first described in 1868 by Falconer. The succession also contains molluscs, ostrocodes, charophytes, spores, pollens and plant remains. On the evidence of the vertebrate fauna, Pilgrim (1913) divided the Siwalik succession into three units assigning them Middle Miocene, Late Miocene to Early Pliocene and Late Pliocene to Early Pleistocene ages (Table 11.3).

Table 11.3: Classification of the Siwalik Group
(Based on Pilgrim, 1913; Colbert, 1934, 1942)

Sub-Group	Formation	Lithology	Characteristic fossils
	OLDER ALLUVIUM		
	----- gradational -----		
Upper (Upper Pliocene to Lower Pleistocene)	Boulder Conglomerate	Coarsse boulder conglomerates, clays, sands, grits	*Elephas nomadicus, Equus, Camelus, Buffelus palaeindicus*
	Pinjor Formation	Conglomerates, sandstones, clays	*Elephas planiferous, Hemibos, Stegodon*
	Tatrot Formation	Sandstones, clays conglomerates	*Hyppophys, Leptobos*
Middle (Upper Miocene to Lower Pliocene)	Dhokpathan Formation	Sandstones, shales, clays, pebbly at the top	*Stegodon, Mastodon, Giraffoid, Sus*
	Nagri Formation	Massive sandstones shales, red clays	*Mastodon, Hipparion, Prostegodon, Zamapithicus*
Lower (Middle Miocene)	Chinji Formation	Nodular shales, clays, sandstones	*Listriodon, Amphicyon, Giraffokery, Tetrabelodon*
	Kamlial Formation	Dark hard sandstones red and purple shales	*Aceratherium, Telemastodon, Tetrabelodon, Anthropoids, Hypoboops*

----- Conformable and gradational -----
KASAULI FORMATION (UPPER MURREES)

Colbert (1934, 1942) further subdivided the lower unit into the Kamlial and Chinji "Zones", the middle unit into the Dhokpathan and Nagri "Zones" and the upper into Tatrot, Pinjor and Boulder Conglomerate "Zones". The three sub-groups of the Siwalik Group have also shown a definite pattern of variation of heavy mineral composition (Dehadrai, 1958; Sinha, 1970 and Chaudhri, 1972). Staurolite dominates the heavy mineral components of the Lower Siwaliks, whereas the Kyanite and Hornblende are the dominant heavy minerals in the sediments of the Middle and the Upper Siwaliks respectively.

Warm and humid climate seems to have prevailed during the greater part of the deposition of the Siwalik sediments. The Lower Siwaliks comprising intercalations of fine and coarse sandstones were deposited dominantly in lacustrine environments of fluctuating climates. The fine sediments of dry season were presumably derived from the northern parts of the Peninsula whereas the coarse sediments transported by the rivers of relatively high velocities were derived from the northern regions of emerging Himalayas during the wet seasons. The Middle Siwalik sediments were laid in marshy environments. High energy fluviatile influence attained the predominance during the deposition of the Upper Siwaliks.

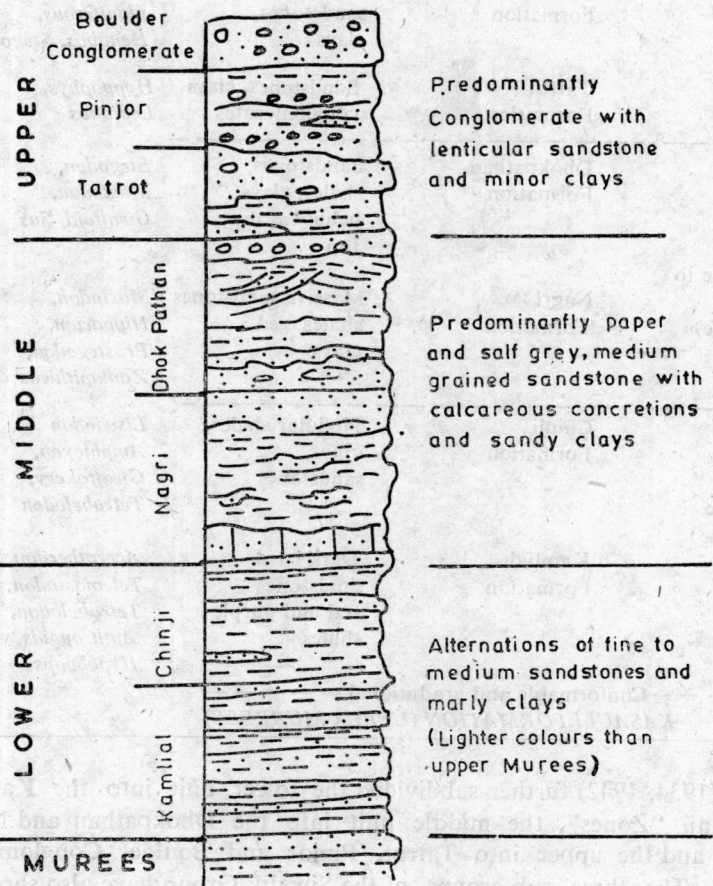

Fig. 11.5: Stratigraphy of the Siwaliks in the Punjab Himalaya (based on Gansser, 1964).

The **Kamlial Formation** in the type area near Khaur oil field in Pakistan comprises about 600 to 1000 m thick succession of dark coloured hard sandstones, purple shales and pseudo-conglomerates. The Kamlial Formation exposed in the Jawalamukhi area in Himachal Pradesh consists of green sandstones. The **Chinji Formation** named after the type area in Pakistan comprises 400 to 1800 metre thick bright red nodular shales, clay, grey sandstones and conglomerates. The formation is exposed in Ramnagar, Tawi Valley and Poonch areas of Jammu.

The **Nagri Formation** first described from the village Nagri in Attock district (Pakistan) consists of intercalated massive micaceous sandstones, clays, shales and conglomerates. The formation is well exposed in the Tawi Valley (Jammu) and Hartalyangar areas (H.P.). The overlying **Dhokpathan Formation** named after a village of that name in Pakistan consists of a succession of variegated coloured sandstones, shales and clays with some beds of conglomerates. The formation exposed in the Hartalyangar area laterally grades into the upper parts of the Nagri Formation.

The transition between the Middle and the Upper Siwaliks is marked with a phase of folding. The **Tatrot Formation** well exposed in the north of Chandigarh and Narayangarh Tehsil of the Ambala District comprises soft grey, brown to red massive sandstones, silts, variegated clays and conglomerates. The faunal assemblage of the formation contains a number of genera common to the underlying Dhokpathan Formation. The fauna is suggestive of a Late Pliocene age for the formation. The Pinjor and the Boulder Conglomerate Formations comprising the upper part of the Upper Siwalik has been assigned a Quaternary (Pleistocene) age. The Neogene-Quaternary boundary defined on the faunal evidence either lies in the basal part of the Pinjor Formation or at the contact of the Pinjor and Tatrot Formations.

The **Pinjor Formation** named after a town near Kalka in Haryana comprises a succession of coarse grit, siltstones, clay and claystones intercalated with several beds of conglomerates. The red and pink colours are characteristics of the fine sediments of the formation. The conglomerates become frequent towards the top of the succession. Halstead and

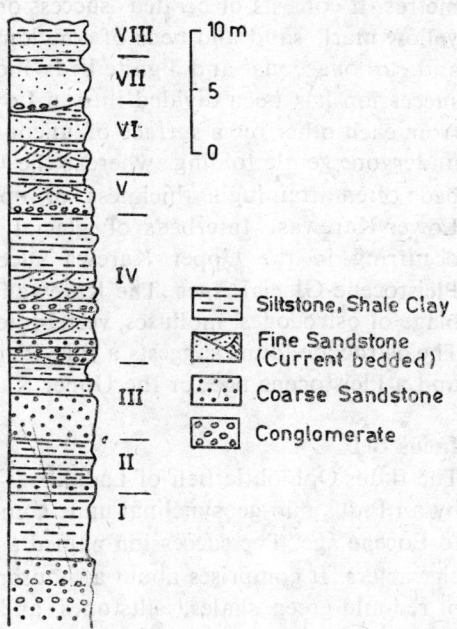

Fig. 11.6 : Cyclothems in the Pinjor Formation (based on Halstead and Nanda, 1973).

Nanda (1973) have shown the cyclic nature of the Pinjor succession deposited in an environment of flat flood plain with meandering rivers. Eight cyclothems representing eight major cycles of deposition have been recorded from a section near Chandigarh (Fig. 11.6). The overlying **Boulder Conglomerate Formation** contains boulders, pebbles and cobbles of granites, quartzites, slates and limestones derived from the fast eroding Lesser and Higher Himalayas. The formation generally occurring at the point of emergence of large rivers grades laterally and upwards into the Holocene alluvial sediments of the Indo-Gangetic Plain.

Karewa Formation: The term "Karewa" in colloquial language of the Kashmir Valley means terraces. The formation consists of lacustrine and fluvial sediments intercalated with glacial tills of Pliocene-Pleistocene age. The basin of deposition originally covered an area of about 3600 square kilometres. The sediments were derived from the northern Higher Himalayan and the southern Pir-Panjal Ranges that had emerged as the dominating orographic features at the onset of the filling of the molassic basin.

The Karewa Formation attains a maximum thickness of about 2000 metres. It consists of bedded succession of grey and buff coloured silt, yellow marl, sand and beds of conglomerate, varved clays of glacial origin and carbonaceous and lignite beds indicating warm humid climate. The succession has been divided into a Lower and an Upper Units separated from each other by a surface of unconformity. The Lower Karewas have undergone gentle folding whereas the Upper Karewas lie flat. The lignite beds often attaining a thickness of up to 3 metres are frequent in the Lower Karewas. Interbeds of glacial tills and varved clays frequently occurring in the Upper Karewa succession represent the relics of the Pleistocene Glacial Ages. The Karewa Formation has yielded a rich assemblage of ostrocodes, molluscs, vertebrates, plant fossils, pollens and spores. The ostrocode fauna suggests a Plio-Pleistocene age for the Lower Karewas and a Pleistocene age for the Upper Karewas (Bhatia, 1969; Singh, 1971).

Indus Belt

The Indus Ophiolitic Belt of Ladakh (Chapter 9) is bounded in the north by an Indus mio-geosynclinal unit comprising rocks of Late Cretaceous to Eocene age. The succession named as **Indus Formation** has a flyschoidal characters. It comprises about a thousand metre thick marine succession of red and green shales, siltstones and conglomerates. The rocks show well preserved sedimentary structures such as graded bedding, sole marks, ripple marks, flute casts and cross-bedding. The lower beds of the succession have yielded *Orbitolina parma* and *O. discoidea*, gastropods and ostrocods indicating a Late Cretaceous age for the beds. Presence of *Nummulites* spp. and *Assilina* spp. from the upper beds suggest an Eocene age for the upper parts of the succession.

The Indus Formation is disconformably overlain by molassic sediments deposited in intra-deep and back deep molassic basins of Ladakh. The succession named as **Karoo Formation** comprises about 800 m thick succession of intercalations of shales, sandstones and conglomerates. The upper part of the formation known as **Hemis Conglomerate** has a transgressive contact over the Ladakh Batholith. The Karoo Formation has yielded a fossil assemblage comprising *Unio, Melania, Viviparous* sp., *Planorbis* sp., *Hypoboops* and a fan palm. The assemblage indicates a Miocene age for the formation.

Deccan Traps

The Deccan Traps extend over 500,000 square kilometres in the northern and western parts of the Peninsula and attain a maximum thickness of about 2000 metres in Western Ghats. The traps lie flat as horizontal sheets much like the Tertiary lavas of Iceland, Skye and Mull. They are believed to have erupted sub-aerially through the fissures in the earth's crust. Such fissures now exposed as dikes are mostly seen in Satpuras, Konkan and Gujarat. Central type of volcanic activity that took place during the Deccan lava emission is evidenced by the presence of volcanic vents and related features. Lonar lake situated about 120 km from Aurangabad presumably represents the remnant of such a volcanic crater.

The tectonic disturbances related with the fragmentation of the Gondwanaland at the close of the Jurassic Period were associated with volcanic activity represented by the Rajmahal and Sylhet traps of Eastern India. The Deccan volcanism is regarded as the continuation of this tectonic disturbance during the Early Palaeocene Epoch. Radiometric dating tends to establish that the Deccan Trap volcanism took place in a relatively shorter span of time. The fissures along which the lava found its way on to the earth's surface trend parallel to the major lineaments of the Precambrian Basement. Several dike systems and fissure zones trending parallel to the ENE-WSW Son-Narmada-Godawari lineament, N-S Konkan lineament and NNE-SSW Cambay Graben lineament are known to occur in thick pile of lava flows. Deep seismic studies have shown that some of the major fractures of the lineaments extend down to the earth's mantle.

The results of the palaeomagnetic studies made on the Deccan Traps suggest that the geomagnetic field reversed its polarity several time during the eruption of the lavas (Fig. 11.7). The reversed flows dominate the lower parts of the succession whereas the upper parts of the lava succession contain the relics of the normal polarity. The palaeomagnetic data also seem to suggest that the continental drift occurred over a protracted period of time punctuated by a comparative cessation of the movement when the lava extrusion took place (Pal and Bhimsankaran, 1971).

The lower and upper parts of the Deccan Trap succession can also be differentiated on the basis of their differing chemical characters (Ghosh,

1976). The lower unit exposed in the eastern and southern parts of the Deccan country is composed of uniform horizontal thoelitic flows representing the quiet type of eruptions. The upper unit exposed in the northern parts of the Deccan country is characterised by an explosive activity.

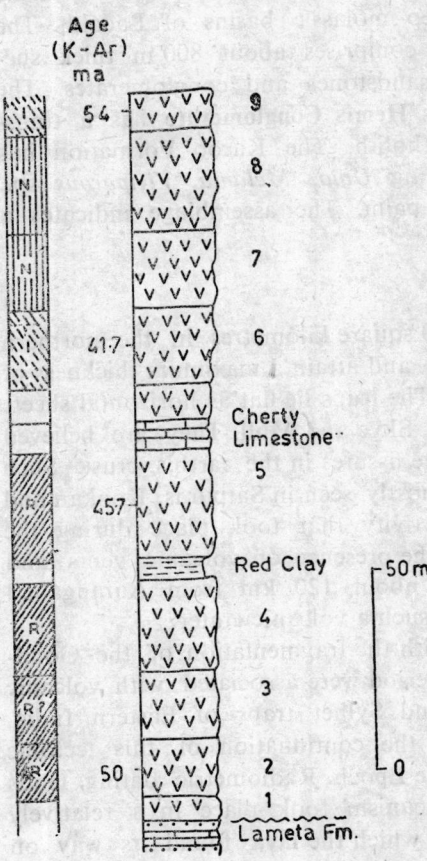

Fig. 11.7 : Reversed (R) and normal (N) magnetic polarity in Deccan Trap succession (based on Alexander, 1981).

The trap succession has been generally classified into three stratigraphic units. The lower about 150 m thick succession of traps exposed in parts of Madhya Pradesh and eastern areas are associated with numerous inter-trappean beds. The middle unit comprising about 1200 m thick lava flows and ash beds is practically devoid of inter-trappean beds. This unit is exposed in Central India and Malva region of the northern Peninsula. The upper unit of the succession consisting of about 450 m thick sequence of lava flows with numerous inter-trappean beds is exposed in the northwestern parts of the Peninsula.

The intertrappean beds of the trap succession consist of black, cherty detritus and carbonate deposits of lacustrine and fluviatile origin. The individual beds usually 1 to 3 m thick extend laterally for 5 to 8 km. The beds contain a rich record of faunal and floral remains. The flora is rich in palm (*Palmoxylon*). The invertebrate fauna include *Physa* (*Bullinus*) *princepi*, *Lymnaea*, *Unio*, *Paludina*, *Valvata*, *Melania*, *Natica*, *Vicarya*, *Gerithium*, *Turritella*, *Pupa* and *Cypris*. The vertebrates are represented by some insects and fragments of fishes, frogs and tortoise.

The Deccan Trap overlie the Bagh and Lameta Beds that have been assigned Turonian age (Chapter 9). The *Cardita beaumonti* Beds of Sind (Pakistan) which are interstratified with a few flows of traps of Deccan Trap affinity have been assigned a Danian age. The Deccan Traps of Saurashtra region have been unconformably overlain by the Nummulitic rocks of Early Eocene age. These stratigraphic relationships establish a Palaeocene age for the volcanic episode.

Assam-Arakan Region

The Assam-Arakan mountain belt of the Eastern India and its adjoining regions of Shillong-Mikir Plateau (Fig. 11.8) expose a vast thickness of Tertiary and Quaternary rocks. The thickness of individual stratigraphic units vary in different areas of the region. Larger thicknesses are generally observed in the geosynclinal areas of the Arakan mountains. Some stratigraphic units while having conformable relationships with the overlying

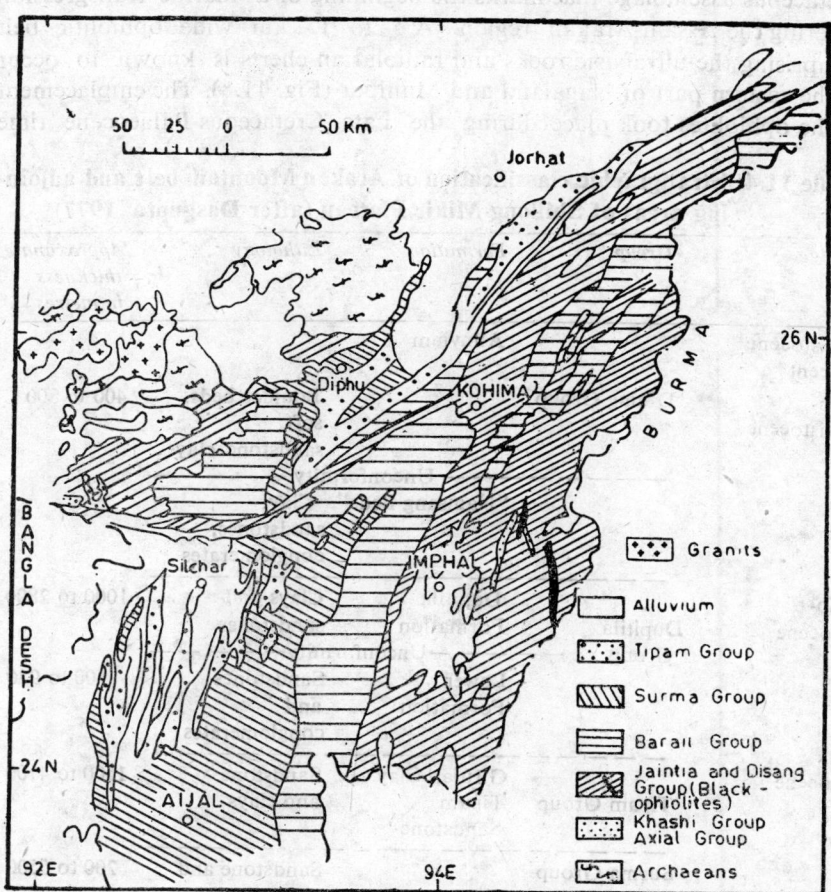

Fig. 11.3: Geological map of the northern parts of Assam-Arakan Region (based on Dasgupta, 1977; political boundaries are approximate).

and underlying units in the geosynclinal part of the region have unconformable contacts in the shelf areas of Assam and Shillong plateau. The Cenozoic rocks of the region were laid over a basement comprising Precambrian rocks and Lower Gondwana sediments. The basement in the core of the Arakan mountains is believed to include the representatives of

Upper Triassic. A group of low to medium grade quartzites, schists, limestones with sheared granites and minor serpentinites known as Naga Metamorphics is exposed in a linear overthrust belt along the Indo-Burma border (Brunnschweiler, 1966; Chattopadhyay et al., 1983). This sequence is unfossiliferous and it has been correlated with the Precambrian or Lower Palaeozoic rocks of Burma.

The oldest fossiliferous sequence of the region contain an Upper Cretaceous assemblage that marks the beginning of a marine transgression covering the Assam-Arakan region. A 5 to 15 km wide ophiolitic belt comprising the ultrabasic rocks and radiolarian cherts is known to occur in the eastern part of Nagaland and Manipur (Fig. 11.8). The emplacement of the ophiolites took place during the Late Cretaceous-Palaeocene time

Table 11.4: Stratigraphic classification of Arakan Mountain belt and adjoining areas of Shillong-Mikir Plateau (after Dasgupta, 1977)

Age	Group	Formation	Lithology	Approximate thickness in metres
Pleistocene Recent		Alluvium		
Plio-Pleistocene	Dihing Group		Boulder beds, soft sandstone clay	400 to 800
	————————Unconformity————————			
		"Namsang Beds"	Clay, sandstones, conglomerates	
Mio-Pliocene	Dupitila Group	Dupitila Formation	Clays and sandstones	1000 to 2800
		————Unconformity————		
		Lower Formation	Sandstones and conglomerates	500 to 930
Miocene	Tipam Group	Girujan Clay Tipam Sandstone	Sandstone and clays	1500 to 4100
	Surma Group		Sandstone and shales	200 to 5500
Oligocene	Barail Group		Sandstones and shales	1200
Palaeocene-Eocene	Jaintia Group (Disang Group in Naga-Paktoi hills)	Kopili Shales Sylhet Limestone	Shales Limestone & Sandstones	380 to 800
Cretaceous		Eastern Disang Group		

(Chattopadhyay et al., 1983). The emplacement of the ophilites also coincided with a phase of regional metamorphism that took place prior to the deposition of Upper Eocene rocks (Ranga Rao, 1983). During the Eocene time, a general shallowing the Assam-Arakan basin has been noticed and the close of the Eocene is marked by the uplift of the axial zone of the Indo-Burma region represented by the ophilites and associated rocks. The sea gradually regressed during the Miocene time as the Plio-Pleistocene sequence of the region consists of predominantly deltaic and channel facies.

Upper Cretaceous-Eocene: The Cretaceous-Eocene sequence exposed in the southern and southeastern margin of the Shillong Plateau consists of Mahadek Formation overlain by Langpar Formation. The 230 m thick succession of the **Mahadek Formation** consists of boulder beds overlain by some hard massive cliff forming gritty glauconitic sandstones with a thin bed of softer sandstone at the top. The soft sandstone has yielded Maastrichtian fauna. The micro-faunal assemblage of the Mahadek Formation includes the species of *Bolivine, Buliminella, Clavulinoides, Globotruncana, Gumbelina, Gumblitria, Orbitoides, Pseudolextularis, Siderolites* and *Ventilabrella*. The faunal assemblage persists in the lower parts of the overlying **Langpar Formation** that comprises a 200 m thick succession of impure sandy limestones and shales containing molluscan shells of Danian age. The micro-faunal assemblage of the Langpar Formation include *Angulogerina, Anomalina, Cibicides, Coleites, Globigerina, Trifarina, Veginulina* and *Valvulineria*. The faunal assemblages of Mahadek and Langpar Formations indicate shallow water open sea environments of deposition.

The **Jaintia Group** that conformably overlies the Langpar Formation has been subdivided into two formations. The **Sylhet Limestone** comprising the lower formation consists of alternating four limestone sequences separated by beds of sugary sandstones. The four carbonate sequences have been correlated with Lower Ranikot (Lower Palaeocene), Upper Ranikot (Upper Palaeocene), Laki (Lower Eocene) and Kirthar (Middle Eocene) successions of the western India. The limestone succession has yielded microfaunal assemblages indicating warm clear shallow water marine environment. The assemblages include species of *Discocyclina, Opertorbitolites, Alveolina, Orbitolites, Assilina, Astercyclina, Nummulites,* etc. The limestone-sandstone succession of the Sylhet Limestone indicates alternating phases of peneplanation and uplift of the Shillong Plateau region that formed the source area for the supply of the sediments. In Mikir Hills, the Sylhet Limestone is underlain by a typical transgressive sequence of sandstones and shales with coal beds known as Tura Formation. The formation has yielded a faunal and floral assemblage indicating a Palaeocene to Early Eocene age (Ranga Rao, 1983).

The Sylhet Limestone is conformably overlain by 380 to 800 m thick **Kopili Formation** in the southeastern part of the Shillong Plateau. The formation consists predominantly of sandstones in the lower parts overlain by sand shale alternations. In Upper Assam, the Kopili Formation consists predominantly of shales with a few fossiliferous limestone beds near the top of the succession. The fossil assemblage includes the mixture of calcareous and arenaceous micro-forms indicating a Late Eocene age. The assemblage comprises the species of *Asterocyclina, Cancris, Globigerina, Rotalia* and others.

The Upper Cretaceous-Eocene sequence of the Arakan mountains comprises about 3000 m thick succession of geosynclinal sediments known as **Disang Group**. The succession named after the Disang river (Mallet, 1876) has yielded scanty fossil record suggesting Late Cretaceous-Eocene age range for the Group. The Disang Group consists of fine clastics derived from a source other than the Shillong Plateau. The provenance for these sediments were presumably located in the Mishmi Hills or Burma. In the western and central parts, the Disang Group is composed of splintery shales with some thin bands of hard flaggy sandstones. Eastward, the rocks progressively acquire metamorphic characters changing into slates, phyllites and schists. The eastern outcrops also include beds of limestones, conglomerates, grits, tuffs and intrusives of ultrabasics and serpentinous rocks. The upper part of the Disang succession has yielded a foraminiferal assemblage that includes *Ammobaculites, Ammodiscus, Bathysiphon, Gaudryina, Haplophragmoides* and *Cyclammina* (Nagappa, 1959). The assemblage suggests an age "not younger than Upper Eocene".

Oligocene: The Jaintia and Disang Groups are conformably overlain by about 1200 m thick succession of alternating hard sandstones and shales. The succession has been named as **Barail Group** after the Barail Range skirting southeastern margin of the Shillong Plateau. The succession of the Barail Range represents a shelf facies whereas the Barails in the north-east are primarily coarse sandstones with some coal shale of coastal facies. The succession in the 'Kohima Syncline' in the south comprises interbedded shales and sandstones containing some thick coal seams in the upper part of the succession. The Barails of the Garo Hills have yielded micro-faunal assemblage comprising *Cyclammina, Miliammina, Haplophragmoides,* and *Trochammina* with rare Rotalids and globular indeterminate bodies. The arenaceous foraminifers are not considered as age diagnostic (Ranga Rao, 1983). The Barail succession has been assigned an Oligocene age as this succession overlies the Upper Eocene rocks comprising the Kopili Formation.

Miocene: The Miocene succession of the region has been subdivided into Surma and Tipam Groups. The succession is exposed in a large area extending southward from the centre of the Kohima Syncline through Surma Valley, Tripura and Chittagong to Arakan mountains of Burma. The **Surma Group** ranging in thickness from 2900 m to more than 4000 m comprises a succession of alternating sandstones and shales. The rocks unconformably overlie the Barail Group. The Surma succession has yielded shallow brackish water micro-faunal assemblage indicating a Miocene age. The assemblage includes *Ammodiscus, Clavulina, Gadryina, Haplophragmoides, Textularia, Angulogerina, Anomalina, Bolivina, Globogerina* and *Rotalia*.

The **Tipam Group** that conformably overlies the Surma Group consists of 3000 m to 4000 m thick succession of massive sandstones with subordinate clays and shales. The Tipam succession that was first described from the Tipam river section in Upper Assam (Mallet, 1876) has been subdivided into two formations. The lower formation known as **Tipam Sandstone** consists of massive sandstones with subordinate clays and shales. The overlying formation known as **Girujan Clay** comprises variegated and mottled clays with some impersistent interbeds of argillaceous sandstones. The Tipam and overlying successions are virtually barren of microforms although some beds have yielded useful spores and pollens. The Girujan clay has yielded some plant fossils that have been identified as belonging to *Terminalia* sp. (Ranga Rao, 1983).

Mio-Pliocene: The **Dupi-Tila Group** that conformably overlies the Girujan Clay has been assigned Mio-Pliocene age on indirect evidence. The Dupi-Tila rocks are exposed in the northern Cachar district of Assam. The succession has been subdivided into a lower and an upper formations. The lower formation consists of sandstones and conglomerates whereas the upper formation comprises a succession of mottled clays and variegated sandstones with occasional pebbles of quartz. The two formations are separated from each other by an angular unconformity that has been attributed to a folding phase prior to the deposition of the Upper Dupi-Tila Formation. The angular discordance becomes progressively less marked over the limbs of the major anticlinal folds.

Plio-Pleistocene: The Girujan Clay is unconformably overlain by a sequence of mottled clay, sandstone and numerous conglomerate beds exposed between Langting and Dihing valleys. The succession named as **Namsang Beds** reflect a fresh phase of uplift in Naga-Paktoi hiterlands. The conglomerates contain pebbles of coal suggesting that the Barail succession was uplifted above the erosion level during deposition of the Namsang Beds.

The Namsang Beds are unconformably overlain by a succession of well rounded boulder and pebble beds with interspersed lenses of soft sands and clays. The succession best exposed in the Dihing Valley has been named as **Dihing Group**. The rocks of the group have undergone the last phase of folding of Arakan mountains acquiring often steep dips. The succession has been unconformably overlain by a series of boulder terraces of Pleistocene to Holocene age.

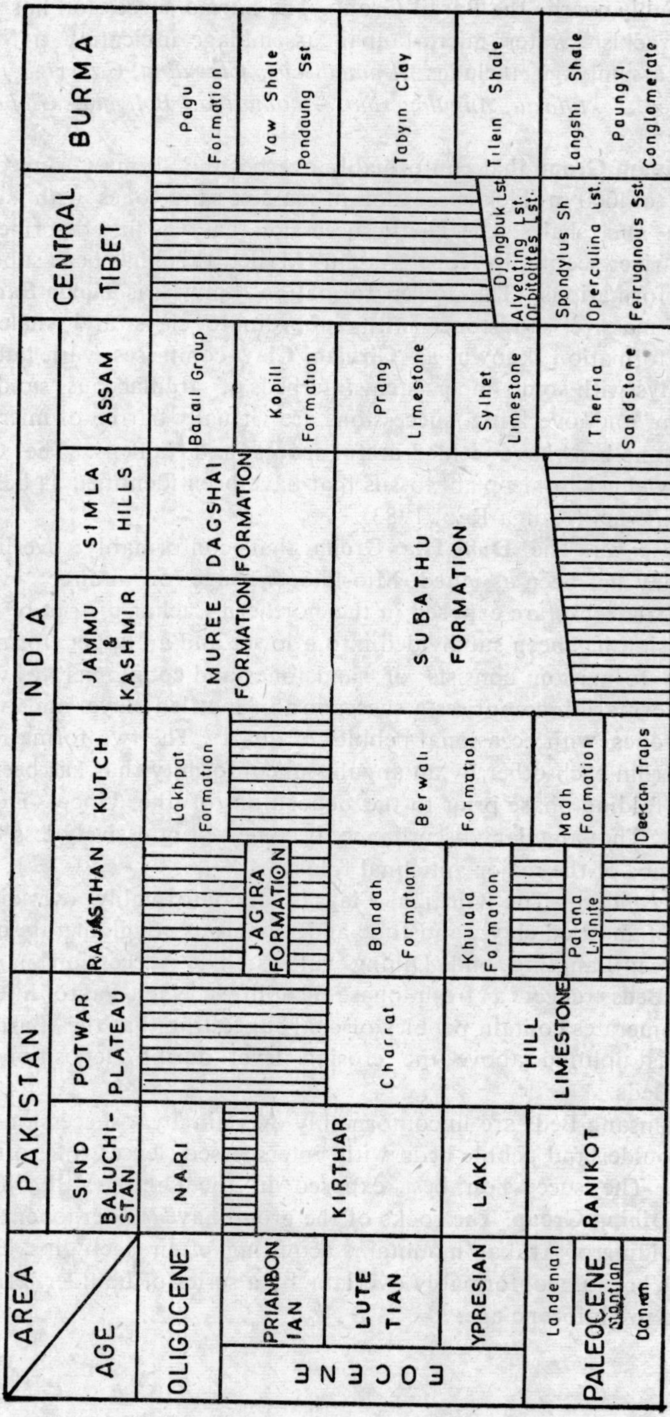

218 Stratigraphy of India

AGE \ AREA	PAKISTAN			INDIA					CENTRAL TIBET	BURMA
	SIND BALUCHISTAN	POTWAR PLATEAU	RAJASTHAN	KUTCH	JAMMU & KASHMIR	SIMLA HILLS	ASSAM			
OLIGOCENE PRIABONIAN	NARI	═══	═══		MUREE FORMATION	DAGSHAI FORMATION	Barail Group	═══		Pagu Formation
			JAGIRA FORMATION	Lokhpat Formation						Yaw Shale
EOCENE LUTETIAN	KIRTHAR	Charrat	Bandah Formation	═══			Kopill Formation	═══		Pondaung Sst
	═══			Berwali Formation			Prang Limestone	═══		Tabyin Clay
YPRESIAN	LAKI	HILL LIMESTONE	Khuiala Formation		SUBATHU FORMATION		Sylhet Limestone		Djongbuk Lst. Alyeating Lst. Orbitolites Lst.	Tilein Shale
			Palana Lignite	Madh Formation					Spondylus Sh	Langshi Shale
PALEOCENE Landenian	RANIKOT		═══	Deccan Traps			Therria Sandstone		Operculina Lst.	Paungyi Conglomerate
Montian Danian									Ferruginous Sst	

Fig. 11.9: Correlation of the Palaeogene sediments of the Indian sub-continent (based on Sahni and Kumar, 1974).

Andaman-Nicobar Islands

The Cenozoic rocks of the Andaman-Nicobar islands belong to the Arakan folded belt of Eastern India and Burma (Fig. 4.3). The Cenozoic succession conformably overlying the Cretaceous rocks has been divided into Upper Cretaceous-Palaeogene Group and Archipelago Group (Table 11.4).

Upper Cretaceous-Palaeogene Group: The oldest rocks exposed on the Andaman-Nicobar islands consist of an Upper Cretaceous succession known as **Port Meadow Formation**. The formation comprises a succession of radiolarian chert, jaspers, quartzites, dark shales and clays associated with ophilitic suite of ultrabasic and basic lava flows. The Port Meadow Formation is conformably overlain by the **Baratang Formation** which consists

Table 11.5: Classification of Cenozoic succession of Andaman-Nicobar Islands (based on Srinivasan and Azmi, 1976 and Roy, 1983)

Group	Formation	Lithology	Age
	Shell limestone, beach sand	Coral rag and sand	Sub-recent to Recent
	— — — — — Unconformity — — — — —		
	Neil West Coast Formation	Bio-calcarenite	Late Pliocene to Pleistocene
	Guitar Formation	Calcareous mudstone	Early to Late Pliocene
	Sawai Bay Formation	Calcareous mudstone	Late Miocene to Early Pliocene
Archipelago Group	Long Formation	Calcareous siltstone and silty mudstone	Middle to Late Miocene
	Inglis Formation	Calcareous Chalk	Early Miocene
	Strait Formation (Round Formation)	Shell sandstone, limestone and conglomerates	Early to Middle Miocene
	— — — — Unconformity — — — —		
Upper Cretaceous-Palaeogene Group	Port Blair Formation	Conglomerate, grit and greywacke	Middle Eocene to Oligocene
	Bartang Formation	Shale-greywacke	Late Cretaceous to Eocene
	Port-Meadow Formation	Radiolarian chert, quartzites, shale with ophiolites	Late Cretaceous

of about 1000 m thick turbidite sequence of dark grey shales, greywackes and some limestones. The lower beds of the succession have yielded *Globotruncana* assemblage of Late Cretaceous age whereas the upper beds contain Late Eocene foraminifera (Pandey, 1972).

A flyschoidal succession known as **Port Blair Formation** conformably overlies the Bartang Formation. The group exposed in the Andaman islands consists of about 700 m thick sequence of conglomerates, grits and greywackes with a few interbeds of shales and limestones. The major upper part of the Port Blair Formation is unfossiliferous. The underlying conglomerates and grits have yielded Middle Eocene microfauna suggesting a Middle Eocene to Oligocene age for the Port Blair succession.

Archipelago Group: The Palaeogene phase of sedimentation was terminated by a phase of folding and uplift during the late Oligocene time (Roy, 1983). The Neogene succession named as the **Archipelago Group** after the Rithchie's Archipelago in the Andaman sea (Oldham, 1885) consists of about 1650 m thick sequence of sandy limestone, marl, siliceous and calcareous chalk, silt and silty mudstone. The base of group though not well exposed is presumed to unconformably overlie the Oligocene rocks. The Archipelago succession has been divided into six formations (Srinivasan and Azmi, 1976). The formations have yielded a rich micro-faunal assemblage containing abundant planktonic foraminifera which help in assigning stratigraphic ages to the formations (Table 11.5).

Northwestern Peninsula

After a brief phase of marine regression towards the close of the Cretaceous Period, the northwestern Peninsula was once again submerged by a shelf sea during the Tertiary Period (Fig. 11.1). The basin extended to Sind and Baluchistan regions of Pakistan. The Tertiary rocks exposed in the hill ranges of Sind and Baluchistan have been classified into six formations (Table 11.6). The fossil assemblages characterising the six formations have also been identified in the Tertiary formations of the northwestern and other parts of India. The stratigraphic units of Sind and Baluchistan have acquired a significance similar to that of the units of the Standard Stratigraphic Scale for the purpose of correlation of the rocks of the Indian subcontinent.

Western Rajasthan: The Palaeogene succession exposed in Bikaner, Barmer and Jaisalmer districts have been classified into Palana, Khuiala, Bandah and Jogira Formations (Fig. 11.9). The **Palana Formation** comprises a clay predominant succession intercalated with friable and ferruginous sandstone of continental origin. A thin band of lignite encountered in the bore holes occurs in the lower part of the succession. The Palana Formation is unfossiliferous except for the lignite bed which contain fresh water alga *Botryococcus braunii* and pollen comparable to Potamogeton natans (Vimal, 1953) indicating a Palaeocene age.

Table 11.6: Classification of Tertiary succession of Sind, Pakistan (based on Krishnan, 1968)

Formation	Member	Lithology	Age
Manchar Formation (3050 m)	Upper Manchar	Grey sandstone and conglomerate	Pliocene
	Lower Manchar	Conglomerates	Middle to Late Miocene
Gaj Formation (455 m)		Yellow limestone and shales	Early Miocene
Nari Formation (1830 m)	Upper Nari	Fluviatile sandstone	Late Oligocene
	Lower Nari	Marine limestone	Early Oligocene
Kirthar Formation (900 to 2800 m)		Massive nummulitic limestone, shale, sandstone	Middle Eocene
Laki Formation (150 to 250 m)		Shales with coal, *Aveolina* limestone	Early Eocene
	Unconformity		
Ranikot Formation (610 m)	Upper Ranikot	Brown limestone & Shale	Late Palaeocene
	Lower Ranikot	Fluviatile shales and sandstones with gypseous and carbonaceous beds	Early Palaeocene
	Cardita beaumonti Beds	Shales and sandstones	
Basement of Conformable Cretaceous Rocks			Late Cretaceous

The Palana Formation is unconformably overlain by the **Khuiala Formation** that marks the first marine transgression in the Western Rajasthan during the Early Eocene. The formation consists of about 100 m thick fossiliferous succession of calcareous shales, limestones, fuller's earth, red shales and sandstones. The succession has been subdivided into six zones named as *Assilina granulosa* Zone, *Assilina deviesi* Zone, *Assilina lacunata* Zone, *Venericardia mutabilis* Zone, *Assilina* sp. Zone and Berren Zone (Khosla, 1977). The faunal assemblages indicate a warm shallow transgressive marine environment of deposition. The Barren Zone occurring at the top of the formation marks the regressive phase of the shelf sea.

The marine conditions were re-established during the deposition of the **Bandah Formation** of Middle Eocene age. The formation comprises an arenaceous limestone succession that becomes ferruginous in the upper parts. Two zones, namely, *Discocyclina sella* Zone and *Nummulites* *macclatus/Discocyclina dispansa* Zone, have been recognised in the Bandeh succession (Khosla, 1977). The microfaunal record suggests a warm shallow infra-neritic marine environment of deposition.

The **Jogira Formation** comprises 13 to 20 m thick unfossiliferous succession of greyish white grit and friable sandstone in the lower part overlain by calcareous and ferruginous nodular beds in the upper part. In the Jaisalmer area, the formation rests unconformably over the Bandeh Formation which is completely overlapped in Bikaner area bringing the Jogira Formation to rest directly over the Khuiala Formation. The Jogira Formation has been tentatively assigned a Late Eocene age.

Saurashtra-Kutch: Marine Tertiary rocks are exposed in the Surat Broach region of the Southern Gujarat. The Tertiary rocks have also been encountered in the drill cores of the oil producing Cambay region. The Tertiary succession exposed in the western and southern parts of Kutch attains a maximum thickness of about 1000 m. The Palaeogene rocks of the succession have been grouped into Madha, Berwati and Lakhpat Formations (Fig. 11.9), and the Neogene rocks are classified into Khari and Kankawati Formations (Fig. 11.10).

The **Madh Formation** consists of volcano-clastic sequence deposited in continental fluviatile and marine littoral basins. The volcanic sediments were presumably derived from the vaning phase of volcanism associated

AGE \ AREA			PAKISTAN: SIND & BALUCHI-STAN	PAKISTAN: POTWAR PLATEAU	INDIA: SIMLA & GARHNAL	INDIA: JAMMU & KASHMIR	INDIA: KUTCH	INDIA: KATHIA-WAR	INDIA: ASSAM	BURMA
Pleistocene			Older Alluvium	Alluvium / Boulder Congl. Fm. / Pinjor Fm.	Alluvium / Boulder Congl. Fm. / Pinjor Fm.	Karewa Formation	Pobanda Limeston	Pobanda Limeston	Older Alluvium / Dihing Group	Alluvium
Pliocene		Piacenzian	Upper Manchar	Tatrot Fm. / Dhok Pathan Fm	Tatrot Fm. / Dhok Pathan Fm	Karewa Formation	Kankaneti Formation	Dwarka Beds	Dihing Group	Irravady Group
Pliocene		Zanclean	Upper Manchar	Nagri Formation	Nagri Formation	Nagri Formation	Kankaneti Formation	Preram Beds	Tipam Group	Irravady Group
MIOCENE	MIDDLE & UPPER	Messinian	Lower Manchar	Chingi Formation	Chingi Formation	Chingi Formation	Khari Formation	Gogha Beds	Tipam Group	Irravady Group
MIOCENE	MIDDLE & UPPER	Toronian	Lower Manchar	Chingi Formation	Chingi Formation	Chingi Formation	Khari Formation	Gogha Beds	Tipam Group	Irravady Group
MIOCENE	MIDDLE & UPPER	Serravallia	Lower Manchar	Kamlial Formation	Kamlial Formation	Kamlial Formation	Khari Formation	Gogha Beds	Tipam Group	Upper Pegu Group
MIOCENE	MIDDLE & UPPER	Langhian	Lower Manchar	Kamlial Formation	Kamlial Formation	Kamlial Formation	Khari Formation	Gogha Beds	Surma Group	Upper Pegu Group
MIOCENE	LOWER	Burdigalia	Upper Gaj	Muree Formation	Kasauli Formation	Murree Formation	Khari Formation	Upper Gaj	Surma Group	Upper Pegu Group
MIOCENE	LOWER	Aguitanian	Lower Gaj	Muree Formation	Kasauli Formation	Murree Formation		Lower Gaj	Surma Group	Upper Pegu Group

Fig. 11.10: Correlation of the Neogene sediments of the Indian sub-continent (based on Sahni and Mitra, 1980).

with the Deccan Traps which conformably underlie the Madh Formation. The formation has yielded well preserved dicot leaf impressions, spores and pollens indicating a Palaeocene age for the formation. The marine beds have yielded a rich micro-faunal assemblage that includes *Globoconusa, Globigerina, Globigerinoides, Globorotalia, Subbotina,* etc.

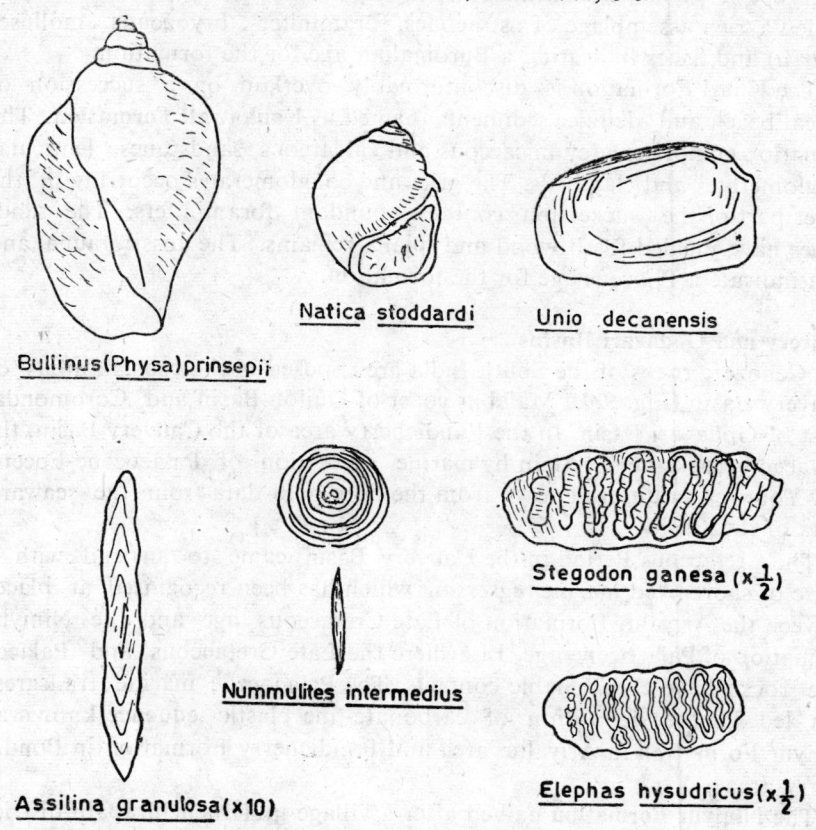

Fig. 11.11: Tertiary fossils from India.

The **Berwali Formation** consists of gypseous and ocherous clay, lignite, oolitic sandstone and marl containing *Assilina granulosa* and molluscs in the lower part overlain by dense fossiliferous fragmental limestone. The micro-faunal assemblage of the lower part of the succession is suggestive of littoral to epi-neritic environment of deposition and an Early Eocene age for the lower part of the formation. The fragmental limestone occurring in the upper part of the Berawali succession contains a basal bed of calcareous clay yielding a rich assemblage of Middle Eocene ostracodes.

The **Lakhpat Formation** consists of greenish grey marl and argillaceous limestone succession with a basal bed of bouldery clayey marl. The bouldery bed characterises the unconformity between the Lakhpat Formation and the underlying Berwali Formation. The Lakhpat succession has yielded foraminifers and ostrocodes of Oligocene age.

The Khari Formation resting unconformably over the Lakhpat Formation comprises variegated siltstones and grey gypseous marl deposited in quiet epi-neritic marine basin. The upper part of the Khari Formation has yielded a rich assemblage of ostracodes, foraminifers, bryozoans, molluscs (*Ostrea*) and fishes indicating a Burdigalian age for the formation.

The Khari Formation is disconformably overlain by a succession of typical beach and deltaic sediments named as **Kankawati Formation**. The formation comprises grey micaceous and calcareous sandstones, lenticular conglomerates and clay beds. The grits and conglomerates occurring in the upper part of the succession contain abundant foraminifers. The sandstones have yielded fossil wood and plant remains. The fossil fauna and flora indicate a Pliocene age for the formation.

Cauvery and Godavari Basins

The Cenozoic rocks of the South India are exposed in the coastal areas of Cauvery Basin (Fig. 9.9), Malabar coast of Quilon Basin and Coromondal coast of Godavari Basin. In the Pondicherry area of the Cauvery Basin, the Cretaceous rocks are overlain by marine succession of Palaeocene-Eocene age. Younger strata are known from the bore-hole data from the seaward parts of the Basin.

The Cretaceous Period in the Cauvery Basin came to an end with a phase of short-lived marine regression which has been recognised at places between the Ariyalur Formation of Late Cretaceous age and the Niniyur Formation of Palaeocene age. Elsewhere the Late Cretaceous and Palaeocene rocks have conformable contacts. The Palaeocene marine transgression led to the deposition of carbonate-fine clastic sequence known as Niniyur Formation in Ariyalur area and Pondicherry Formation in Pondicherry area.

The **Niniyur Formation** named after a village in Tiruchchirapalli district in Tamil Nadu is mostly composed of about 230 m thick succession of purple and grey ocherous clayey sandstones and cream and buff coloured fossiliferous limestone with abundant algal structures. The algal limestone were deposited in algal banks and shelf lagoons. The formation has yielded cephalopods *Nautilus* (*Hercoglossa*), lamellibranchs *Tellina, Lucina* (*Codakia*) and *Cardita*, gastropods *Psedoliva, Euspira, Solarium, Lyria* and *Turritella*, corals and algae. The formation is also known to contain ostracodes and foraminifers. The cephalopod fauna and the planktonic foraminifers indicate a Danian age for the beds yielding the fauna.

The **Pondicherry Formation** consists of about 70 m thick succession of marls and algal foraminiferal limestones. The formation conformably overlies the Cretaceous rocks and it is unconformably overlain by the rocks of Miocene age (Cuddalore Sandstone). The Pondicherry Formation is regarded as a facies equivalent of the Niniyur Formation. The formation has yielded Danian forminifers, ostracodes, bryozoans and algae.

Lower Eocene rocks have been recognised in the Cauvery Basin on the evidence of the occurrence of certain Nummulitic limestones in the Pondicherry area (Rama Rao, 1939). Middle and Upper Eocene rocks have also been identified on the evidence of the presence of *Discocyclina* and other typical fossils (Gowda, 1964). Part of the Cauvery Basin seems to have undergone erosion during the Middle Eocene-Oligocene regression.

Estuarine conditions were established in parts of the Cauvery and Godavari Basins during the Miocene Epoch. The **Cuddalore Sandstone** of the Cauvery Basin that was deposited in these estuaries consists of soft red yellow and mottled ferruginous sandstones, sandy clays, sands, clays and pebble beds. The Cuddalore Sandstone rests unconformably over the rocks of different ages, namely, the Precambrian Gneisses, the Upper Gondwana, the Cretaceous and the Palaeocene rocks. The sandstone formation has yielded algal, foraminiferal and molluscan remains indicating a Late Miocene to Pliocene age for the formation. The molluscan genera include *Ostrea, Fusus, Terebra, Oliva* and *Conus*. The Cuddolore Sandstone also contains large silicified trunks of angiosperm trees and lignite beds. The thick lignites seam at Neyveli about 35 km west of Cuddalore (Fig. 9.9) has a large economic potentiality. The sandstone formation is also known to contain artesian aquifers. In the Godavari Basin, the equivalent formation has been named as **Rajahmundri Sandstone**.

In the Godavari Basin, the Tertiary rocks consist of a trap sequence (stratigraphic equivalent of Deccan Trap) overlain by about 600 m thick sequence of coarse friable sandstones, grits and conglomerates. The later sequence known as **Rajahmundry Sandstone** was deposited primarily in continental basins. Offshore subsurface data have shown the presence of about 2000 m thick marine sequence of shales, mudstones and sandstones that laterally grades into the Rajahmundri Sandstone (Kumar, 1983). The sandstone succession of the Godavari Basin has been correlated with the Cuddalore Sandstone of the Cauvery Basin.

Bibliography of Selected References

Chapter 1

Adams, F.D.; 1938, *The Birth and Development of Geological Sciences*; Reprinted, Dover Publication Inc., N.Y., 506 pp.

Darwin, Charles; 1859, *On the Origin of Species by Means of Natural Selection*; 6th Edition, the Wath, London, viii+434 pp. (reprinted 1972, Murray, London).

Lyell, Charles; 1830-33, *Principles of Geology*; 3 volumes, various editions, John Murray, London.

Pivsoon, L.V. and Schuchert, C.; 1924, *A Text-book of Geology*; Willey, New York.

Ray, S.; 1963, *Fifty Years of Science in India*; Indian Science Congress Association, Calcutta, 194 pp.

———; 1972, *A Decade of Science in India*; Indian Science Congress Association, Calcutta, 130 pp.

Records of the Geological Survey of India; 1971, *Review Papers on Stratigraphy of India*; volume 101, Govt. of India Publication, Calcutta.

Schuchert, C.; 1924, *Historical Geology*; New York, Wiley.

Suess, E.; 1904, *The Face of the Earth*; Oxford, Clarendon Press.

Weller, J.M.; 1960, *Stratigraphic Principles and Practice*; Indian reprint, Universal Book Stall, Delhi, 725 pp.

Chapter 2

Chamberlin, T.C.; 1909, Diastrophism as the ultimate basis of correlation; *Jour. Geol.*, *17*, 685-693.

Gilluly, J.; 1949, Distribution of mountain building in geologic time; *Bull. Geol. Soc. Am.*, *60*, 561-590.

Grabau, A.W.; 1913, *Principles of Stratigraphy*; Seiler, New York (reprinted 1924), 1185 pp. (in two volumes).

Hedberg, H.D. (Editor); 1976, *International Stratigraphic Guide*; John Wiley & Sons, New York, 200 pp.

Indian Committee on Stratigraphic Nomenclature; 1971, Code of Stratigraphic nomenclature of India; *Geol. Surv. Ind. Misc. Publ.*, *20*, 28 pp.

Sheriff, R.E., 1982, *Seismic Stratigraphy*; (Indian Reprint), The English Book Depot Dehradun, 227 pp.

Simpson, G.G.; 1967, *The Meaning of Evolution*; (Indian Edition), Oxford & IBH Publishing Co., New Delhi, 368 pp.

Srinivasan, M.S. and Shrivastava, S.S.; 1975, Late Neogene bio-stratigraphy and planktonic foraminifers of Andman-Nicobar islands, Bay of Bengal; *Late Neogene Epoch Boundaries*; *Micropalaeontology Press Special Publication No. 1*, pp. 124-161.

Symposium; 1916, General considerations of palaeontological criteria used in time relations; *Bull. Geol. Soc. Am.*, *27*, 451-530.

Teichert, C., 1958, Some biostratigraphic concepts; *Bull. Geol. Soc. Am.*, *69*, 99-120.

Weller, J.M.; 1960, *Stratigraphic Principles and Practices*; (Indian reprint) Universal Book Stall, Publishers, Delhi, 725 pp.

Wheeler, H.E. and Mallory, V.S.; 1953, Designation of stratigraphic units; *Bull. Am. Assoc. Petrol. Geol.*, *37*, 2407-2421.

Williams, J.S.; 1954, Problems of boundaries between geologic systems; *Bull. Am. Assoc. Perol. Geol.*, *38*, 1602-1605.

Chapter 3

Bouma, A.H.; 1962, *Sedimentology of Some Flysch Deposits*; Elsevier, Amsterdam, 168 pp.

228 Bibliography of Selected References

Bouma, A.H. and Brouwer, A. (Editors); 1964, *Turbidites*: *Developments* in *Sedimentology*, v. 3; Elsevier, Amsterdam, 264 pp.

Grabau, A.W.; 1913, *Principles of Stratigraphy*; (Reprinted 1960), Dover, New York, 1185 pp. (in two volumes).

Hedgepeth, J.W. (Editor); 1957, *Treatise on Marine Ecology and Palaeoecology: Volume I—Ecology*; Mem. Geol. Soc. Am. 67, 1296 pp.

Hough, J.L. and Menard, H.W. (Editors); 1955, *Finding Ancient Shore Lines*; Soc. Econ. Palaeont. Min., Special Publ. no. 3, 129 pp.

Krumbein, W.C.; 1952, Principles of facies map interpretation; *Journ. Sed. Petr.*, 22, 200-211.

Krumbein, W.C. and Sloss, L.L.; 1951, *Stratigraphy and Sedimentation*; Freeman, San Fransisco.

Ladd, H.S. (Editor); 1957, *Treatise on Marine Ecology and Palaeoecology: Volume II—Palaeoecology*; Mem. Geol. Soc. Am., 67, 1077 pp.

Mathur, L.P., Rao, K.L.N. and Chaube, A.N.; 1980, Tectonic framework of Cambay Basin of India; *Bull. Oil Nat. Gas. Comm.*, 17(2), 202-222.

Moore, R.C.; 1949, Meaning of facies; *Mem. Geol. Soc. Am.*, 30, 1-34.

Pettijohn, H.J.; 1962, Palaeocurrents and palaeogeography; *Bull. Am. Assoc. Petr. Geol.*, 46, 1468-1493.

Schuchert, C.; 1927, Unconformities as seen in disconformities and diastems; *Am. Jour. Sc., Sr. 5*, 13, 260-262.

Sloss, L.L.; 1958, Palaeontological and lithological associations; *Jour. Pal.*, 32, 715-729.

Sloss, L.L., Krumbein, W.C. and Dapples, E.C.; 1949, Integrated facies analysis; *Mem. Geol. Soc. Am.*, 30, 92-123.

Sudhakar, R. and Basu, D.N.; 1973, A reappraisal of the Palaeogene Stratigraphy of southern Cambay Basin; *Bull. Oil Nat. Gas. Comm.*, 10 (1-2), 55-76.

Teichert, C.; 1958, Concept of facies; *Bull. Am. Assoc. Petr. Geol.* 42, 2718-2744.

Weller, J.M.; 1960, Stratigraphic Principles and Practices; (Indian reprint), Universal Book Stall, Delhi, 725 pp.

Wheeler, H.E.; 1958, Time in Stratigraphy; *Bull. Am. Associ. Petr. Geol.*, 42, 1047-1063.

Chapter 4

Aswathanarayanan, K.; 1954, Absolute ages of Archaean orogenic cycles of India; *Am. Jour. Sc.*, 254, 19.

———; 1964, Isotopic ages from the Eastern Ghats and Cuddapah of India; *Jour. Geophys. Res.*, 69, 3479-3486.

Balasundaram, M.S. and Balasubramanyam, M.N.; 1973, Geochronology of the Indian Precambrian; *Bull. Geol. Soc. Malayasia*, 6, 213-226.

Beloussov, V.V.; 1962, *Basic Problems in Geotectonics*; McGraw Hill, New York, 809 pp.

Berthelsen, A.; 1951, A geological section through the Himalayas; *Medd. Dansk. Geol. Foren*, 12, 102-104.

Evans, P.; 1964, The tectonic framework of Assam; *Jour. Geol. Soc. Ind.*, 5, 80-96.

Gansser, A.; 1964, *Geology of the Himalayas*; Interscience Publishers, London, 289 pp.

Holmes, A.; 1955, Dating of Precambrian of Peninsular India and Ceylon; *Proc. Geol. Assoc. Can.*, 7, 81-106.

Krishnan, M.S.; 1953, *The Structure and Tectonics of India*; Mem. Geol. Surv. Ind., 81, 109 pp.

Pitchamuthu, C.S.; 1971, Precambrian geochronology of Peninsular India; *Jour. Geol. Soc. Ind.*, 12 (3), 262-273.

Sarkar, S.N.; 1980, Precambrian stratigraphy and geochronology of Peninsular India: a review; *Ind. Jour. Earth Sc.*, 7 (1), 12-26.

Sundaram, P.K., Swaminath, J. and Venkatesh, N.; 1964, Tectonics of Peninsular

India; *Proceedings 22nd Int. Geol. Cong.*, 4, 557-565.
Tectonic Map of India; 1963, Geological Survey of India, Govt. of India Publication, map on the scale 1 : 2 million.
Tectonic Map of India; 1968, Oil and Natural Gas Commission Publication; map on the scale 1 : 2 million; Principles of preparation in *Bull. Oil Nat. Gas Comm.*, 6(1) 1-11.
Valdiya, K.S.; 1973, Tectonic framework of India: a review and interpretation of recent structural and tectonic studies; *Bull. Geophys. Res.*, *11*(2), 79-114.
Wadia, D.N.; 1942, The Making of India; *Procceedings of Indian Science Congress, Presidential Addresses, Pt. II*, 91-118.

Chapter 5
Ball, V.; 1877, Geology of the Mahanadi basin; *Rec. Geol. Surv. Ind. 10*, 167-185.
Banerji, A.K.; 1964, Structure and stratigraphy of part of northern Singhbhum, south of Tatanagar; *Proc. Nat. Inst. Sc. Ind.*, *30*A (4), 486-510.
Beckinsale, R.D., Dury, S.A. and Holt R.W.; 1980, 3360 m.y. old gneisses from the South Indian craton; *Nature*, *283*, 469-470.
Crawford, A.R.; 1969, Reconnanaissance Rb-Sr dating of the Precambrian rocks of southern Peninsular India; *Jour. Geol. Soc. Ind.*, *19*, 531-549.
Crookshank, H.; 1963, Geology of southern Bastar and Jeypore from Bailadila range to the Eastern Ghats; *Mem. Geol. Surv. Ind.*, *87*, 1-150.
Dunn, J.A.; 1929, The geology of the north Singhbhum including parts of Ranchi and Manbhum districts; *Mem. Geol. Surv. Ind.*, *54*, 1-166.
———; 1940, The stratigraphy of south Singhbhum; *Mem. Geol. Surv. Ind.*, *63* (3), 303-369.
Dunn, J.A. and Dey, A.K.; 1942, Geology and petrology of eastern Singhbhum and surrounding areas; *Mem. Geol. Ind.*, *69*(2), 281-456.
Fermor, L.L.; 1926, in Pascoe, E.H.: General Report for 1925; *Rec. Geol. Surv. Ind.*, *59*, 75-80.
———; 1936, An attempt at the correlation of the ancient schistose formations of Peninsular India; *Mem. Geol. Surv. Ind.*, *70*, 1-217.
Foot, R.B.; 1888-89, The Dharwar System, the chief auriferous rock series in South India; *Rec. Geol. Surv. Ind.*, *21*, 40-56, and *22*, 17-39.
Ghosh, P.K.; 1941, Charnockite series of Bastar State and Western Jaypore; *Rec. Geol. Surv. Ind.*, *75*, 1-55.
Heron, A.M.; 1935, Synopsis of the pre-Vindhyan geology of Rajputana; *Trans. Nat. Inst. Sc. Ind.*, *1*, 17-33.
———; 1953, Geology of Central Rajputana; *Mem. Geol. Surv. Ind.*, *79*, 1-389.
Holland, T.H.; 1900, The Charnockite Series, a group of Archaean hypersthene rocks in Peninsular India: *Mem. Geol. Surv. Ind.*, *28*(2), 119-249.
Iyengar, S.V.P. and Murty, Y.G.K.; 1982, The evolution of the Archaean-Proterozoic crust in parts of Bihar and Orissa, Eastern India; *Rec. Geol. Surv. Ind.*, *112*(3), 1-6.
Jones, H.C.; 1934, The iron-ore deposits of Bihar and Orissa; *Mem. Geol. Surv. Ind.*, *63*(2), 167-302.
Kanungo, D.N. and Mahalik, N.K.; 1967, Structure and stratigraphic position of Gangpur Series in the Archaeans of Peninsular India; *Proc. Symp. Upp. Mantle*, Hyderabad, India; 458-478.
King, W.; 1885, Sketch of the geological work in the Chattisgarh division of the Central Provinces; *Rec. Geol. Surv. Ind.*, *18*, 169-200.
Krishnan, M.S.; 1935, The geology of Gangpur state, Eastern States; *Mem. Geol. Surv. Ind.*, *71*, 1-181.
Maisonnave, J.; 1982, The composition of the Precambrian ocean waters; *Sed. Geol.*, *31*, 1-11.
Mukhopadhyay, D.; 1976, Precambrian stratigraphy of Singhbhum: The problems and a prospect; *Ind. Jour. Earth Sc.*, *3*(2), 308-319.

Murty, V.N. and Acharya, S.; 1975, Lithostratigraphy of Precambrian rocks around Koina, Sundergarh district, Orissa; *Jour. Geol. Soc. Ind.* 16(1), 55-68.

Naha, K. and Ghosh, S.K.; 1960, Archaean paleogeography of eastern and northern Singhbhum, Eastern India; *Geol. Mag.*, 97, 436-439.

Newbold; 1850, Summary of the geology of the southern India; *Jour. Roy. Asiatic Soc.,* 12, 78.

Nisbet, E.G.; 1982, Definition of Archaean: comments and proposal on the recommendation of the International sub-commission on Precambrian stratigraphy; *Precamb. Res.,* 19, 111-118.

Pitchamuthu, C.S.; 1965, Regional metamorphism and charnockitisation in Mysore State, *Ind. Min.*, 6(1), 119-126.

Radhakrishnan, B.P.; 1964, Symposium on stratigraphy, age and correlation of Archaean provinces of India; *Summary of Procceedings, Andhra University*, 33 pp.

Radhakrishnan, B.P. and Vasudev, V.N.; 1977, The Early Precambrian of southern Indian Shield; *Jour. Geol. Soc. Ind.*, 18(10), 525-541.

Rai, K.L., Sarkar, S.N. and Paul, P.R.; 1980, Primary depositional and diagenetic features in the Banded Iron Formations and associated iron ore deposits of Noamandi, Singhbhum district, Bihar, India; *Mineral Deposita*, 15, 189-200.

Rama Rao, B.; 1936, Recent studies on the Archaean complex of Mysore; *Presidential Addresses, 23rd Ind. Sc. Cong.*, 215-244.

———; 1964, The Archaean Provinces of India and their comparison with those of other continental shields; *Jour. Geol. Soc. Ind.*, 5, 56-71.

Saha, A.K.; 1975, The Mayurbhanj Granite: A Precambrian batholith in eastern India; *Jour. Geol. Soc. Ind.*, 16(1), 37-43.

Sarangi, S.K. and Acharya, S.; 1975, Stratigraphy of the Iron Ore Group around Kandadhar, Sundergarh district, Orissa; *Ind. Jour. Earth Sc.*, 2(2), 182-189.

Sarkar, A., Trivedi, J.R., Gopalan, K., Singh, P.N., Das, A.K. and Paul, D.K., 1984 Rb-Sr Geochronology of the Bundelkhand Granitic Complex in the Jhansi-Babina-Talbehat sector, U.P.; (Abstracts) Seminar on Crustal Evolution of the Indian Shield and its Bearing on Metallogeny, Calcutta, 26-28.

Sarkar, S.N.; 1957-58, stratigraphy and tectonics of Dongargarh System: a new system in the Precambrian of Bhandara, Drug, Balaghat area, Bombay and Madhya Pradesh; *Jour. Sc. Ind. IIT, Kharakpur*, 1, 237-268, 2, 143-160.

———; 1968, *Precambrian Stratigraphy and Geochronology of Peninsular India*; Dhanbad Publishers, India, 33 pp.

———; 1980, Precambrian Stratigraphy and geochronology of Peninsular India: A review; *Ind. Jour. Earth Sc.*, 7(1), 12-26.

Sarkar, S.N., Gopalan, K. and Trivedi, J.R.; 1981, New data on the geochronology of the Precambrian of Bhandara, Drug, Central India; *Ind. Jour. Earth Sc.*, 8(2) 131-151.

Sarkar, S.N., Polkanov, A.A., Gerling, E.K. and Chukrov, F.V; 1967, Precambrian geochronology of Nagpur-Bhandara-Drug, India; *Geol. Mag.*, 104, 525-549.

Sarkar, S.N. and Saha, A.K.; 1962, A revision of the Precambrian stratigraphy and tectonics of Singhbhum and adjacent regions; *Quart. Jour. Geol. Min. Met. Soc. India*; 34 (2 & 3), 97-136.

——— and ———; 1963, On the occurrence of two intersecting Precambrian orogenic belts in Singhbhum and adjacent areas, India; *Geol. Mag.*, 100, 69-92.

——— and ———; 1977, The present status of precambrian stratigraphy, tectonics and geochronology of Singhbhum-Keonjhar-Mayurbhanj region, Eastern India; *Ind. Jour. Earth. Sc., S. Ray volume*, 37-65.

——— and ———; 1982, Structure and tectonics of the Singhbhum-Orissa Iron Ore Craton, Eastern India; *Rec. Res. Geol.*, 10, 1-25.

Sarkar, S.N., Saha, A.K., Boelrijk, N.A.I.M. and Hebeda, E.H.; 1979, New data on the geochronology of the Older Metamorphic Group and the Singhbhum Granite of

Singhbhum-Keonjhar-Mayurbhanj region, Eastern India; *Ind. Jour. Earth. Sc.,* 6(1), 32-51.
Sims, P.K.; 1980, Subdivision of the Proterozoic and Archaean Eons: recommendations and suggestions of the International Sub-commission on Precambrian Stratigraphy; *Precamb. Res., 13,* 379-380.
Smeeth, W.; 1916, Outline of the geological history of Mysore; *Deptt. Mines & Geol., Mysore State, Bull.,* 6, 1.
Straczek, J.A., Subramanyam, M.R., Narayanaswamy, S., Shukla. K.D., Vemban N.A., Chakravarty, S.C. and Yenkatesh, V.; 1956, Manganese ore deposits of Madhya Pradesh, India; *XX Int. Geol. Cong., Mexico, Tomo IV,* 63-96.
Swaminath, J. and Ramakrishnan, M. (Editors); 1981, Early Precambrian supracrustals of Southern Karnataka; *Mem. Geol. Surv. Ind., 112,* 1-329.
Vemban N.A.; 1961, Structure and tectonics of the manganese ore belts of Madhya Pradesh and adjoining parts of Bombay; *Proc. Ind. Acad. Sc., 53(3), Section B,* 125-134.
Viswanathaiah, M.N., Venkatachalapathy, V. and Mahalakchnamma, A.P.; 1976, Primitive microstructures in Sargur Schists (Precambrian) near Mysore, Karnataka; *Jour. Geol. Soc. Ind., 17* (1), 112-113.
West, W.D.; 1936, Nappe structure in the Archaean rocks of Nagpur District; *Trans. Nat. Inst. Sc. Ind., 1* (6), 93-102.

Chapter 6

Aswathanarayana, U.; 1964, Isotopic ages from the Eastern Ghats and Cuddapahs of India; *Jour. Geophys. Res.,* 69 (16), 3479-3486.
Auden. J.B.; 1933, Vindhyan sedimentation in Son valley, Mirjapur district; *Mem. Geol. Surv. Ind.,* 62, 141-150.
Basumallick, S.; 1967, Purana sedimentation in parts of the Godawari valley; *Jour. Geol. Soc. Ind.,* 8, 130-141.
Chandrashekhara Gowda, M.J.; 1981-82, A litho-stratigraphic classification of the Kaladagi Group; *Karnataka Sc. Jour., Univ. Mysore,* 28, 126-134.
Crawford, A.R.; 1970, The Precambrian geochronology of Rajasthan and Bundelkhand, northern India; *Canad. Jour. Earth. Sc.,* 7, 91-110.
Crawford, A.R. and Compston, W.; 1970, The age of the Vindhyan System of Peninsular India; *Quart. Jour. Geol. Soc. Lond.,* 125, 351-371.
Davies, A. and Crawford, A.R.; 1971, Petrography and age of the rocks of Bullard Hill, Kirana Hills, Sargoda district, West Pakistan; *Geol. Mag., 108* (3), 235-246.
Dutt, N.V.B.S.; 1964, Suggested succession of the Purana formations of Chattisgarh; *Rec. Geol. Surv. Ind., 93* (2), 143-148.
Hacket, C.A.; 1870, Geology of Gwalior and vicinity; *Rec. Geol. Surv. Ind., 3* (2), 33-42.
————; 1877, Note on the Aravalli Series in north-eastern Rajputana; *Rec. Geol. Surv. Ind., 10* (2), 84-92.
Heron, A.M.; 1917, Geology of northeastern Rajputana; *Mem. Geol. Surv. Ind., 45* (1), 1-128.
————; 1953, Geology of Central Rajputana; *Mem. Geol. Surv. Ind.,* 79, 1-389.
Jones, H.C.; 1909, General report of the geological survey of India; *Rec. Geol. Surv. Ind.,* 38, 66.
King, W.; 1872, On the Cuddapah and Kurnool formations in Madras Presidency; *Mem. Geol. Surv. Ind.,* 8, 1-313.
Kochhar, N.; 1982, Petrochemistry and petrogenesis of the Malani Igneous Suite, India: Discussion and Reply; *Bull. Geol. Soc. Am.,* 93, 926-928.
Krishnan, M.S. and Swaminath, J.; 1959, The great Vindhyan basin of northern India; *Jour, Geol. Soc. Ind., 1,* 10-30.
Mahadevan, C.; 1946, The Bhima Series in the Gulbarga district, Hyderabad; *Jour. Hyderabad, Geol. Surv.,* 5 (1).

Medlicott, H.B.; 1859, Vindhyan rocks and their associates in Bundelkhand; *Mem. Geol. Surv. Ind.*, *2* (1), 1-

Misra, R. C.; 1969, The Vindhyan System; *Presidential Address, 56th Ind. Sc. Cong.*, 1-32.

Narayanaswamy, S.; 1966, Tectonics of the Cuddapah Basin; *Jour. Geol. Soc. Ind.*, 7, 33-50.

Oldham, R.D., Vredenberg, E. and Datta, P.N.; 1901, Geology of the Son valley in Rewah state and parts of the adjoining districts of Jubbalpore and Mirzapur; *Mem. Geol. Surv. Ind.*, *31* (1), 1 -

Pareek, H.S.; 1981, Petrochemistry and petrogenesis of the Malani Igneous Suite, India: summary; *Geol. Soc. Am. Bull.*, *92*, 67-70.

Sahni, M.R.; 1977, Vindhyan Palaeobiology, stratigraphy and depositional environments: a critical review; *Jour. Pal. Soc. Ind.* 20, 289-384.

Sarkar, S.N.; 1980, Precambrian stratigraphy and geochronology of Peninsular India: a review; *Ind. Jour. Earth. Sc.*, 7 (1), 12-26.

Vinagradov, A.P., Tugarinov, A.I., Zhukov, C., Staprikova, N., Bibikova, E. and Khorre, K., 1964, Geochronology of Indian Precambrian; *Proc. 22nd Int. Geol. Cong., Section 10*, 553-567.

Viswanathiah, M.N.; 1968, Badami Series, a new post-Kaladgi formation, Mysore State; *Bull. Geol. Soc. Ind.*, *5*, 94-97.

Vredenberg, E.; 1906, Suggestions for the classification of the Vindhyan System; *Rec. Geol. Surv. Ind.*, 33, 255-260.

Chapter 7

Arita, K., Ohita, Y., Akiba, Ch. and Maruo, Y.; 1973, Kathmandu Region; in S. Hashimoto et al. (Editors): *Geology of Nepal Himalaya*; Saikon, Tokyo, 99-145.

Auden, J.B.; 1934; The Geology of the Krol Belt; *Rec. Geol. Surv. Ind.*, *67* (4), 357-454.

Bhargava, O.N.; 1972, A reinterpretation of the Krol Belt.; *Him. Geol.*, *2*, 47-81.

Bordet, P.; 1961, *Researches Geologiques dans l'Himalaya du Nepal, Region du Makalu*; Ed. C.N.R.S., Paris, 275 pp.

Gansser, A.; 1964, *Geology of the Himalayas*; Interscience Publishers, London, 289 pp.

Griesbach, C.L.; 1891, Geology of the Central Himalayas; *Mem. Geol. Surv. Ind.*, *23*, 1-232.

Kumar, R.; 1980, Outline of the stratigraphy of Central Nepal; in Valdiya, K.S. and Bhatia, S.B. (Editors): *Stratigraphy and Correlation of Lesser Himalayan Formations*, Hindustan Publishing Corporation (India), Delhi, 180-190.

Kumar, R. and Gupta, V.J.; 1981, Stratigraphy of Nepal Himalaya; in Sinha, A.K. (Editor) : *Contemporary Geoscientific Researches in Himalaya*, Volume I (Nautiyal Volume), Bishen Singh Mahendra Pal Singh Publishers, Dehradun, 161-176.

Kumar, R. and Kapila, S.P.; 1980, Occurrence of angular unconformity at the base of Nagthat Formation near Khairna, Kumaun Himalaya; *Bull. Ind. Geol. Assoc.*, *13* (1), 83-88.

Kumar, R., Shah, A.N. and Bingham, D.K.; 1978, Positive evidence of a Precambrian tectonic phase in Central Nepal Himalaya; *Jour. Geol. Soc. Ind.*, *19*(1), 519-522.

Lombard, A., 1953, Presentation d'un profil geologique du Mt. Everest a la plaine du Gange (Nepal Oriental) ; *Bull. Soc. Belg. Geol. Pal. Hydrogeol.*, Brussels, *62* (1), 123-129.

Mallet, F.R.; 1875, On the geology and mineral resources of the Darjeeling district and the Western Duars; *Mem. Geol. Surv. Ind.*, *11*(1), 1-50.

Medlicott, H.B.; 1864, On the geological structures and relations of the southern portion of the Himalayan ranges between the rivers Ganges and Ravee; *Mem. Geol. Surv. Ind. 3* (2), 1-212.

Misch, P.; 1949, Metasomatic granitisation of batholithic dimensions; *Am. Jour. Sc. 247*, 209-245, 372-406 and 673-705.

Oldham, R.D.; 1888, Some notes on the geology of the north-west Himalaya; *Rec. Geol. Surv. Ind.*, *21* (4), 149-157.

Pascoe, E.H.; 1950, *A manual of Geology of India and Burma*; 3rd Edition, Volume I, Government of India Press, Calcutta, 483 pp.

Pilgrim, G.E. and West, W.D.; 1928, The structure and correlation of the Simla rocks; *Mem. Geol. Surv. Ind.*, *53*, 1-140.

Raha, P.K.; 1980, Stratigraphy and depositional environments of Jammu Limestone, Udhampur district, Jammu; in Valdiya, K.S. and Bhatia, S.B. (Editors): *Stratigraphy and Correlation of Lesser Himalayan Formations*, Hindustan Publishing Corporation (India), Delhi, 145-151.

Raha, P.K. and Sastry, M.V.A.; 1973, Stromatolites from Jammu Limestone, Udhampur district, J & K State, and their stratigraphic and palaeogeographic significance; *Him. Geol.*, *3*, 135-147.

Sinha, A.K. ; 1977, Riphean stromatolites from the Western Lower Himalaya, H.P., India; in Flugel E. (Editor): *Fossil Algae*, Springer Verlag, Berlin, 86-100.

Srikantia, S.V. and Sharma, R.P. ; 1971, Simla Group: a reclassification of the "Chail Series", "Jaunsar Series" and "Simla Slates" in the Simla Himalaya; *Jour. Geol. Soc. Ind.*, *12*, 234-240.

———— and ————; 1976, Geology of the Shali Belt and adjoining areas; *Mem. Geol. Surv. Ind.*, *106* (1), 31-166.

Stocklin, J.; 1980, Geology of Nepal and its regional frame; *Jour. Geol. Soc. Lond.*, *137*, 1-34.

Valdiaya, K.S.; 1967, Occurrence of magnesite deposits and time controlled variation of stromatolites in the Shali Series, district Mahasu, H.P.; *Bull. Geol. Soc. Ind.*, *4*, 125-128.

————; 1980, Stratigraphic scheme of the sedimentary units of the Kumaun Lesser Himalaya; in Valdiya, K.S. and Bhatia, S.B. (Editors) : *Stratigraphy and Correlation of Lesser Himalayan Formations*, Hindustan Publishing Corporation (India), Delhi, 1-48.

Wadia, D.N.; 1934, The Cambrian-Trias sequence of North-western Kashmir (parts of Muzzafarabad and Baramula districts); *Rec. Geol, Surv. Ind.*, *68* (2), 121-176.

Chapter 8

Auden, J.B.; 1934, Geology of the Krol Belt; *Rec. Geol. Surv. Ind.*, *67*, 357-434.

Azmi, R.J. and Pancholi V. P.; 1983, Early Cambrian (Tommotian) conodonts and other shelly microfauna from the Upper Krol of Mussoorie Syncline, Garhwal Lesser Himalaya with remarks on Precambrian-Cambrian boundary; *Him. Geol.*, *11*, 360-372.

Bhargava, O.N. ; 1972, A reinterpretation of the Krol Belt; *Him. Geol.* *2*, 47-81.

Bhatia, S.B.; 1975, The anatomy of the Blainis; *Bull. Ind. Geol. Assoc.*, *8* (2), 243-258.

Bhatt, D. K., Mamgain, V.D., Misra, R.S. and Srivastava, J. P.; 1983, Shelly microfossils of Tommotian age (Lower Cambrian) from the chert-phosphorite member of Lower Tal Formation, Maldeota, Dehradun district, Uttar Pradesh; *Geophytology*, *13* (1), 116-123.

Bordet, P., Colchen, M., Le Fort, P., Mouterde, R. and Remy, J.M.; 1967, Donnees nouvelles sur la geologie de la Thakhola (Himalaya du Nepal); *Bull. Soc. Geol. France* S-7, T-9. 883-896.

Bordet, P., Colchen, M., Krummencacher, D., Le Fort, P., Mouterde, R. and Remy, J.M.; 1971, *Researches Geologiques dans l'Himalaya du Nepal, Region de la Thakhola*, Paris Ed. C.N.R.S., 279 pp.

Bordet, P., Colchen, M. and Krummenacher, D.; 1975, *Researches Geologiques dans l 'Himalaya du Nepal, Region du Nyi Shang*; Paris Ed. C.N.R.S., 138 pp.

Cowie, J.W.; 1978, IUGS/IGCP Project 29. Precambrian-Cambrian Boundary Working Group in Cambridge; *Geol. Mag.*, *115*, 151-152.

Diener, C.; 1915, The anthrocolithic fauna of Kashmir, Kinnaur and Spiti; *Pal. Indica*

Bibliography of Selected References

Geol. Surv. Ind., N.S., 5 (2), 1-38.

Egeler, C.G., Boudenhousen, J.W.A., De Booty, T. and Nijhuis, H.G.; 1964, On the geology of the Central-West Nepal—a preliminary note; *Proc. 22nd Int. Geol. Cong., 11* 101-122.

Fuchs, G.; 1967, Zum bau des Himalaya; *Osterr. Acad. Wis. Nat. Kl. DenKschr, 113,* 1-211.

———; 1968, The geological history of the Himalayas; *23rd Int. Geol, Cong, Prague, Proceedings vol. 3,* 1-61.

———; 1977, The geology of the Karnali and Dolpo regions, Western Nepal; *Jahrb. Geol. B., 120* (2), 165-217.

Fuchs, G. and Frank, W.; 1970, The geology of the West Nepal between the rivers Kali Gandaki and Thulo Bheri; *Jahrb. Geol. Budesan, 18,* 1-103.

Ganesan, T.M.; 1972, Fenestellid bryozoans from the boulder slate sequence of Garhwal; *Him. Geol., 2,* 431-451.

Ghosh, A.K. and Srivastava, S.K.; 1962, Microfloristic evidence on the age of the Krol Beds and associated formations; *Proc. Nat. Inst. Sc. Ind., 28* (A), 710-717.

Griesbach, C.L.; 1891, Geology of the Central Himalaya; *Mem. Geol. Surv. Ind., 23,* 1-232.

Gupta, V.J.; 1966, *Palaeontology, stratigraphy and structure of the Palaeozoic rocks of Parts of the Kashmir Himalaya*; Unpublished Ph.D. thesis, Panjab University, Chandigarh, 239 pp.

———; 1969, Palaeozoic stratigraphy of the area south-east of Srinagar; *Res. Bull.* (N.S.), *P.U., 20* (1 & 2), 1-14.

———; 1971, Silurian-Devonian boundary in the Indian sub-continent; *Jour. Geol. Soc. Ind., 12* (3), 274-279.

———; 1973. *Indian Palaeozoic Stratigraphy*; Hindustan Publishing Corporation, Delhi, 207 pp.

Gupta, V.J. and Jain, S.P.; 1972 (for 1967), Lower Palaeozoic fossils from Lossar, Spiti; *Res. Bull.* (N.S.), *P.U., 18* (1 & 2), 5-12.

Gupta, V. J. and Chetri, V. S. ; 1977, Geology of the area around Phulchauki, Kathmandu, Nepal ; *Chayanika Geologia, 3* (2), 133-146.

Gupta, V.J. and Stocklin, J.; 1978, Stratigraphy and structure of Phulchauki-Chandragiri area, Nepal; *Rec. Res. Geol.,* 7, 263-275.

Hayden, H.H.; 1904, The geology of Spiti; *Mem. Geol. Surv. Ind., 36* (1), 1-129.

Heim, A. and Gansser, A.; 1939, Central Himalaya : geological observations of the Swiss Expedition-1936; *Mem. Soc. Helv. Sci. Nat., 73* (1), 1-245.

Holland, T.H.; 1908, On the occurrence of the striated boulders in the Blaini formations of Simla with a discussion on the geological ages of the beds; *Rec. Geol. Surv. Ind., 37,* 129-135.

Jain, S.P., Bhatia, S.B. and Gupta, V.J.; 1972, Carboniferous ostrocods from near Losar, Spiti valley; *Him. Geol., 2,* 168-187.

Kumar, G., Raina, V.K., Bhargava, O.N., Maithy, P.K. and Babu, R., (In Press), The Precambrian-Cambrian boundary problem and its prospects, north-western Himalaya, India; *Geol. Mag.,121* (3), 1984, 211-219

Kumar, R. and Gupta, V. J.; 1981, Stratigraphy of Nepal Himalaya; in Sinha, A. K. (Editor): *Contemporary Geoscientific Researches in Himalaya,* Volume I (Nautiyal volume), Bishen Singh Mahendra Pal Singh Publishers, Dehradun, 161-176.

Lyddekar, R.; 1883, Geology of Kashmir, Kistwar and Chamba; *Mem. Geol. Surv. Ind., 22,* 1-344.

Mathews, S. C. and Missarzhevsky V. V.; 1975, Small shelly fossils of late Precambrian and early Cambrian age: a review of recent work; *Jour. Geol. Soc. Lond., 131* (3), 289-304.

Mehdi, H. S., Kumar, G. and Prakash, G.; 1972, Tectonic evolution of the Eastern Kumaun Himalaya: a new approach; *Him. Geol., 2,* 481-501.

Medlicott, H.B.; 1864, On the geological structure and relation of the southern portion of the Himalayan ranges between the rivers Ganges and Ravee; *Mem. Geol. Surv. Ind.,* 3, 1-207.
Middlemiss, C.S.; 1910, A revision of the Silurian-Trias sequence in Kashmir; *Rec. Geol. Surv. Ind.,* 40, 206-260.
Nakazawa, K., Kapoor, H.M., Ishi, K., Bando, Y., Maegoya, T., Shimizu, D., Nogami, Y., Tukuoka, T. and Nohda, S.; 1970, Preliminary report on the Permo-Trias of Kashmir; *Mem. Fac. Sci. Kyoto University.,* 37 (1), 163-172.
Oldham, R.D.; 1883, Note on the geology of Jaunsar and Lower Himalayas; *Rec. Geol. Surv. Ind.* 16 (4), 193-198.
———; 1888, Some notes on the geology of the North-West Himalayas; *Rec. Geol. Surv. Ind.,* 21 (4), 149-157.
Pande, I.C.; 1967, Palaeotectonic evolution of the Himalayas; *Publ. Cent. Adv. Stud. Geol. Panjab Univ.,* 3, 107-116.
Pilgrim, G. E. and West, W. D.; 1928, The structure and correlation of Simla rocks; *Mem. Geol. Surv. Ind.,* 53, 1-140.
Powar, K. B. and Phansalkar, V. G.; 1971, On the occurrence of plant remains in the phyllites of Bhikhiasen area, Almora district, U.P.; *Curr. Sc.,* 40, 377-378.
Reed, F.R.C.; 1912, Ordovician and Silurian fossils of the Central Himalayas; *Palaeont. Indica, Geol. Surv. Ind., Ser.,* 15, 7 (2), 1-168.
———; 1932, New fossils from the Agglomeratic Slates of Kashmir; *Palaeont. Indica, N.S., Geol. Surv. Ind.,* 20 (1), 1-77.
———; 1934, Cambrian and Ordovician fossils of Kashmir; *Palaeont. Indica, N.S., Geol. Surv. Ind.,* 21 (2), 1-38.
Rupke, J.; 1968, Note on the Blaini Boulder Beds of Tehri Garhwal, Kumaun Himalayas; *Jour. Geol. Soc. Ind.,* 9, 171-177.
Sah, S.C.D., Venkatchala, B.S. and Lakhanpal, R.N., 1968, Palynological evidence on the age of the Krols; *Publ. Cent. Adv. Stud. Geol.; Panj. Univ.,* 5, 115-120.
Sastry, M.V.A. and Shah, S.C.; 1964, Permian marine transgression in Peninsular India; *Proc. 22nd Int. Geol. Cong., Delhi, part* 9, 139-150.
Saxena, M. N.; 1971, Geological classification and tectonic history of the Himalaya; *Proc. Ind. Nat. Sc. Acad.,* 37 (A), (1), 28-54.
Schindewolf, O.H. and Seilacher, A.; 1955, Beitrage zur Kenntnis des Kambriums in der Salt Range (Pakistan); *Abh. Akad. Wiss Litt. Manz,* 10, 1-446.
Srikantia, S.V. and Bhargava, O.N.; 1974, The Jaunsar problem in the Himalaya—a critical analysis and elucidation; *Jour. Geol. Soc. Ind.,* 15, 115-136.
Srivastava, R.N. and Venkataraman, K.; 1975, Palynostratigraphy of the Blaini Formation; *Bull. Ind. Geol. Assoc.,* 8 (2), 196-199.
Stocklin, J., Termier, G. and Bhattarai, K.; 1977, A propos des roches fossiliferous de Chandragiri (Mahabharat du Nepal); *Bull. Geol. Soc. Tran. Ser.* 7, 19 (2), 367-373.
Suneja, I.J., 1971, A note on the Ordovician fossils from the Baramula district, Kashmir; *Curr. Sc.,* 40 (17), 466.
Tewari, B.S. and Singh, R.Y., 1981, The Late Palaeozoic fossils from the Kumaun Himalayas and their stratigraphic significance; *Proc. IX Ind. Coll. Micropal. Strat.,* 206-219.
Valdiya, K.S.; 1980, Stratigraphic scheme of the sedimentary units of the Kumaun Lesser Himalaya; in Valdiya, K.S. and Bhatia (S.B. (Editors): *Stratigraphy and Correlation of Lesser Himalayan Formations*; Hindustan Publishing Corp. (India) Delhi, 7-48.
Valdiya, K.S. and Gupta, V.J.; 1972, A contribution to the geology of the north-eastern part of Kumaun Himalaya; *Him. Geol.,* 2, 1-31.
Varadrajan, S.; 1977, Potassium-argon age of the metabasics from the Bhimtal-Bhowali area, Nainital district, Kumaun Himalaya and its significance; *Rec. Res. Geol.,* 3, 233-243.
Wadia, D.N.; 1928, The geology of the Poonch State (Kashmir) and adjacent portions of Punjab; *Mem. Geol. Surv. Ind.,* 51 (2), 185-370.

———; 1934, Cambrian-Trias sequence of north-western Kashmir (parts of Muzzafarabad and Baramula districts); *Rec. Geol. Surv. Ind.*, 68 (2), 121-176.
———; 1957, *Geology of India*; 3rd Edition, McMillan, London, 536 pp.
Waterhouse, J.B. and Gupta, V.J.; 1978, Early Permian fossils from Bijni tectonic unit, Garhwal Himalaya; *Rec. Res. Geol.*, 4, 410-437.
Wynne, A.B.; 1978, On the geology of the Salt Range in the Punjab; *Mem. Geol. Surv. Ind.*, 14, 1-314.

Chapter 9

Agarwal, S.K.; 1957, Kutch Mesozoic: A study of the Jurassic of Kutch with special reference to the Jhura Dome; *Jour. Pal. Soc. Ind.*, 2, 119-130.
Ahluwalia, A.D.; 1978, Discovery of Upper Palaeozoic fossils (Foraminifera or Porifera?) from Mussoorie phosphorite, Lower Tal Formation, Kumaun Lower Himalaya, India; *Contr. Him. Geol.*, 1, 20-24.
Auden, J.B.; 1934, The geology of the Krol-Belt; *Rec. Geol. Surv. Ind.*, 67, 357-454.
Azmi, R.J., Joshi, M.N. and Juyal, K.P.; 1981, Discovery of the Cambro-Ordovician conodonts from the Mussoorie Tal phosphorite: its significance in correlation of the Lesser Himalaya; in Sinha, A.K. (Editor): *Contemporary geoscientific Researches in Himalaya*, Volume I (Nautiyal Volume), Bishen Singh Mahendra Pal Singh, Publishers, Dehradun, 245-250.
Balagopal, A.T. and Srivastava, V.K.; 1975, A study of the palaeocurrents and the provenance of the Jurassic rocks of Central Kutch, Gujrat State; *Ind. Jour. Earth. Sc.*, 2, 62-76.
Banerjee, P.K.; 1973, Foraminiferal biostratigraphy and geological evolution of the Thanjavur sub-basin, South India; *Jour. Geol. Soc. Ind.*, 14 (3) 257-274.
Bhalla, S.N.; 1983, India; in Mullade, M. and Nairn, A.E.M. (Editors): *The Phanerozoic Geology of the World II: The Mesozoic, B*; Elsevier Amsterdam, 305-352.
Bhatia, S.B.; 1975, The anatomy of the Blainis; *Bull. Ind. Geol. Assoc.* 8 (2), 243-258.
———; 1980, The Tal tangle; in Valdiya, K.S. and Bhatia, S.B. (Editors), *Stratigraphy and Correlation of Lesser Himalayan Formations*; Hindustan Publishing Corporation (India), Delhi, 79-96.
Bhatia, S.B. and Jain, S.P.; 1969, Dalmiapuram Formation : a new Lower Cretaceous horizon in South India; *Bull. Ind. Geol. Assoc.*, 2, 105-108.
Bhatt, D.K., Joshi, V.K. and Arora, A.K.; 1981 (for 1980), Conodonts of the *Otoceras* Bed of Spiti; *Jour. Pal. Soc. Ind.*, 25, 130-134.
Bhattacharya, S.C. and Niyogi, D.; 1971, Geological evolution of the Krol Belt in Simla Hills, H.P.; *Him. Geol.*, 1, 178-212.
Biswas, S.K. and Deshpande, S.V.; 1970, Geologic and tectonic map of Kutch; *Bull. Oil Nat. Gas Comm.*, 7(2), 115-123.
———and———; 1983, Geology hydrocarbon prospects of Kutch, Saurashtra and Narmada basins; *Petroleum Asia Journal*, Nov.; 83, 111-126.
Chiplonkar, G.W.; 1972, Newer observations on the stratigraphy of Bagh Beds; *Jour. Geol. Soc. Ind.*, 13(1), 92-95.
Dasgupta, S.K.; 1975, A revision of the Mesozoic-Tertiary stratigraphy of the Jaisalmer basin, Rajasthan; *Ind. Jour. Earth Sc.*, 2(1), 77-94.
Diener, C.; 1912; The Trias of the Himalaya; *Mem. Geol. Surv. Ind.*, 36, 1-176.
Ganesan, T.M.; 1972, Fenestellid bryozoans from the Boulder Slate Sequence of Garhwal; *Him. Geol.*, 2, 431-451.
Ghosh, A.K. and Srivastava, S.K.; 1962, Microfloristic evidence on the age of Krol Beds and associated formations; *Proc. Nat. Inst. Sc. India*, 28A (5), 710-717.
Gupta, V.J.; 1975, *Indian Mesozoic Stratigraphy*; Hindustan Publishing Corporation (India), Delhi, 267 pp.
Jain, S.P.; 1971, *Contribution to the study of marine Cretaceous ostrocodes of the Peninsular India*; unpublished Ph.D. thesis of the Panjab University, Chandigarh,

396 pp.
Kalia, P.; 1972, Upper Permian Fusulinids from Garhwal Himalaya; *Proc. 2nd Ind. Colloq. Micropal. Strat.*, Lucknow, 107-110.
Kumar, R. and Gupta, V.J.; 1982, Indus Ophiolite Belt of Ladakh and Himalayan Orogeny; *Rec. Res. Geol.*, 8, 494-509.
Lakhanpal, R.N., Sah, S.C.D. and Dube, S.N.; 1958, Further observations on plant microfossils from a carbonaceous shale (Krols) near Nainital with a discussion on the age of the beds; *Palaeobotanist*, 7(2), 111-120.
La Touche T.H.D.; 1911, Geology of Western Rajasthan; *Mem. Geol. Surv. Ind.* 35, 1-116.
Medlicott, H.B.; 1964, On the geological structure and relations of the southern portions of the Himalayan ranges between the rivers Ganges and Ravee; *Mem. Geol. Surv. Ind.*, 3, 1-212.
Middlemiss, C.S.; 1885, A fossiliferous series in the Lower Himalaya, Garhwal; *Rec. Geol. Surv. Ind.*, 18, 73-77.
———; 1910, A revision of the Silurian-Trias sequence in Kashmir; *Rec. Geol. Surv. Ind.*, 40 (3), 206-260.
Mithal, R.S. and Chaturvedi, R.S.; 1969, Possible algal structures in the Upper Krol Limestone of the Mussoorie area, U.P.; *Bull. Ind. Geol. Assoc.*, 2(3&4), 89-90.
Patwardhan, A.M.; 1978, First Moravamminids from the Himalaya; *Nat. Acad. Sc. Letters*, 1(1), 7-8.
Poddar, M.C.; 1964, Mesozoics of western India—their geology and oil possibilities; *Proc. 22nd Int. Geol. Cong.*, 1, 126-143.
Rajnath; 1932; A contribution to the stratigraphy of Kutch; *Quart. Jour. Geol. Min. Met. Soc. Ind.*, 4, 161-174.
Ramanathan, S.; 1968, Stratigraphy of Cauvery Basin with reference to its oil prospects; *Mem. Geol. Soc. Ind.*, 2, 153-167.
Rama Rao, L.; 1956, Recent contributions to our knowledge of the Cretaceous rocks of South India; *Proc. Ind. Acad. Sc.*, 44 B(4). 185-245.
Robinson, P.L.; 1967, Indian Gondwana formations: a review; *Proc. 1st Symp. Gond. Strat. IUGS*, 201-268.
Sarbadhikari, T.R.; 1979, Some problems of the Triassic Gondwana of India; in Laskar, B. and Raja Rao, C.S. (Editors): *Proceedings of the Gondwana Symposium*, Calcutta, 470-477.
Sastry, M.V.A. and Mamgain, V.D.; 1971, The marine Mesozoic formations of India; *Rec. Geol. Surv. Ind.*, 101(2), 162-177.
Sastry, M.V.A. and Rao, B.R.J.; 1964, Cretaceous-Tertiary boundary in South India; *Proc. 22nd Int. Geol. Cong.*, 3, 92-103.
Sastry, M.V.A., Rao, B.R.J. and Mamgain, V.D.; 1968, Biostratigraphic Zonation of the upper cretaceous formations of Trichinopoly district, South India; *Geol. Soc. Ind. Mem.*, 2, 10-17.
Singh, P. and Shukla, S.D.; 1981, Fossils from the Lower Tal: their age and its bearing on the stratigraphy of Lesser Himalaya; *Geosc. Jour.*, 2(2), 157-176.
Sinha, A.K.; 1975, Calcareous nannofossils from Simla Hills (Himalaya, India) with discussion on their age in the tectono-stratigraphic column; *Jour. Geol. Soc. Ind.*, 16(1), 69-77.
Spath, L.F.; 1933, Revision of Jurassic cephalopod fauna of Kutch (Cutch), *Geol. Surv. Ind., Palaeont. Indica, N. Ser.*, 9(2), 1-6, 1-945.
Srivastava, J.P.; 1972, Geology of parts of Almora and Nainital districts lying between Bhatranjkhan and Ramnagar; *Misc. Publ. Geol. Surv. Ind.*, 15, 131-138.
Teichert, C., Kummel, B. and Kapoor, H.M.; 1970, Mixed Permian-Triassic fauna, Guryul ravine, Kashmir; *Science*, 167(3915), 174-175.
Tewari, B.S.; 1969, Nannofossils from the Krols; *Bull. Ind. Geol. Assoc.* 2 (3 & 4), 22.
Tewari, B.S. and Kumar, R.; 1967, Foramininfera and Nummulitic Beds of Nilkanth

and organic remains from Tal Limestone, Garhwal Himalayas; *Publ. Cent. Adv. Stud. Geol. Panj. Univ.*, 3, 33-42.

Valdiya, K.S.; 1980, Stratigraphic scheme of the sedimentary units of the Kumaun Lesser Himalaya; in Valdiya, K.S. and Bhatia, S.B. (Editors): *Stratigraphy and Correlation of Lesser Himalayan Formations*, Hindustan Publishing Corp. (India), Delhi, 1-48.

Wadia, D.N.; 1957, *Geology of India*; 3rd Edition, McMillan, London, 536 pp.

Wynne, A.B.; 1872, Geology of Kutch; *Mem. Geol. Surv. Ind.*, 9(1), 1-289.

Chapter 10

Acharya, S.K.; 1971, Ranjit Pebble Slate—a new formation from Darjeeling Hills; *Ind. Min.* 25, 61-64.

Acharya, S.K., Ghosh, S.C., Ghosh, R.N. and Shah, S.C.; 1975, The continental Gondwana Group and associated marine sequence of Arunachal Pradesh (NEFA), Eastern Himalaya; *Him. Geol.*, 5, 6082.

Ahmad, F.; 1961, Palaeogeography of the Gondwana period and Gondwana land with special reference to India and its bearing on the theory of Continental Drift; *Mem. Geol. Surv. Ind.*, 90, 1-142.

———; 1964, The Permian basin of Peninsular India; *Proc. 22nd Int. Geol. Cong.*, 9, 123-138.

Ball, V.V.; 1877, Geology of Rajmahal Hills; *Mem. Geol. Surv. Ind.*, 13(2), 155-248.

Banarjee, I.; 1963, Trends of sedimentary differentiation in Barakar sandstones of south Karanpura coal field, India; *Jour. Sed. Petr.*, 33, 320-332.

Bhatia, S.B.; 1959, Additional microfossils from the Umaria marine beds, Central India; *Jour. Geol. Soc. Ind.*, 1, 116-125.

Bhatia, S.B. and Singh, S.K.; 1959, Carboniferous (Uralian) foraminifers from Mahendragarh, Central India; *Micropalaeontology*, 5, 127-134.

Cashyap, S.M.; 1979, Patterns of sedimentation in Gondwana basins; in Laskar, B. and Raja Rao, C.S. (Editors): *Proceedings of the IV International Gondwana Symposium*, 1977, Calcutta, 525-551.

Crookshank, H.; 1936, The geology of the northern slope of Satpuras between the Morand and Sher rivers; *Mem. Geol. Surv. Ind.*, 66, 173-381.

Datta, N.R., Mitra, N.D. and Bandyopadhyay, S.K.; 1983, Recent trends in the study of Gondwana basins of Peninsular and Extra-peninsular India; *Petroleum Asia Journal*, Nov. 1983, 159-69.

Feismantal, O.; 1880-82, The flora of Damuda and Panchet divisions; *Palaeont. Indica, Geol. Surv. Ind.*, Ser. 12, 3, 1-149.

Fox, C.S.; 1931, The Gondwana System and related formations; *Mem. Geol. Surv. Ind.*, 58, 1-113.

—; 1932, The natural history of Indian coal; *Mem. Geol. Surv. Ind.*, 57, 1-283.

Ghosh, S.; 1954, Discovery of new locality of marine Gondwana formations near Manendragarh in M.P.; *Science and Culture*, 19 (12), 620.

Hughes, T.H.W.; 1881, Notes on the South Rewa Gondwana basin; *Rec. Geol. Surv. Ind.*, 14, 126-138.

King, W., 1881, The geology of the Prahnita-Godawari valley; *Mem. Geol. Surv. Ind.*, 18 (30), 150-311 (reprinted 1930).

Krishnan, M.S.; 1958, Some problems on the geology of Gondwana formations; *Rec. Geol. Surv. Ind.*, 85 (4), 409-569.

Kumar, R. and Gupta, V.J.; 1981, Stratigraphy of Nepal Himalaya; in Sinha A.K. (Editor): *Contemporary Geoscientific Researches in Himalayas:* Volume I (Nautiyal volume), Bishen Singh Mahendra Pal Singh, Publishers, Dehradun, 161-176.

Kutty, T.S.; 1969, Some contributions to the stratigraphy of the Upper Gondwana formations of Pranhita-Godavari valley, Central India; *Jour. Geol. Soc. Ind.*, 10(1), 33-48.

Medlicott, H.B.; 1872, Note on the Lameta or Infra-Trappean formation in Central

India; *Rec. Geol. Surv. Ind.*, 5, 115-120.
Mehta, D.R.S.; 1964, *Gondwanas in India*; 22nd Int. Geol. Cong. (Brochure), 21 pp.
Misra, J.S., Shrivastava, B.P. and Jain, S.K.; 1961, Discovery of marine Permo-Carboniferous in Western Rajasthan; Curr. Sc., 30(7), 262-263.
Oldham, R.D.; 1861, On the geological relations and probable geological age of several systems of rocks in Central India and Bengal; *Mem. Geol. Surv. Ind.*, 2, 209-235.
—; 1886, Prospects of finding coal in Bap Boulder Beds, Western Rajasthan; Rec. Geol. Surv. Ind., 19(3).
—; 1893, *Geology of India*; Govt. India Publication, Calcutta, 543 pp.
Pareek, H.S.; 1979, The Permian-Mesozoic-Palaeogeography of the Rajasthan and Gujrat Shelf and correlation with that of the Indus basin; in Laskar, B. and Raja Rao, C.S. (Editors): *Fourth International Gondwana Symposium: Papers (Volume I)*; Hindustan Publishing Corporation (India), Delhi, 23-36.
Ranga Rao, A., Dhar, C.L. and Obergfell, F.A.; 1979, Badhaura Formation of Rajasthan—its stratigraphy and age; in Laskar, B. and Raja Rao, C.S. (Editors): *Fourth International Gondwana Symposium: Papers (Volume I)*; Hindustan Publishing Corporation (India), Delhi, 481-490.
Reed, F.R.C.; 1928, A Permo-Carboniferous marine fauna from Umaria Coal field; *Rec. Geol. Surv. Ind.*, 60, 367-398.
Robinson, P.L.; 1967, The Indian Gondwana formations: a review; *Proceedings First Symposium on Gondwana Stratigraphy*; I.U.G.S., 201-268.
Sastry, M.V.A., Achary, S.K. Shah, S.C., Satsangi, P.P., Ghosh. S.C. and Singh G.; 1979, Classification of Indian Gondwana Sequence—a reappraisal; in Laskar, B. and Raja Rao, C.S. (Editors): *Fourth International Gondwana Symposium: Papers (Volume I)*; Hindustan Publishing Corporation (India), Delhi, 502-510.
Sengupta, S; 1966, Palaeocurrents and depositional environments of the Gondwana rocks around Bheemaram—a preliminary study; *Bull. Geol. Soc. Ind.*, 3(1), 5-8.
—; 1970, Gondwana sedimentation around Bheemaram (Bhimaram), Pranhita-Godavari valley, India; *Jour. Sed. Petr.*, 40 (1), 140-170.
Vredenberg, E., 1914, The classification of the Gondwana System; *Proc. 1st* Ind. Sc. Cong.
Wadia, D.N.; 1957, *Geology of India*; 3rd Edition, McMillan, London, 536 pp.

Chapter 11

Alexander, P.O., 1981, Age and duration of Deccan volcanism-K-Ar evidence; *Mem. Geol. Soc. Ind.*, 3, 244-258.
Bhandari, L.L., Fuloria, R.C. and Sastry, V.V.; 1973, Stratigraphy of Assam Valley, India; *Bull. Am. Assoc. Petrol. Geol.*, 57, 642-654.
Bhatia, S.B.; 1969, Some ostracodes from the Lower Karewas near Nichahom, Kashmir; *Bull. Ind. Geol. Assoc.*, 2(1&2), 69.
Brunnschweiler, R.O.; 1966, On the geology of the Indo-Burman Ranges; *Jour. Geol. Soc. Australia*, 13(1), 137-194.
Chattopadhyay, B., Venkataramana, P., Roy D.K., Ghosh, S. and Bhattacharya, S.; 1983, Geology of Naga Hills ophiolites; *Rec. Geol. Surv. Ind.*, 112(2), 59-115.
Chaudhri, R.S. 1968, Stratigraphy of the Lower Tertiary formations of Punjab Himalaya; *Geol. Mag.*, 105 (5), 421-430.
—————; 1972, Heavy minerals from the Siwalik formations from the north-western Himalayas; *Sed. Geol.*, 8, 77-82.
Colbert, E.H.; 1934, A new rhinoceros from the Siwalik beds of India; *Am. Mus. Nov.*, 635, 1-9.
—————; 1935, Siwaliks mammals in the American museum of natural history; *Trans. Am. Phil. Soc., New Series*, 26, 1-401.

————; 1942, The geological succession of Proboscidae; in Orborn, H.F. (Editor): *Proboscidea*, 2, 1421-1521.

Dasgupta, A.B.; 1977, Geology of Assam-Arakan region; *Oil Commentary*, 14(17), 4-35.

Dehadrai, P.V.; 1958, Boundary between Lower Siwaliks and Middle Siwaliks of Jwalamukhi area; *Quart. Jour. Geol. Min. Met. Soc. Ind.*, 30, 211-214.

Dutta, S.K.; 1976, Tertiary palaeogeography of north-eastern India; *Geophytology*, 6(1), 65-74.

Eames, F.E.; 1963, Discussion on "Cretaceous-Tertiary Boundary"; *Proc. Int. Geol. Cong,, Copenhagen, pt. 27,* 60-61.

Falconer, H.; 1868, *Palaeontological Memoirs* (Edited by R.I. Murchusion), volumes I & II, 556 pp.

Ghosh, N.C.; 1976, Composition and origin of Deccan basalts; *Lithos*, 9, 65-73.

Gowda, S.; 1964, The foraminifera of South Indian Cretaceous-Eocene; *Ecol. Helv.*, 57(1), 299-313.

Gupta, V.J.; 1976, *Indian Cenozoic Stratigraphy*; Hindustan Publishing Corporation (India), Delhi, 344 pp.

Halstead, H.B. and Nanda, A.C.; 1973, Environ of deposition of the Pinjor Formation, Upper Siwaliks near Chandigarh; *Bull. Ind. Geol. Assoc.*, 6(1) 63-70.

Hofker, J.; 1966, *Maastrichtian, Danian and Palaeocene Foraminifera*; Paleontographica Suppl., Band 10, Stuttgart.

Jacob, K.; 1949, Land connections between Ceylon and Peninsular India; *Proc. Nat. Inst. Sc. Ind.*, 15(8), 341-343.

Khosla, S.C.; 1977, Palaeoecology of the Eocene beds of Rajasthan; *Publ. Cent. Adv. Stud. Geol. Panj. Univ.*, 11, 1-12.

Krishnan, M.S.; 1968, *Geology of India and Burma*; Higginbothams (Pvt.) Ltd., Madras, 5th Edition, 536 pp.

Krynnine, P.D.; 1937, Petrography and genesis of Siwalik series; *Am. Jour. Sc.*, 34, 423-446.

Kumar, R. and Gupta, V.J.; 1982, Indus Ophiolite Belt of Ladakh and Himalayan orogeny; *Rec. Res. Geol.*, 8, 494-509.

Kumar, S.P.; 1983, Geology and hydrocarbon prospects of Krishna, Godavari and Cauveri basins; *Petroleum Asia Journal*, Nov. 1983, 57-65.

Loeblich, A.R. and Tappan, H.; 1957, Correlation of the Gulf and Atlantic coastal plain Palaeocene and Lower Eocene formations by means of planktonic foraminifers; *Jour. Palaeont.*, 31(6), 1109-1137.

Mallet, F.R.; 1876, On the coal fields of Naga Hills bordering the Lakhimpur and Sibsagar districts, Assam; *Mem. Geol. Surv. Ind.*, 12(2), 286 pp.

Mathur, N.S.; 1969, *Biostratigraphy of the Subathu beds (Eocene), Simla Hills*; Unpublished Ph.D. thesis, Panj. Univ., Chandigarh, 314 pp.

Nagappa, Y.; 1959, Foraminiferal biostratigraphy of the Cretaceous-Eocene succession in India, Pakistan and Burma region; *Micropalaeontology*, 5(2), 145-192.

————; 1963, The Cretaceous-Tertiary boundary in India, Pakistan subcontinent; *Proceedings 21 st. Int. Geol. Con.*, 15, 41-49.

Nandi, B.; 1975, Palynostratigraphy of Siwalik Group of Punjab; *Him. Geol.*, 5 411-424.

Oldham, R.D.; 1885, Geology of the Andman islands; *Rec. Geol. Surv. Ind.*, 18(3), 135-145.

Pal, P.C, and Bhimshankaran, V.L.S.; 1971, Palaeomagnetism and the Deccan Trap volcanism; *Bull. Volcanologique, Tom XXXV-3*, 766-789.

Pandey, J.; 1972, Depositional, environmental and geological history of Baratang Formation, Andman Islands; *Proc. 2nd Colloq. India Micropal. Strat.*, 66-76.

Pascoe, E.; 1919, Early history of the Indus, Brahmaputra and Ganges; *Quart. Jour. Geol. Soc., London*, 75, 138-159.

Pilgrim, G.E.; 1913, Correlation of Siwaliks with the mammal horizons of Europe; *Rec. Geol. Surv. Ind.*, *43*, 262-376.

Prasad, K.N.; 1972, Observations on the Siwalik System of India; *Rec. Geol. Surv. Ind.*, *99*(2), 17-26.

Raiverman, V., Ganju J.L. and Misra, V.N.; 1979, A new look into the stratigraphy of Cenozoic sediments of the Himalayan foothills between the Ravee and Yamuna rivers; *Misc. Publ. Geol. Surv. Ind.*, *41*(5), 233-246.

Raiverman, V. and Raman, K.S.; 1971, Facies relations in the Subathu sediments, Simla Hills, north-western Himalaya; *Geol. Mag.*, *108*(4), 329-341.

Rama Rao, L.; 1939, On the occurrence of an Eocene bed in the Pondicherri area; *Curr. Sc. 8*, 1.

───────── ; 1956, Recent contributions to our knowledge of the Cretaceous rocks of South India; *Proc. Ind. Acad. Sc., Ser. B*, *44*(4), 1.

Ranga Rao, A.; 1983, Geology and hydrocarbon potential of a part of Assam-Arakan basin and its adjacent regions; *Petroleum Asia Journal*, Nov. 1983, 127-158.

Ranga Rao, A. and Obergfell, F.A.; 1973, *Harchyus asiaticus*, a new species of an Upper Eocene tapirod (mammalia, Perissodactyla) from Kalkot, India; *Oil Nat. Gas. Comm., special paper 3*, 1-8.

Roy, T.K.; 1983, Geology and hydrocarbon prospects of Andaman-Nicobar basin; *Petroleum Asia Journal*, Nov. 1983, 37-50.

Sah, S.C.D. and Dutta, S.K.; 1967, Palynostratigraphy of the Tertiary sedimentary formation of Assam; *Palaeobotanist*, *16*(2), 177-194.

Sahni, A. and Khare, S.K.; 1973, Additional Eocene mammals from the Subathu Formation of Jammu and Kashmir; *Jour. Palaeont. Soc. Ind.*, *17*, 31-49.

Sahni, A. and Kumar, V.; 1974, Palaeogene palaeogeography of the Indian subcontinent; *Palaeogeog. Palaeoclim. Palaeoecol.*, *15*(3), 209-226.

Sahni, A. and Mitra, H.C.; 1980, Neogene palaeogeography of the Indian subcontinent with special reference to fossil vertebrate; *Palaeogeog. Palaeoclim. Palaeoecol.*, *31*, 39-62.

Sastry, M.V.A. and Rao, B.R.J.; 1964, Cretaceous-Tertiary boundary in South India; *Proceedings 22nd Int. Geol. Cong.*, *3*, 92-103.

Sikka, D.B., Saxena, M.N., Bhatia, S.B. and Jain, S.P.; 1961, Occurrence of graywackes in the Lower Siwaliks, Simla Hills; *Nature*, *192*, 62.

Singh, D.; 1971, *The Quaternary and Recent Ostracode Fauna from Kashmir*; Unpublished Ph.D. thesis, Panj. Univ., Chandigarh, 300 pp.

Sinha, R.N.; 1970. Heavy mineral investigations in the Siwaliks of Mohand district, S.V. Saharanpur; U.P., India; *Jour. Geol. Soc. Ind.*, *11*(2), 163-167.

Srikantia, S.V. and Bhargava, O.N.; 1967, Kakara Series—a new Palaeoeocene formation in Simla Hills; *Bull. Geol. Soc. Ind.*, *4*(4), 114-116.

Srinavasan, M.S. and Azmi, R.J.; 1976, New developments in the Late Cenozoic lithostratigraphy of Andman-Nicobar islands, Bay of Bengal; *Proceedings VI Ind. Colloq. Micropal. Strat.*, 302-327.

Vimal, K.P.; 1953, Tertiary pollen from lignite of Palana (Bikaner); *Proceedings 40th Ind. Sc. Cong. (Abstracts)*, *3*, 27.

Vita-Finzi, C.; 1973, *Recent Earth History*; Macmillan, London, 138 pp.

Subject Index

Aalenian 147, 148
Abor Volcanics 186
Abur Formation 164, 169-170
Abyssal zone 31
Acadian 122
Aceratherium 207
Acrothele 98
Age 21
Agglomeratic Slate "Series" 135
Agnostus 131
Ajabgarh Group 88
Albian 148, 158, 173
Alectryonia ungulata 175
Alveolina 215
Alwar Group 87
Amblypterus 189
Amgaon Group 72-73
Ammodiscus 217
Amplexus 143
Amri Tectonic Unit 146
Anceps Beds 165
Anchignathodus 159
Ancyrodella 141
Angulogerina 215, 217
Anisian 148, 157
Anodontophoria 155
Anomalina 215, 217
Anomia 155
Anomocare 133
A. hundwanse 135
Antactosaurus septentrinales 172
Anthropoids 207
Aporthophylla 142
Aptian 148, 159, 167, 170, 172
Aquitanian 195
Arancorites 193
Aravalli Group 79-80, 84, 87, 100
Arca 158
Archaean 59
Archaeofavosia 100

Archaeozoic 19, 20
Archipelago Group 219-220
Arenigian 122
Arialur Formation 164, 175, 202
Arid plains 32
Arthrophyens 140
Arthyris 134
Artinskian 122
Asaphus 138
Ashgillian 122
Assilina 204, 211, 215
A. deviesi Zone 221
A. granulosa 223
A. lacunata Zone 221
Assilina sp. Zone 221
Astercyclina 215
Asthenosphere 46
Astian 195
Astrate 166
Atactoporella 138
Atdabanian 127
Athleta Beds 165
Attock Slates 112, 131
Auk Shale Member 101
Aulacaosphinctes 166
Auloosphinctes 166
Aulopora 143
Australliceras 165, 167
Aviculopecten 181
Azoic 19

Bababudan Group 67
Badami Group 102
Badesar Formation 164, 169
Badhaura Formation 167, 182
Bagh Beds 164, 170-172, 212
Bagra Conglomerate 187, 190
Baicalia 81
Baikalika 116
Bairakonda Quartzite 91, 93

244 Subject Index

Bairath Granite 89
Baisakhi Formation 164, 169
Bajocian 148
Bandah Formation 221
Banded Gneissic Complex 79-80, 84, 87, 100
Banganpalle Formation 101
Banjar Group 111, 118
Banjar Thrust 111
Bap Formation 182
Barail Group 214, 215
Barakar Formation 180, 183, 184
Baratang Formation 219
Barmer Sandstone 169
Barrenian 148, 169
Barren Zone 221
Basal Formation 97
Basaltic layer 41
Basantpur Formation 115, 117
Bastar Depression 52, 94, 101, 102
Bathonian 148, 166
Bathyal Zone 30
Beacon rocks 180
Bed 23
Belemnites 165
Belemnites marl of Jurum 165
Belemnite Shale
Belemnopsis gerardi 158
Bellerophon 138
Bellanella 81
Benighat Slates 118, 119
Benthonic fauna 29
Berriasian 148
Berwali Formation 223
Beufort Beds 154, 180
Bhainsedoban Marble 107, 109, 110
Bhander Group 97, 99
Bhander Limestone 97
Bhander Sandstone 97
Bhareli Formation 186
Bhima Group 84, 101, 102
Bhimasaram Formation 187, 192
Bhimphedi Group 104, 107-110, 130, 141
Bhimtal Volcanics 144
Bhuj Formation 165, 167
Biberian 194
Bijagarh Shales 97, 98
Bijawar Group 84, 87, 89, 98
Bijawar Shelf Sea 94
Bijli Rhyolites 73, 74
Bijni Tectonic Unit 145-146
Bijori Formation 183, 185
Biocoenose 14, 29, 33
Biotope 15, 29

Blanfordiceras 158
Blaini Boulder Bed 135, 144
Blaini Formation 118, 144-145
Blaini tillites 144
Bleaching Shales 97
Boileaugunge Formation 110
Bolivina 217
Bortryococcus braunii 220
Boundary problems:
—Basement-cover transition 83
—Cretaceous-Tertiary 200-203
—Neogene-Quaternary 203
—Permian-Triassic 151-155
—Precambrian-Cambrian 126-130
Buckaye Tillite 180
Budhaites 155
Budhi Schists 137, 140
Buffelus planiferus 207
Bundelkhand (Gneiss) Gneissic Complex 50, 79, 80, 98
Bundelkhand Massif 79, 80, 87, 89
Bunter (Bunstanstein) 147
Burdigalian 195
Buxa Dolomite 120
Buxa Group 120

Cainozoic see Cenozoic
Calceola 134
C. sandelina 135
Caledonian Orogeny 123, 144
Callovian 148, 166, 169
Calycoceras newboldi Zone 174
Calymene 134, 138, 140
C. nivalis 139
Calvulina 217
Camelus 207
Cameroboechia 140
Cambrian 20, 21, 121, 122
—Lower (Early) 129, 131, 133, 138
—Middle 138
—Upper (Late) 133, 138
Cambro-Ordovician 133
Cambro-Silurian 105
Campanian 148, 159, 175
Canina 140
Caninia 143
Caninophyllum 140
Carboniferous 20, 21, 122, 123
—Lower (Early) 134, 138, 139, 140, 143
—Upper (Late) 129, 134, 136, 139
Cardita 224
C. beaumonti Beds 202
C. (Venericardia) beaumonti 202

Subject Index 245

C. (V.) jaquinoti 203
Carnian 148, 156, 157
Caryocrinites 141
Cathophyllum (Thamnophyllum) multizonatum 139
Cenomanian 148, 159, 167, 170
Cenozoic (Cainozoic) 16, 19, 20, 194
Cenozoic life 199-200
Central Gneisses 104, 111
Ceratites 157
Cerethium 203, 210
Cerinopora dispar 171
Cerurus 138
Chaibasa Formation 77-78
Chail (Series) Group 111, 118, 144
Chail Thrust 111
Chakrata Formation 116, 117
Chamalwas Slates 134
Chandpur Formation 116, 117, 118, 143
Chandragiri Limestone 137, 141
Chandrapur Sandstone 102
Chaosa Formation 115
Chattian 195
Chari Formation 164, 165, 166
Charnockite Series 68-69
Chattisgarh Depression 52, 94, 101, 102
Chattisgarh Group 84, 102
Chaugaon Beds 187, 190-191
Cheloniceras 167
Cheyyair Group 91, 93
Chidambu "Stage" 158
Chikkim Limestone 159
Chilpi Ghat Group 71
Chinji Formation 207, 208, 209
Chisapani Formation 109, 110
Chitlang Formation 137, 141
Chitradurga Group 67
Chonetes 140
Chor Granite 110, 111, 112
Cibicides 215
Clararia 153, 154, 156
Clay line 28
Closepet Granite 64, 68, 80
Clypeina 162
Codacystes 141
Coilopoceras bossei 171
C. scindiae 171
Coleites 215
Collenia sp. 98
C. baicalica 114
C. columnaris 94, 98, 114
C. compacta 94
C. Symmetrica 114

Colleniella 81
Compensated sequence 38
Compsosuchus solus 172
Condensed horizons 31
Coniacian 148, 170
Coniopteris 193
Conocoryphe 133
Conophyton cylendricus 98
Continental shelf 27
Continental slope 27
Conularia 155
Conus 225
Coraline Limestone 170-172
Corbis 158
Corbicula 172
Corbula 166
C. lyrata 165, 169
Cornulites 138
Costiferina 140
Couvinian 122
Craton 41, 44
Cretaceous 20, 21, 106, 132, 147-148
Cretaceous-Tertiary Boundary 200-203
Crinoidal Limestone Formation 137
Cryptozoan proliferum 94
Cucullaea 158
Cuddalore Sandstone 225
Cuddapah Depression 50, 85, 89, 100
Cuddapah Supergroup 7, 84, 89-94, 100
Cumbum Shales 91, 93
Cuttack Chelf 52
Cyathophyllum 138
Cyclemina 216
Cyclobus 139
Cyclodendron 185
Cyclopteris 189
Cylendroporella-Johnson 162
Cypris 212
Cyretella 186

Dagshai Formation 205
Dalbhum Formation 77, 78
Daling Group 104, 112, 120
Dalma Lava 75, 76, 78
Dalmiapurum Formation 164, 172-173
Damatha Group 116-117
Damuda Group 182-185
—of Eastern Himalaya 185-186
Danian 148, 163, 175, 195, 201, 202
Danubian 194
Daonella 157
Daonella Limestone 157
Daonella Shales 157

246 Subject Index

Darjeeling Gneiss 112
Dark Band Formation 137, 142
Deccan Syneclise 52, 101
Deccan Traps 6, 52, 211, 212
Delhi Geosyncline (orogen) 94
Delhi Orogeny 89
Delhi Supergroup 79, 84, 87-89, 100
Delhi Synclinorium 84, 87
Dendrocystites 141
Denwa Clays 187, 190
Deoban Formation 116-117, 118, 120
Deola Marl 171
Devonian 20, 21, 121, 122
—Lower (Early) 138, 140, 142, 144
—Middle 138, 140, 142
—Upper (Late) 138, 141
Dhading Dolomite 118, 119-120
Dhanjori Group 75, 76, 78
Dhanjori-Simlipal Lava 76
Dharangadhara Formation 164, 167
Dharmaram Formation 187, 193
Dharmshala Group 205-206
Dharwar Supergroup 7, 64, 66-68
Dhauladhar Gneiss 112
Dhaulagiri Limestone 142
Dhimpu Gneiss 142
Dhokpathan Formation 207, 208, 209
Dhosa Oolites 165
Dicrodium-Noegerathiopsis Zone 187
Dictyoconoides 204
Dictyzamites-Pterophyllum Zone 187, 191
Dicynodonts 189
Dihing Group 214, 217
Disang Group 214, 217
Discocyclina 199, 215, 225
D. dispansa Zone 222
D. ranikotensis 202
D. sella Zone 222
Discovery Edge Formation 180
Dogger Series 147
Dogra Slates 104, 105, 112, 131-134, 137
Dolikephalites 165
Dongargarh Granite 73
Downtonian 122
Dras Volcanics 105, 159, 196
Dubrajpur Formation 187, 191
Dupi-Tila Group 214, 217
Dupi-Tila Formation 214
Dwyka Shales 180
Dwyka Tillites 180

Ecca Series 180
Elatocladus 193

Elephus 203, 207
E. nomadicus 207
Emsian 122
Entoleum 158
Eocambrian 81
Eocene 16, 20, 194, 195
Eomiodon 166
Eoserranus 172
Eparchaean Unconformity 63, 83
Epimayaites 165
Epoch 21
Episageceras 151
Equisetites 193
Equus 203, 207
Era (Erathem) 18
Erinpura Granite 87, 88-89
Estrata nova 180
Euapidoceras 165
Eubaculus Vagina 175
Euomphalus 138, 155
Euspira 224
Euspirifer 134
Eurydesma 135, 181, 182, 184
E. rotundata 135
Euryhaline 31

Facies 26, 33
Facies map 35, 37, 39
Fagfog Quartzite 118, 119
Femennian 122
Favosites 138
F. spitiensis 139, 140
Fenestella 134, 140, 186
Fenestella Shales 134, 137
Fermoria 98
Flawn Limestone 97, 98
Folded Mountain Belt 41, 44-45
Foredeep 32
Formation 10, 11, 23
Fossil assemblage 14-16
Frasnian 122
Fundamental Gneiss 48, 66, 80
Fussella mucronata 143
Fusus 225

Gadryina 217
Gaj Formation 221
Gangamopteris 179
G. cyclepteroides 179, 186
Gangamopteris Zone 183
Gangpur Formation 187, 193
Gangpur (Series) Group 7, 78
Gangparh Shales 97

Subject Index

Garbayang Formation 133, 137, 140
Gauran Beds 133, 134
Gedinnian 122
Geochronology 4, 23
Geological Time 17, 18
Georgian 122
Geosyncline 45
Giraffoid 207
Girujan Clay 214, 217
Giumal Sandstone 158
Givetian 122
Glacial zone 32
Glauconitic Member 97
Globigerina 199, 202, 215, 216, 217, 223
Globigerinoides 223
Globorotalia 216, 223
G. tosaensis fenuitheca Zone 203
G. truncatulinoides Zone 203
Globotruncana 158, 175
G. stuarti 175
Glossopteris 179, 182, 189, 190
Glossopteris Assemblage Zone 183
G. communis 186, 192
G. conspicua—G. retifera Zone 183
Glossopteris flora 154, 176, 178, 185, 188
G. indica 179, 185, 186, 192
G. walkomiella Zone 183
Godavari Graben 52, 101, 176
Godavari Shelf 52
Golden oolites 165
Gondolella 159
Gondwana Sequence 6, 52, 123, 154, 176
—classification 176-177
—Lower Gondwana 112, 120, 167, 178-186
—marine intercalations 180-181
—sedimentation 177-178
—Upper Gondwana 186-193
Gondwanaland 124, 149, 179
Gondwanidium-Buriada Zone 183
Goniaceras 138
Gonioglyptus 189
Goniophera 138
Gothlandian 122
Graben 44
Granitic layer 41
Granomarginata 100
Group 11, 22
Gryphea 158
Gryphoceras 155
Guembilina 202
Guitar Formation 219
Gulcheru Quartzite 91-92
Gunzian 195

Gwalior Group 79, 84, 87, 89, 95, 98
Gymmites 155, 157
Gymnosolen 81

Haimanta Group 106, 118, 133, 136, 137
Halobia 157
Halobia Limestone 157
Halysites 138
Hansmanica 193
Haploceras 166
Haplophragmoides 216, 217
Hauterivian 148, 158, 171
Hazara Slates 131
Hedenstroemica Zone 155
Hemiaster fourteavi 171
Hemibos 207
Hemis Conglomerate 211
Hercynian Orogeny 123, 126
Hettangian 148
Hildoglochiceras 165
Himalayan Geosyncline 145
Himalayan Orogeny 196-199
Himalayicalix 141
Hipparion 203, 207
Hirpur Formation 203
Holeostephanus 158
Holmla 126
Holocene 20, 194
Homo sapiens 200
Homotaxial 9
Hoplites 158
Hormotoma 138
Hornstone Breccia 88
Hostimella 140
Hubertoceras 165
Humid plains 32
Hungarites 155
Hypoboops 207, 211
Hyppophys 207

Ichthyostegids 125
Icriodus 141
Idicycloceras singulare 169
Illaenus 134, 138
Index fossil 14
Indo-Brahm river 206
Indo-Gangetic Plain 95
Indocephalites diadematus 166
Indosphinctes 165
Indosuchus raptorius 172
Indus Formation 210
Indus (Ophiolitic) Belt 155, 159, 210
Indus Suture 53, 56

Subject Index

Infra-krol Formation 145, 162
Inglish Formation 219
Inseria 81
Intermontane basin 32
Intertrappean Beds 7, 52, 212
Irati Shales 180
Irlakonda Quartzite 91, 93
Iron Ore Group 75, 76-77, 78, 89
Iron Ore Orogeny 74, 75
Irravadi "System" 222
Isculites 155
Isopach maps 37-39
Itarare Series 180

Jabalpur Group 167, 187, 190-191
Jabalpur Beds 187, 191
Jainti Quartzite 120
Jaintia Group 214, 215
Jaisalmer Formation 164, 169
Jamalamadugu Formation 101
Jammu Limestone 113, 114, 115, 117, 120
Jaunsar Group 117, 118, 143-144
Jhiri Shales 97
Joannites thenamensis 157
Jodhpur Sandstone 99
Jogira Formation 222
Jomsom Limestone 160
Jurrassic 20, 21, 130, 147, 148
Jurusania 115
Jutogh Group 104, 110-112, 113, 118
Jutogh Nappe 111, 112
Jutogh Thrust 111, 118

Kaimur Group 97, 98-99
Kaimur Sandstone 97
Kajrahat Limestone 97
Kakra "Series" 204-205
Kaladgi Basin 85, 94, 102
Kaladgi Group 84, 94, 102
Kalhel Limestone 156
Kali Formation 137, 140
Kalitar Formation 107, 109, 110
Kamlial Formation 207, 208-209
Kamthi Formation 183, 185, 187
Kankawati Formation 224
Kantkote Sandstone 165
Karewa (Group) Formation 203, 210
Karharbari (Member) Formation 183, 184
Karoo Formation 211
Karsog Granite 112
Kasauli Formation 205
Kashmir Nappe 104
Kathmandu Nappe 107, 108, 141

Kathmandu (Mahabharat) Thrust 107
Katrol Formation 164, 165, 166
Katroliceras 165
Kazanian 122
Keuper Series 147
Khaira Member 113
Khamisara Member 167
Kheinjua Formation 97, 98
Khelong Formation 186
Khondalite Series 69-70
Khuiala Formation 221
Kimmeridgian 148, 158, 169
Kinkeliniceras 165
Kioto Limestone 158
Kirthar Formation 215, 221
Kistna Group 91, 93
Kohima Syncline 216
Koilkundla Limestone Member 101
Kolamnala Shales 91, 93
Kolhan Group 78, 84, 89
Kopili Formation 214, 215-216
Kossamaticeras theoboldianum Zone 175
Kota Formation 187, 193
Krinkrong Formation 137, 141

Ladenian 195
Ladinian 148, 149, 157
Laevisuchus indicus 172
Lagoons 31
Lagynocystis 141
Lakhpat Formation 223
Laki Formation 215, 221
Lamellibranch Zone 136
Lameta Beds 164, 172, 212
Langpar Formation 175, 203, 215
Larji Group 111
Larjung Formation 137, 142
Lathi Formation 164, 167
Lenian 127
Lenticulina 203
Lepidocyclina 199
Lepidoobitoides 203
Lepidosteus 172
Leplatosaurus medagascariensis 172
Leptaena 133, 138
L. trachealis 139
Leptelloidea 133
Leptobos 203, 207
Lesser Himalayan Sea 113, 143
Lewesiceras vaju Zone 175
Lias Series 147
Lichas 138
Lilang Group 157

Subject Index   249

Lima 158
Limestone Formation 202
Linella 81
Lingula 182
Linoproductus 140, 186
L. pollex 143
Lipak Formation 138-139, 140
Listriodon 207
Lithosphere 6, 46
Littoral Zone 29
Llandeillian 122
Llandoverian 122, 142
Llanvirnian 122
Lochambal "Stage" 158
Lokhartia 204
L. haemei 202
Lolab Formation 129
Long Formation 219
Lophocarionophyllum 140
Lophospiral 138
Lucian percrassa 203
Lucina (Codakia) 224
Ludlovian 122
Lumachelle Formation 160
Lutenian 195
Lymnaea 212
Lyria 224
L. crassicostata 175
Lystosaurus 154, 189

Maastrichtian 148, 159, 163, 175, 203
Machaeraia 140
Macrocephalites 165
M. triangularis 166
M. macrocephalus 166
Macrocephalus Beds 165
Madh Formation 222-223
Magnesian Sandstone 127-129
Mahabharat Thrust see Kathmandu Thrust
Mahadek Formation 175, 203, 215
Mahadeva (Formation) Group 154, 187, 190
Mahanadi Graben 52, 176
Main Boundary Thrust 53, 118, 198
Makrana Marble 84
Malachimata Member 167
Malani Igneous Suite 99-100
Malani Volcanics 84, 94, 98, 99-100
Malania 172
Malekhu Limestone 118, 119
Maleri Formation 187, 192-193
Malm Series 147
Mammites concilatum Zone 174

Manchar Formation 221
Mandhali Formation 116-117, 118, 143
Mandi Granite 112
Mahendragarh Marine Beds 180-181
Mangli Beds 187
Marginifera 141
M. himalaensis 139
M. himalaensis Zone 136
Marine transgression/regression 12, 36
Marker horizon 10
Markhu Formation 109, 110, 137
Martoli Formation 129, 137, 140
Maslovielia 116
Mastodon 207
Mastodonosaurus indicus 190
Mayaites 158, 165
Medlicotia 151
Meekoceras Zone 155
Megalodon 156
Megalodon Limestone 158
Melania 211, 212
Member 10, 23
Mesosaurus 180
Mesozoic 7, 19, 20, 147
Metapeltoceras 165
Metasandstone Formation 118, 119
Michelina megstoma 143
Microcanthoceras 166
Microdiscus 131
Micropalaeontology 17
Mid-oceanic ridge 40, 45
Mikir Hill Massif 52
Miliammina 216
Mindelian 194
Minjaria 116
Miocene 16, 20, 194, 195
Miscellina Miscella 202
Mississipian 20, 122, 123
Modiola 155
Mollassic deposit 33
Monograptus 134
Monophyllites 157
Monotis salinaria 160
Monotrypa 134
Montian 195
Morar Formation 89
Moscovian 122
Motur Formation 183, 185
Mount Glossopteris Formation 180
Mukut Limestone 160
Murree Group 105, 206
Murree Thrust 132
Muschelkalk Series 147, 157

250 Subject Index

Mussoorie Group 145
Muth Quartzite 133, 134, 137, 138, 140, 142
Myophoria 155

Naga Metamorphics 214
Nagri Formation 207, 208, 209
Nagri Quartzite 91, 93
Nagrota Formation 203
Nagthat Formation 117, 118, 143
Nahan Formation 206
Nallamalai Group 91, 93
Namikla Flysch 159
Namsang Beds 214, 217
Namurian 122
Nandayal Shale Member 101
Nandgaon Group 73
Nanga Parbat Massif 104, 106
Nari Formation 221
Narj Limestone Member 101
Narmada-Son-Damodar Graben 52, 176
Natica 212
Naubag Beds 134
Nautilus (Hercoglossa) 224
Nautilus indicus 203
Nawakot Group 118-120
Neil West Coast Formation 219
Neobolus Shales 127-129
Neocomian 148, 169
Neocomites 158
Neogene 20, 21, 111, 132, 194, 195
Neospathodus 153
Neospirifer 182
Nerita (Ostostoma) divaricata 175
Neritic Zone 30
Newer Greenstone Belt 48
Nilgiri Carbonate Group 142
Nilgiri Limestone 137, 142
Nimar Sandstone 170-171
Nimbahar Limestone and Shales 97
Niniyur Formation 202, 224
Niobe 126
Nipania Plant Beds 187, 191
Nisusia depsaensis Zone 138
Nodular Limestone 170, 171
Noeggerathiopsis-Paranocladus Zone 183
Nonian 203
Norian 148, 160
North Face Quartzite 137, 142
Nourpur Formation 118, 119
Novella kingi 157
Nucleela 81
Nucula 166
Nummulites 199, 204, 210, 215

N. maculatus Zone 221-222
N. (Ranikothalia) nuttali 202
N. (R.) sindensis 202

Obolella 98
Obolus 104
Oceanic trench 27, 45, 46
Old Red Sandstone 123
Older Greenstone Belt 48, 64
Older Metamorphic Group 74, 75-76
Olenus 126
O. haimantaensis Zone 138
Oligocene 20, 194, 195
Oliva 225
Olive Shales 97
Ontogeny 16
Opekina 142
Operculina 203
Opertorbitolites 215
Ophiceras Bed 153, 154
Ophiceras Zone 155
Opis 158
Oppelia 166
Orbitoides macropora 202
O. media 202
Orbitolina parma 210
O. discoidea 210
Orbitolites 215
Ordovician 20, 21, 121, 122
—Lower (Early) 142
—Upper (Late) 133
—fauna 138
Orthambonites 142
Orthis 104, 133, 134, 138, 140, 141
O. (Plectoorthis) spitiensis 139
O. pustilifera 135
Orthoceras 138, 155
Orthotetis 138
Orusia 126
Ostrea 158, 225
Otoceras Bed 153, 154
Otoceras woodwardi 153
Otoceras Zone 155
Otozamites 193
Oxfordian 148, 158, 166, 169

Pab Sandstone 202
Pachygonia 189
Pachysphinctes aff. bathylocus 169
Pacopteris 190
Pagiophyllum 193
Pagiophyllum-Brachyphyllum Zone
   (Assemblage) 187, 191

Pakhal Group 84, 94, 101
Palaeocene 20, 194, 195
Palaeogene 20, 21, 194, 195
Palaeogeography 26
Palaeogeographic maps 4, 34-36
Palaeontology 3
Palaeotectonic map 37
Palaeozoic 19, 20, 121
—Lower (Early) 104, 107, 121, 143, 144
— subdivisions 122
—Upper (Late) 120, 121, 144
Palana Formation 220
Pali Formation 187, 192
Palmatolepsis 141
Paludina 172, 212
Panchet Formation 154, 155, 180, 187, 188-189
Panchmarhi Sandstone 187, 190
Pangaea 149
Panian Quartzite 101
Panjal Thrust 132
Panjal Traps 105
Panjal Volcanics 136, 137, 152, 155, 157
Papaghani Group 91-93
Par Formation 89
Para "Stage" 158
Paraconularia 182, 186
Paradoxides 126
Paranautiloids 155
Parastantoceras mintoi 171
Parihar Formation 164, 169
Paramandal Sandstone 203
Parsora Formation 187, 192
Patcham Formation 164, 165-166
Pecopteris 189
Pecten 158
Pegu "System" 222
Pelagic fauna 29
Peltoceras 165
Peltura 126
Peninsular (Gneiss) Gneissic Complex 48, 64, 66, 94
Pennsylvanian 122, 123
Pentamerus 138
Period 20-21
Perisphinctes 158
Permian 20, 21, 122, 123
Permo-Carboniferous 112
Phanerozoic 18, 19, 20, 44, 46, 112, 136
Phulchauki Group 141
Phylogeny 16
Physa (*Bullinus*) 172
P. (*B.*) *princepi* 212

Phyllotheca 186
Pinacoceras 155
Pinjor Formation 207, 208, 209
Pinnackled Quartzite Member 101, 102
Pitepani Volcanics 73, 74
Pitted Calc-schist Formation 137, 142
Placenticeras tamulicum Zone 175
Plaisanchian 195
Planitrochus 138
Planorbis 211
Plate Tectonics 5, 8, 45-46, 198-199
Plateform cover 14
Pleistocene 20, 194
Plesiosaurus indica 167
Pliensbachian 148
Pliocene 16, 20, 194, 195
Po Formation 137, 139, 141
Polygnathus 141
Pondicheri Formation 224-225
Porcellinite Formation 97, 98
Porcellinic Shales 97
Port Blair Formation 219, 220
Port Meadow Formation 219
Portlandian 148, 166
Posidonia 162
Potsdamian 122
Prasopora 134
Precambrian 8, 19, 20, 25, 59-63
Precambrian-Cambrian Boundary 126-130
Precambrian Basement 41, 48, 59, 61-63
Priabonian 195
Priscogalea 100
Proceraties 165
Productus 135, 139, 141, 180, 186
P. (*Dictyoclostus*) *genuinus* 135
Productus Limestone 11
P. semireticulatus Zone 136
Productus Shale 11, 137, 139, 141, 151, 152
P. spitiensis 139
Proplacenticeras santoi var. *bolli* 171
Prostegodon 207
Proterosuchus 154, 189
Proterozoic 19, 44, 46, 59, 84-85, 111
Protobolella 98
Protocardia 165, 166
Protoretepora 141, 145, 181
P. cf. *ampla* 135, 186
Protoretepora Zone 136
Psedoliva 224
Pseudomonatis 158
Pseudotrapezium 165, 166
Psilophyton princepi 138
Pterinea thanamensis 138

252 Subject Index

Ptychites 155, 157
Ptychoparia 133
P. admissa Zone 138
P. civica Zone 138
P. pervulgata Zone 138
P. spitiensis Zone 138
Ptychophylloceras 166
Ptyllophyllum flora 154, 177, 178, 188
Pupa 212
Purana 7, 84, 132
—Lower 84-94
—Upper 94-102
Purple Sandstone 99, 127-129
Pycnactis mitrata 133
Pycnodus 172

Quaternary 21, 44, 47, 132, 194

Radua Formation 107, 109, 110
Refinesquina 133, 138, 140, 141
Raialo Group 83, 84, 87, 100
Raipur Formation 102
Rajahmundri Sandstone 225
Rajasthan Shelf 52
Rajmahal Formation 187, 191
Rajmahal Plant Beds 187, 191
Rajmahal Traps 187, 191
Ralam Conglomerate 129, 137
Ramgarh Quartz Porphyry 117
Rangea 83
Rangit Pebble Slate Formation 185
Ranibennur Group 67
Raniganj Formation 154, 180, 183, 185
Ranikot Formation 202, 215, 221
Ranikothalia 204
Ranipat Member 167
Rautgarha Formation 116, 117
Raymondella 133
Red Shale Member 161, 162
Redlichia 127, 133
R. noetlingi Zone 138
Rehmani Beds 165
Reineckia 165
R. tyranniformis 165
R. rehmani 165
Reitzia himaica 157
Resserella 133
Reticularia 140, 180
Rewa Group 97, 99
Rewa Sandstone 97
Rhacopteris flora 139
Rhaetian 148, 157, 158
Rhynochonella 155

R. griesbachi 157
Riasi Limestone 113
Rilu Formation 186
Riphean 81, 84, 85, 98, 113, 114, 116, 129
Rissian 194
Robustoschwagerina 145
Rohtas Formation 97
Rotalia 203
Rubulus 203
Rudistes 159

Sabal major 205, 206
Sabellidites 82
Sakmarian 122, 146, 186
Sakoli Group
Saline "Series" 115, 127-129
Salkhala Group 104-105, 112, 131, 137
Salopina 140, 144
Salt Pseudomorph Beds 127-129
Samorpsis 189
Sanjauli Formation 115
Santa Maria Formation 180
Santonian 148
Sargur Schist Complex 64-66
Sarmatian 195
Satpura (Orogeny) Orogen 51, 85, 87, 101
Saurashtra-Kutch Shelf 52
Sausar Group 70-71
Sawai Formation 219
Schellwienaella 140
Schizoneura 186, 189
Schizophoria 134
Schloenbachia inflata Zone 174
Scythian 147, 148, 159
Semri Group 89, 95, 97, 98
Senonian 148, 159, 174, 175
Series 21
Shali Group 111, 113-115, 116, 118, 120
Shelf 44
Shelly Limestone 165
Shiala Formation 137, 140
Shield 41
Shillong Massif 52
Siberites 155
Siderolites 203
S. calcitrapoids 202
Siegenian 122
Silurian 20, 21, 121, 122
Silurian fauna 138
Simla Group 111, 113-115, 117, 118, 133
Sinchua Formation 120
Sindeites sindensis 169
Sinemurian 148

Singhbhum Granite 77, 89
Singhbhum Group 77-78
Singhbhum Orogeny 51, 74
Singhbhum Shear Zone 74, 75
Siphogenerina 199
Siphonodella 141
Sirmur Group 117
Sivajiceras 165
S. congener 166
Siwalik Group 53, 206-210
Solarium 224
Solenopleura 133
Sopyang Formation 141
Sowerbyella 133, 138
Sphenophyllia 143
Sphenopteris 193
Spirifer 135, 139, 141, 181
Spiriferellah rajah Zone 136
Spiriferina 155, 157, 158, 181
S. stracheyi 156
Speirigerina 157, 158
Spiti Shales 158
Spiticeras 158
Srisailam Quartzite 91, 93
Stage 21
Stampian 195
Stegodon 207
Stenohaline 31
Stephanian (Uralian) 122
Steptelasma 138, 140
Stigmatopygus elatus 175
Strait Formation 219
Straparolus 138
Stratigraphy 4
—biostratigraphy 16
—chronostratigraphic units 18
—classification and correlation 9-17
—lithostratigraphic classification 22-23
—Standard Stratigraphic 18-22
Strophalosia 141
Stropheodonta interstrialis 138
Strophomena 133, 138, 140
Sturia 157
Subathu Formation 113, 114, 204
Subbotina 223
Suket Shales 97, 98
Sullavai Group 84, 101-102
Sundernagar Formation 116, 118
Supergroup 22
Surajdewal Member 167
Surma Group 214, 216
Sus 207
Sylhet Limestone 214, 215

Syneclise 44
Syntaxial bend 53
System 20-21
Syringonautiloids 155
Syringothyris 134, 135
S. curzoni glaber 143
S. cuspidata 135, 139
Syringothyris Limestone 134, 137, 140
Syringothyris Zone 134

Tadpatri Shales 91, 93
Taeniopteris 189
Tagling "Stage" 158
Tal Formation 130, 145, 160, 162-163
Talcher Boulder Bed 135, 178
Talcher Formation 178-180
Talcher tillites 144, 179, 180
Tanawals 133-134
Taramelliceras 165
Tarap Shales 160
Tatarian 122
Tatrot Formation 207, 208, 209
Tectonics 4, 11, 12, 36
—epierogenic/orogenic 37
Tejam Group 111-117
Telemastodon 207
Tellina 158, 224
Tentaculites 142
Terebra 225
Terebratula biplicata 169
Tertiary 7, 20, 21, 194, 195
Tetrabelodon 207
Tethys palaeo-sea
Textularia 217
Thabo Member 139
Than Member 167
Thanjavur Shelf 52
Thenatocoenose 15, 33
Theory of Catastrophy 5
Therria Formation 203
Thini-Chu Formation 137, 143
Tibetan slab 106-107
Tilicho Lake Formation 137, 143
Tilicho Pass Formation 137, 142, 143
Tipam Group 214, 217
Tipam Sandstone 214, 217
Tirodi Biotite Gneiss 71
Tistung Formation 130, 137, 141
Titanosaurus indicus 172
Tithonian 148, 158, 160, 167, 169
Toarcian 148
Tommotian 127, 130
Torquaphinctes 165

Subject Index

Tournasian 122
Trachyceras 157
Transform fault 45, 46
Trappoid Beds 97
Tremadocian 122
Triassic 20, 21, 147, 148
Trichinopolly Formation 164, 175-175, 202
Trifarina 215
Trigonia 165, 166
Trigonia Beds 165
Triplica 134
Tripocium 167
Trochammina 216
Tunda Pathar Limestone 115
Tungussia 81
Tupe Tillite 180
Tura Formation 215
Turonian 148, 159, 172, 174
Turritella 200, 212, 224

Ukra Beds 165, 166-167
Umaria Marine Beds 237-238
Umia Ammonite Beds 165
Umia Formation 164, 165, 166
Umia Plant Beds 165, 167
Umiatites 166
Unconformity 11, 39
Uniformitarianism 5
Unio 205, 211, 212
Uralian 122, 186
Uttatur Formation 164, 174, 202

Vaikrita Group 104, 106-107, 111, 118, 137
Vaikrita Thrust 111
Vailanginian 148, 158
Valvata 212

Valvulinerina 215
Variscan Orogeny 123
Veginulina 215
Vempelle Shales and Limestones 91, 92
Vendian 81, 85, 115, 129
Venericardia mutabilis Zone 221
Vertebraria 179, 189, 190
V. indica 186, 192
Vicarya 212
Vindhyan Supergroup 6, 7, 84, 89, 95-100
—Lower Vindhyans 97-98
—Upper Vindhyans 98-100
Vindhyan sea 94, 113, 123, 143
Vindhyan Syneclise 50, 94
Vindobonian 195
Virgosphinctes 165, 166, 169
Visean 122
Viviparous 211

Wadhwan Formation 164, 167
Walkomiella 185
Weichselia-Onichiopsis Zone 187
Wenlockian 122, 133
Westphalian 122

Xenaspis 139, 151, 157
Xenodiscus 151

Yerappalli Formation 187, 192
Ypresian 195

Zamapithicus 207
Zapherentites 143
Zewan Formation 136, 137, 141, 152, 155
Zone 22

RAYMOND H. FOGLER LIBRARY
DATE DUE

BOOKS ARE SUBJECT TO
AFTER TWO WEEKS